Methods of Experimental Physics

VOLUME 15

QUANTUM ELECTRONICS

PART A

METHODS OF EXPERIMENTAL PHYSICS:

L. Marton and C. Marton, *Editors-in-Chief*

1. Classical Methods
   Edited by Immanuel Estermann
2. Electronic Methods, Second Edition (in two parts)
   Edited by E. Bleuler and R. O. Haxby
3. Molecular Physics, Second Edition (in two parts)
   Edited by Dudley Williams
4. Atomic and Electron Physics—Part A: Atomic Sources and Detectors, Part B: Free Atoms
   Edited by Vernon W. Hughes and Howard L. Schultz
5. Nuclear Physics (in two parts)
   Edited by Luke C. L. Yuan and Chien-Shiung Wu
6. Solid State Physics (in two parts)
   Edited by K. Lark-Horovitz and Vivian A. Johnson
7. Atomic and Electron Physics—Atomic Interactions (in two parts)
   Edited by Benjamin Bederson and Wade L. Fite
8. Problems and Solutions for Students
   Edited by L. Marton and W. F. Hornyak
9. Plasma Physics (in two parts)
   Edited by Hans R. Griem and Ralph H. Lovberg
10. Physical Principles of Far-Infrared Radiation
    L. C. Robinson
11. Solid State Physics
    Edited by R. V. Coleman
12. Astrophysics—Part A: Optical and Infrared
    Edited by N. Carleton
    Part B: Radio Telescopes, Part C: Radio Observations
    Edited by M. L. Meeks
13. Spectroscopy (in two parts)
    Edited by Dudley Williams
14. Vacuum Physics and Technology
    Edited by G. L. Weissler and R. W. Carlson
15. Quantum Electronics (in two parts)
    Edited by C. L. Tang

Volume 15

Quantum Electronics

PART A

Edited by

C. L. TANG

*School of Electrical Engineering
and the Materials Science Center
Cornell University
Ithaca, New York*

1979

ACADEMIC PRESS · **New York** **San Francisco** **London**
A Subsidiary of Harcourt Brace Jovanovich, Publishers

COPYRIGHT © 1979, BY ACADEMIC PRESS, INC.
ALL RIGHTS RESERVED.
NO PART OF THIS PUBLICATION MAY BE REPRODUCED OR
TRANSMITTED IN ANY FORM OR BY ANY MEANS, ELECTRONIC
OR MECHANICAL, INCLUDING PHOTOCOPY, RECORDING, OR ANY
INFORMATION STORAGE AND RETRIEVAL SYSTEM, WITHOUT
PERMISSION IN WRITING FROM THE PUBLISHER.

ACADEMIC PRESS, INC.
111 Fifth Avenue, New York, New York 10003

United Kingdom Edition published by
ACADEMIC PRESS, INC. (LONDON) LTD.
24/28 Oval Road, London NW1 7DX

Library of Congress Cataloging in Publication Data

Main entry under title:

Quantum electronics.

   (Methods of experimental physics ; v. 15)
   Includes bibliographical references.
   1. Lasers. 2. Quantum electronics. I. Tang, Chung
Liang, Date II. Marton, Ladislaus Laszlo,
Date   III. Series.
TA1675.Q36   621.36'6   79–14369
ISBN 0–12–475915–7

PRINTED IN THE UNITED STATES OF AMERICA

79 80 81 82   9 8 7 6 5 4 3 2 1

CONTENTS

CONTRIBUTORS	xi
FOREWORD	xiii
PREFACE	xv
CONTENTS OF VOLUME 15, PART B	xvii
CONTRIBUTORS TO VOLUME 15, PART B	xix

1. Introductory Concepts and Results
by C. L. TANG

1.1. Introduction	1
1.2. The Laser	3
1.2.1. Light Amplification by Stimulated Emission	4
1.2.2. Optical Resonator	9
1.2.3. The Laser Oscillator	12
1.3. Density Matrix	20
1.3.1. Definition and Properties	20
1.3.2. Density Matrix Equation	22
1.4. Special Techniques	26
1.4.1. Q-Switching	27
1.4.2. Mode-Locking	28

2. Atomic and Ionic Gas Lasers
by WILLIAM B. BRIDGES

2.1. Neutral Atom Lasers Excited by Electron Collision	33
2.1.1. Cyclic Lasers	34
2.1.2. Continuous Neutral Noble Gas Lasers	55

CONTENTS

- 2.2. Ions Excited by Electron Collision 59
  - 2.2.1. Spectroscopy of Noble-Gas Ion Lasers 62
  - 2.2.2. Noble-Gas Ion Laser Characteristics and Mechanisms 70
  - 2.2.3. Ion Laser Performance 92

- 2.3. Neutral Atom Lasers Excited by Collisions with Atoms 97
  - 2.3.1. Helium–Neon Lasers 97
  - 2.3.2. Other Neutral Atom Lasers Excited by Atomic Collisions 116

- 2.4. Ion Lasers Excited by Collisions with Atoms or Ions . . 120
  - 2.4.1. Noble-Gas Ion Lasers Excited by Atomic Collisions 121
  - 2.4.2. Mercury Ion Laser 124
  - 2.4.3. Cadmium Ion Laser 128
  - 2.4.4. Zinc Ion Laser 143
  - 2.4.5. Iodine Ion Laser 146
  - 2.4.6. Selenium Ion Laser 149
  - 2.4.7. Other Metal Vapor Ion Lasers 151

3. Solid State Lasers
by M. J. WEBER

- 3.1. Introduction 167
- 3.2. Physical Processes 170
  - 3.2.1. Energy Levels 171
  - 3.2.2. Transition Probabilities 176
  - 3.2.3. Ion–Ion Interactions 179

- 3.3. Laser Materials 181
  - 3.3.1. Laser Ions 181
  - 3.3.2. Host Materials 187
  - 3.3.3. Fluorescence Sensitization 193

3.4. Properties and Comparison of Solid State Lasers	195
3.4.1. Nd:YAG Lasers	196
3.4.2. Nd:Glass Lasers	199
3.4.3. Ruby Lasers	202
3.4.4. Stoichiometric Lasers	203
3.4.5. Other Lasers and Materials	204
3.5. Hazards	205

4. Semiconductor Diode Lasers
by HENRY KRESSEL

4.1. Introduction	209
4.1.1. Laser Topology	211
4.1.2. Vertical Geometry	213
4.2. Injection	216
4.3. Carrier Confinement	219
4.4. Radiation Confinement	220
4.5. Gain Coefficient and Threshold Condition	222
4.6. Temperature Dependence of the Threshold Current Density	224
4.7. Materials	226
4.8. Heterojunction Lasers of Various Materials	230
4.8.1. Near-Infrared Emission Lasers	230
4.8.2. Visible Emission Lasers	231
4.8.3. Infrared Emission	232

4.9.	Performance of Selected Laser Structures	234
4.10.	Radiation Patterns	237
4.11.	Degradation	242
4.12.	Modulation Characteristics	244
4.13.	Distributed-Feedback Lasers	247

5. Dye Lasers
by OTIS GRANVILLE PETERSON

5.1.	Introduction		251
5.2.	Basic Dye Molecule and Dye Solvent Properties		253
	5.2.1.	Energy States of Dye Molecules	253
	5.2.2.	Rate Equations	257
	5.2.3.	Laser Threshold	258
	5.2.4.	Stimulated Emission Cross Section	263
	5.2.5.	Thermal Limitations	264
5.3.	Laser Devices		269
5.4.	Short-Pulse Dye Lasers		269
	5.4.1.	Rate Equation Description	269
	5.4.2.	Amplified Spontaneous Emission	271
	5.4.3.	Short-Pulse Oscillator	276
	5.4.4.	Short-Pulse Amplifiers	281
	5.4.5.	Single-Pass, Short-Pulse Amplifier	281
	5.4.6.	Regenerative Oscillators	287
	5.4.7.	Thermal Limitations	293
5.5.	Steady-State Laser		293
	5.5.1.	General Description	293
	5.5.2.	Steady-State Power Balance	294

5.5.3.	Flashlamp-Excited Laser	297
5.5.4.	Flashlamp Plasma Temperature	298
5.5.5.	Linear Flashlamps	302
5.5.6.	Ablating Flashlamps	305
5.5.7.	Vortex Stabilized Lamps	307
5.5.8.	Linear Flashlamp Reflector Cavities	310
5.5.9.	Dye Cell for Flashlamp Excitation	312
5.5.10.	Free Jet Dye Cell	314
5.5.11.	Shock Wave Effects	318
5.5.12.	Coaxial Flashlamps	319
5.5.13.	Resonators and Tuning Elements	321
5.5.14.	Dye Flow System	324

5.6. CW Laser 325
   5.6.1. Analytical Description 325
   5.6.2. CW Resonator Geometry 336
   5.6.3. Resonator Stability Criteria 338
   5.6.4. Astigmatic Resonator 343
   5.6.5. Dye Cell Astigmatism 344
   5.6.6. Mode Area 347
   5.6.7. Alternative Geometries 351
   5.6.8. Dye System for cw Lasers 353

5.7. Laser Dyes 355

AUTHOR INDEX 361

SUBJECT INDEX 373

CONTRIBUTORS

Numbers in parentheses indicate the pages on which the authors' contributions begin.

WILLIAM B. BRIDGES, *California Institute of Technology, Pasadena, California 91125* (31)

HENRY KRESSEL, *RCA Laboratories, Princeton, New Jersey 08540* (209)

OTIS GRANVILLE PETERSON, *Allied Chemical Corporation, Morristown, New Jersey 07960* (251)

C. L. TANG, *School of Electrical Engineering and The Materials Science Center, Cornell University, Ithaca, New York 14853* (1)

M. J. WEBER, *University of California, Lawrence Livermore Laboratory, Livermore, California 94550* (167)

FOREWORD

Professor C. L. Tang has organized a splendid and broad-based volume on experimental methods in quantum electronics. Because the laser seems to have penetrated nearly every aspect of scientific research, this volume should prove useful to a wide range of researchers. Lasers of various types are discussed, and their principles and characteristics for the experimenter are explored. Each author is quite well known in the area from which he reports, and numerous valuable insights are presented in the succeeding chapters.

With the exception of that on classical methods, the other volumes of "Methods of Experimental Physics" are devoted primarily to a particular field of study. While this volume too is devoted to a particular field, it is also true that one may regard it as dealing with an experimental tool.

We extend our thanks to the volume editor and authors for their efforts.

L. MARTON
C. MARTON

PREFACE

Quantum electronics was the name adopted for the first conference on maser and related physics over twenty years ago. With the advent of the laser shortly afterward, the field has since experienced an explosive growth. Lasers and related devices are now used in many branches of science and technology and have made possible the development of numerous new experimental techniques in physics and chemistry.

There are many excellent review articles and books on numerous specialized topics of quantum electronics and laser physics. The present volume of the "Methods of Experimental Physics" series was prepared with the intent of providing a concise reference or guidebook for investigators and advanced students who wish to do research in the field of quantum electronics or to use laser-type devices and related techniques in their own specialties. As with many of the earlier volumes of this series, it quickly became obvious that to give adequate coverage to the main topics planned, a single volume would be too restrictive. This led to the present two parts of the volume: Volume 15A and Volume 15B. Even with two volumes, a choice still had to be made on whether to emphasize either lasers and related devices as sources of coherent radiation from the IR to the UV or the numerous of experimental techniques that make use of lasers. Although new lasers and related generators of coherent radiation are still being developed, this part of the field appears to have reached a certain degree of maturity. Several basic types of lasers and related devices have emerged and will likely remain as prototypes of sources of coherent radiation from the far IR to the near UV for some time. It is now timely to review and summarize the basic facts and principles of these sources. Therefore, the bulk of Volume 15 deals with these lasers and devices. However, new experimental techniques using lasers are still being rapidly developed. Therefore it is premature as well as extremely difficult to give a comprehensive review that will not become obsolete in a relatively short time. This of course is not to say that laser-related experimental techniques have not already produced important results in physics and chemistry; far from it. A number of such applications have been chosen here as examples of the power and unique potential of laser-related experimental techniques. The time will soon come when a more complete review will be needed.

It has been an honor and a pleasure for me to work with the authors of this volume. Each is an authority and has contributed much to the original research in his specialty. Part 1, however, is an exception. The editor failed in his search for anyone willing to take time off from his busy research to collect and record some of the more basic material. Time was running out and the task unhappily fell upon himself. In addition, as is inevitable with edited works, there is a slight unevenness in the latest literature reviewed in the various articles.

Finally, it is a privilege indeed for me to have worked with Bill Marton, Editor-in-Chief of the "Methods of Experimental Physics" series. I deeply regret that it had not been possible to publish this volume before his passing. I can only hope that this volume has lived up to the high standards of the rest of the volumes of this series.

CONTENTS OF VOLUME 15, PART B

6. Color Center Lasers
   by LINN F. MOLLENAUER

   6.1. Introduction
   6.2. Some Pertinent Color Center Physics
   6.3. Processes for Color Center Formation
   6.4. Optical Gain
   6.5. Laser Cavities With a Highly Concentrated Model Beam
   6.6. Construction and Performance of a cw Laser Using $F_A(II)$ Centers
   6.7. Generation and Performance of a Distributed-Feedback Laser Using $F_A(II)$ Centers
   6.8. Recent Developments in F_2^+ Center Lasers
   6.A. Appendix: Technique for the Piecewise Interferometric Generation of Gratings

7. Molecular Lasers
   7.1. Molecular Infrared Lasers
        by T. MANUCCIA
   7.2. Rare Gas Halide Lasers
        by S. K. SEARLES

8. Chemically Pumped Lasers
   by TERRILL A. COOL
   8.1. Introduction
   8.2. Disequilibrium in Reaction Product Energy States
   8.3. Supersonic Chemical Lasers
   8.4. New Purely Chemical Lasers
   8.5. Rotational Chemical Lasers
   8.6. Problems in the Search for New Chemical Lasers
   8.7. Conclusion

9. Nonlinear Optical Devices
   by F. ZERNIKE
   9.1. Introduction
   9.2. General
   9.3. Parametric Oscillators
   9.4. Generation by Mixing of Two Inputs
   9.5. Upconversion

10. Examples of Laser Techniques and Applications

   10.1. Picosecond Spectroscopy
         by E. P. IPPEN AND C. V. SHANK
   10.2. Vuv Spectroscopy
         by J. J. WYNNE
   10.3. Doppler-Free Laser Spectroscopy
         by P. F. LIAO AND J. E. BJORKHOLM
   10.4. Nonlinear Optical Effects
         by Y. R. SHEN
   10.5. Laser Selective Chemistry
         by JAMES T. YARDLEY

AUTHOR INDEX—SUBJECT INDEX

CONTRIBUTORS TO VOLUME 15, PART B

J. E. BJORKHOLM, *Bell Telephone Laboratories, Holmdel, New Jersey 07733*

TERRILL A. COOL, *School of Applied and Engineering Physics, Cornell University, Ithaca, New York 14853*

E. P. IPPEN, *Bell Telephone Laboratories, Holmdel, New Jersey 07733*

P. F. LIAO, *Bell Telephone Laboratories, Holmdel, New Jersey 07733*

T. MANUCCIA, *Naval Research Laboratory, Washington, D.C. 20375*

LINN F. MOLLENAUER, *Bell Telephone Laboratories, Holmdel, New Jersey 07733*

S. K. SEARLES, *Naval Research Laboratory, Laser Physics Branch, Washington, D.C. 20375*

C. V. SHANK, *Bell Telephone Laboratories, Holmdel, New Jersey 07733*

Y. R. SHEN, *Physics Department, University of California, Berkeley, California 94720*

J. J. WYNNE, *IBM/T. J. Watson Research Center, P.O. Box 218, Yorktown Heights, New York 10598*

JAMES T. YARDLEY,[*] *Department of Chemistry, University of Illinois, Urbana, Illinois 61801*

F. ZERNIKE, *North-American Philips, Briarcliff Manor, New York 10510*

---

[*] Present address: Allied Chemical Corporation, Corporate Research Center, P.O. Box 10212, Morristown, New Jersey 07960

1. INTRODUCTORY CONCEPTS AND RESULTS[†][*]

1.1. Introduction

As a generator of coherent monochromatic electromagnetic radiation ranging from the far infrared to the vacuum ultraviolet, the laser[1-3] as a class of devices has a number of unique properties and is now used in numerous applications in physics, chemistry, and other fields of science.

As is well known, the laser beam can be extremely intense. Field strengths comparable to or greater than, for example, that experienced by electrons in an atom or approximately 10^8 V/cm can be reached. Clearly, at such intensity levels drastic things can happen to all kinds of materials. In fact, many new phenomena can already be observed and new applications become possible at intensity levels considerably below such extreme levels.[4]

The spectral width of the laser can be extremely narrow. For example, linewidths much less than 10^{-3} cm^{-1} can be obtained relatively routinely in gas lasers or tunable dye lasers. This has led to many important new advances in laser-selective chemistry and high-resolution spectroscopy.[5] The laser light can also be highly coherent in the sense that it can closely approach a classical plane wave with a well-defined phase front. Such a

---

[1] A. L. Schawlow and C. H. Townes, *Phys. Rev.* **112**, 1940 (1958); J. P. Gordon, H. J. Zeiger, and C. H. Townes, *ibid.* **95**, 282 (1954); N. G. Basov and A. M. Prokhorov, *Sov. Phys.—JETP (Engl. Transl.)* **27** 431 (1954).

[2] T. H. Maiman, *Nature (London)* **187**, 493 (1960); *Phys. Rev. Lett.* **4**, 564 (1960).

[3] A. Javan, W. R. Bennett, Jr., and E. R. Herriott, *Phys. Rev. Lett.* **6**, 106 (1961).

[4] See, for example, P. A. Franken, A. E. Hill, C. W. Peters, and G. Weinreich, *Phys. Rev. Lett.* **7**, 118 (1961); N. Bloembergen, "Nonlinear Optics." Benjamin, New York, 1965; H. Rabin and C. L. Tang, eds., "Quantum Electronics: A Treatise," Vol. 1, Parts A and B. Academic Press, New York, 1975. Also, the articles by F. Zernike and Y. R. Shen in this book, Part B.

[5] See, for example, H. Walter, ed., "Laser Spectroscopy of Atoms and Molecules." Springer-Verlag, Berlin and New York 1976; K. Shimoda, ed., "High Resolution Laser Spectroscopy." Springer-Verlag, Berlin and New York, 1976. Also, the articles by J. J. Wynne, P. F. Liao, and J. E. Bjorkholm, and J. T. Yardley in this book, Part B.

---

† Supported in part by the Materials Science Center of Cornell University, Ithaca, New York.

* Part 1 is by C. L. Tang.

well-collimated coherent wave can propagate over long distances with minimal spread and can be sharply focused to achieve extremely high light intensity levels. In addition, the availability of coherent monochromatic electromagnetic waves in the spectral range from the far infrared to the ultraviolet[6] has made it possible to observe a whole host of coherent optical phenomena[7] in atoms, molecules, and solids for the first time.

Finally, extremely short pulses of laser light, down to a picosecond or less, can now be generated, thus allowing extremely fast phenomena to be studied.[8]

It must be emphasized, however, that not all these unique properties can be found in the same type of lasers simultaneously. The choice of the lasers to be used obviously depends upon the applications.

There are many types of lasers ranging in size from tiny semiconductor lasers to giant gas-dynamic lasers. Laser media include, for example, free atoms, ions, and molecules, or paramagnetic ions in insulating crystals and glasses, semiconductors, and color centers in crystals. Basic excitation mechanisms include electrical, mechanical, chemical, and optical means or combinations of these. Many of these laser media and excitation mechanisms will be discussed in detail in the following chapters on specific types of lasers. A brief introduction to some of the common features and basic concepts is given in this chapter.

Three features are common to all lasers: There is always a gain medium in which light can be amplified; there is generally an electromagnetic structure or optical resonator to confine the light and to repeatedly feed it back into the gain medium to be amplified; and the light wave interacts with the gain medium via the stimulated emission process. These are discussed in the following section.

To understand some of the dynamic and nonlinear characteristics of the laser, it is often necessary to take into account the atomic coherence, that

---

[6] See, for example, Y. R. Shen, ed., "Nonlinear Infrared Generation." Springer-Verlag, Berlin and New York, 1977; J. J. Ewing and C. A. Brau, *in* "Tunable Lasers and Applications" (A. Mooradian, T. Jaeger, and P. Stokseth, eds.). Springer-Verlag, Berlin and New York, 1976.

[7] See, for example, R. G. Brewer, *in* "Applications of Lasers to Atomic and Molecular Physics" (R. Balian, S. Haroche, and S. Liberman, eds.). North-Holland Publ., Amsterdam, 1977.

[8] S. L. Shapiro, ed., "Ultrashort Laser Pulses." Springer-Verlag, Berlin and New York, 1977.

[9] P. A. M. Dirac, "Quantum Mechanics," Sect. 33. Oxford Univ. Press, London and New York, 1947; R. C. Tolman, "Principles of Statistics Mechanics." Oxford Univ. Press, London and New York, 1939; A. Abragam. "The Principles of Nuclear Magnetism." Oxford Univ. Press, London and New York, 1961.

is, the phase coherence among the wavefunctions of the active atoms or molecules in the gain medium, at least on a statistically averaged basis. The density matrix formalism[9] is a convenient way to do this. This formalism is also particularly useful in describing and analyzing various optically coherent phenomena that can be induced or observed using the laser light. The concept of density matrix and density matrix equation is discussed in some detail in Chapter 1.3.

Finally, two well-known special techniques that are commonly used to modify the output characteristics of lasers, namely, the Q-switching and mode-locking techniques, are described in Chapter 1.4. These are used to generate laser pulses with high peak powers and short pulse durations in a wide variety of lasers.

No attempt is made in either the text or the references of this part to summarize the recent advances in the field of laser physics. The purpose here is to introduce some of the background concepts and results to facilitate reading of the following parts of this volume.

1.2. The Laser

In the case of classical oscillators at radio and microwave frequencies, the basic frequency selection mechanism is provided by the resonance properties of the electromagnetic structure, LC circuits at low frequencies and cavities at microwave frequencies. Consequently, at the high-frequency end of the spectrum the size of the electromagnetic cavity used must be comparable to the wavelength of interest so that the resonances of the cavity are relatively well separated and only the cavity mode nearest the wavelength of interest is excited. A high-frequency limit is eventually reached where one can no longer fabricate cavities of a size comparable to the wavelength of interest. This limits the operating range of conventional generators to the submillimeter region.

In the laser type of oscillators, the basic frequency selectivity generally comes from the active or laser gain medium because the amplification process is due to stimulated emission between relatively well-defined energy levels of atoms, ions, molecules, or solids; only the fine structure of the laser emission is determined by the electromagnetic structure or the optical resonator. The optical cavity nevertheless provides the feedback mechanism that is essential for converting an amplifying medium into an oscillator. Figure 1 shows conceptually the simplest form of a laser. The optical cavity consists simply of a pair of partially reflecting mirrors facing each other. Photons spontaneously emitted by the laser medium will be reflected back and forth between the mirrors and get amplified via the stimulated emission process on each traversal through the laser

1. INTRODUCTORY CONCEPTS AND RESULTS

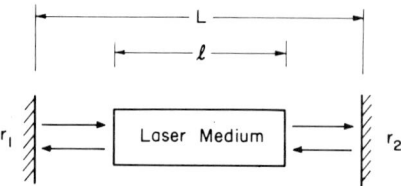

FIG. 1. Schematic of a simple laser. Its structural components consist of a laser medium, in which light can be amplified by the stimulated emission process, and an optical cavity, which provides the feedback for the laser oscillator. The laser cavity consists of a pair of highly reflecting mirrors of reflectivities r_1 and r_2.

medium. If the round-trip gain through the laser medium is greater than the losses at the mirrors and elsewhere in the optical path, laser oscillation will build up until eventually some sort of saturation effect sets in to reduce the gain in the medium to a level just enough to overcome the losses. We discuss first the general conditions that must prevail in the laser medium to achieve sufficient gain and how these conditions are to be realized.

1.2.1. Light Amplification by Stimulated Emission

To simplify the discussion, we consider first a simple two-level system with energy levels E_1 and E_2. Such an atom in the upper level E_2 can spontaneously emit a photon at the transition frequency $\nu_{21} = (E_2 - E_1)/h$ and go to the lower level. Photons spontaneously emitted by an ensemble of such atoms will have random phases or are incoherent. If a coherent electromagnetic wave at ν_{21} is already present and acting on these atoms, stimulated emission will occur instead and the photons emitted will be in phase with the wave present. The rate of emission due to stimulated emission will be proportional to the intensity I of the wave, or the spatial variation of the intensity of the wave due to stimulated emission is

$$\left(\frac{dI}{dz}\right)_{\text{st.em.}} = \left(N_2 - \frac{g_2}{g_1} N_1\right) \sigma_{\text{st}} I, \tag{1.2.1}$$

where N_2 and N_1 are the populations of the upper and lower energy levels, respectively, and g_2 and g_1 are the corresponding degeneracies; $(N_2 - N_1 g_2/g_1)$ is therefore the population inversion; $\sigma_{\text{st.}}$ is the stimulated emission cross section. If there is also distributed loss in the medium characterized by α, the total spatial rate of change of the light intensity in the medium is

$$\frac{dI}{dz} = \left(\frac{dI}{dz}\right)_{\text{st.em.}} + \left(\frac{dI}{dz}\right)_{\text{loss}} = \left(N_2 - \frac{g_2}{g_1} N_1\right) \sigma_{\text{st}} I - \alpha I. \tag{1.2.2}$$

When the intensity is not too strong, or when the laser is below or near the threshold for oscillation, such that the change in the population distribution is relatively small, the population inversion factor can be regarded as a constant. Equation (1.2.1) shows that the intensity varies exponentially due to stimulated emission in the medium with the spatial gain coefficient:

$$g = \left(N_2 - \frac{g_2}{g_1} N_1\right) \sigma_{\text{st}}. \quad (1.2.3)$$

The wave is amplified if g is greater than α or the population inversion is sufficiently large:

$$\left(N_2 - \frac{g_2}{g_1} N_1\right) > \alpha/\sigma_{\text{st}}.$$

If not, the wave will be attenuated.

The stimulated emission cross section σ_{st} describes the nature of the transition, regardless of the condition the laser medium is in. The condition of the medium is specified by the population distribution N_2 and N_1. These parameters g, σ_{st}, α, and the population inversion factor ($N_2 - N_1 g_2/g_1$) are fundamental to the understanding and comparison of all lasers.

To achieve laser action, it is then necessary to find the right kind of laser transition with a large enough stimulated emission cross section and to find the means, or the excitation and relaxation mechanisms, to create a suitable population inversion for that transition. The specific details will obviously differ from laser to laser. Let us examine the basic considerations.

1.2.1.1. Stimulated Emission Cross Section. The stimulated emission cross section obviously depends upon the energy levels involved and the nature of the transition. It also depends on the state of polarization of the incident wave and the symmetry of the active medium. In the case of linearly polarized light and isotropic media consisting of, for example, free two-level atoms and ions, it is related to the spontaneous radiative decay rate $1/\tau_{21}$ (or the Einstein A coefficient $A_{21} = 1/\tau_{21}$) from the upper level of the two-level system to the lower level, the fluorescence line shape $g_f(v_0)$, the relative dielectric constant ϵ_r of the host medium in which the two-level atoms are embedded, and the frequency of the incident wave v_0:

$$\sigma_{\text{st}} = \frac{c^2 g_f(v_0)}{8\pi v_0^2 \epsilon_r \tau_{21}} \quad (1.2.4)$$

The radiative transition rate is in turn related to the matrix element of the transition dipole (say, the z component) between all the degenerate states of the upper and lower levels (e.s.u.):

$$\frac{1}{\tau_{21}} = \sum_{\substack{n=1,2,\ldots,g_1 \\ m=1,2,\ldots,g_2}} \frac{64\pi^4 \nu_{21}^3 |\langle 2m|ez|1n\rangle|^2}{hc^3 g_2}. \quad (1.2.5)$$

If the system is not isotropic, such as in the case where the active "atoms" are impurity ions in anisotropic host crystals, then the stimulated emission cross section is no longer isotropic and simply related to the spontaneous radiative decay rate, which is a scalar; the cross section can be found in terms of the appropriate transition matrix element by substituting Eq. (1.2.5) into Eq. (1.2.4). The detailed nature of the transition is all contained in the appropriate transition matrix element.

Whatever the transition, in the end it is the numerical value of σ_{st} that matters insofar as the laser operation is concerned. Numerically, the stimulated emission cross section can range from 10^{-12} cm² for some of the strongly allowed electronic transitions in free atoms and ions, to 10^{-16} cm² for direct band-gap transitions in semiconductors or electronic transitions in dye molecules, to 10^{-19}–10^{-20} cm² for some of the paramagnetic ions in crystalline solids or glasses. Thus, the required population inversion density for laser oscillation can vary drastically from medium to medium.

1.2.1.2. Population Inversion. The existence of a population inversion in the medium necessarily means that the medium cannot be a closed system in thermodynamic equilibrium. Some external excitation (or "pumping") mechanism must be present to maintain the population inversion or else the medium must be in some transient state in which a population inversion exists temporarily and the laser operates on a self-terminating pulsed basis.

Conceptually, the simplest means of creating a population inversion in an active medium is to mechanically introduce the atoms in the upper laser level into, and remove the atoms in the lower level from, the active region where stimulated emission is to take place. In fact, the very first maser to operate,[1] the ammonium beam maser at 24 GHz, and some of the more recently developed lasers such as the gas-dynamic laser operate this way. In the former case, for example, a beam of ammonium molecules is first sent through some sort of selector, which selects out only those molecules in the upper maser state. The beam is then sent through a hole in the wall of the maser cavity into the interaction region, in which the molecules decay into the lower maser state via the stimulated emission process and exit through a hole on the opposite side of the cavity.

1.2. THE LASER

Generally, more complicated means are used and the population distribution is determined by the detailed balance of various excitation and relaxation processes in the medium. The actual physical processes will naturally vary from system to system and will be discussed in the following chapters in connection with various specific types of lasers. We give here some of the general considerations on the basis of the rate equations.

Consider a nondegenerate n-level system in the absence of any stimulated emission. The rate equation for the population of the ith level is

$$\frac{dN_i(t)}{dt} = R_i - \sum_{j=1,2,\ldots}^{n} [N_i(t)W_{ij} - N_j(t)W_{ji}], \qquad (1.2.6)$$

where R_i is the excitation rate into this level due to some external pumping mechanism and W_{ij} the relaxation rate from the ith to the jth level. The rates for the forward and reverse processes are not independent but are related according to the principle of detailed balancing:

$$W_{ij}/W_{ji} = \exp(E_i - E_j)/kT, \qquad (1.2.7)$$

where T is the temperature of the medium and k the Boltzmann constant. If the total number of atoms over the n levels is conserved, then of course

$$\sum_{i=1,2,\ldots}^{n} N_i(t) = N_0. \qquad (1.2.8)$$

The rate equations give the pumping and relaxation rates required for a given desired population inversion.

In real systems, many levels are usually involved. However, most lasers can be modeled as two-, three-, or four-level systems. Consider first a closed two-level system in the sense that $N_1(t) + N_2(t) = N_0$ and $R_1 = -R_2 = -R$. Assume also that $E_2 - E_1$ is large compared to kT so that the upward reverse relaxation rate W_{12} is negligible compared with the downward forward rate $W_{21} = 1/\tau$, where τ is the lifetime of the upper level. It follows from the corresponding rate equations that the minimum pumping rate $R_{\min}$ required to achieve population inversion between levels 2 and 1 in a steady-state is

$$R_{\min} = N_0/2\tau. \qquad (1.2.9)$$

This means that at least half of the atoms must be pumped up from the lower to the upper level during the lifetime of the atoms in the upper level. It also follows from the rate equations that the steady-state population inversion in the absence of stimulated inversion is

$$\overline{N}_2 - \overline{N}_1 = 2R\tau - N_0 \qquad (1.2.10)$$

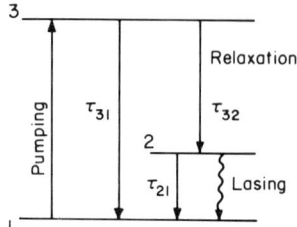

FIG. 2. Schematic of a three-level laser medium with lasing between the lower two levels.

above the minimum pumping rate, Eq. (1.2.9). The fact that at least half of the active atoms must be cycled from the lower to the upper level during the lifetime of the upper state can be a difficult requirement to meet in some cases. The four-level system described later avoids this difficulty.[10]

Few laser systems can be considered true two-level systems. A third level is often involved in either the pumping or relaxation process. Suppose optical or collisional pumping takes place between levels 1 and 3 as shown schematically in Fig. 2. Depending upon the relaxation rates among the levels, lasing can take place either between levels 3 and 2 or levels 2 and 1. In the former case, assuming that relaxation from level 3 proceeds primarily through cascading into level 2, it can be shown on the basis of the rate equations that the basic requirement for achieving population inversion between the top two levels is

$$\tau_{32} > \tau_{21}, \qquad (1.2.11)$$

regardless of the pumping rate. This is a basic requirement on the lifetimes of the laser states for cw lasers when both are excited states. The actual population inversion is, however, directly proportional to the pumping rate. In the latter case, the requirement for achieving population inversion between levels 2 and 1 is the relaxation rate from level 2 into level 1 must be smaller than the pumping rate from level 1 to level 3. If, in addition, the relaxation rate from level 3 to 2 is very much greater than the pumping rate into level 3, the population of level 3 then becomes negligibly small and the three-level system reduces essentially to a two-level system with the ground state becoming the lower laser level. In this case, half of the active atoms must be pumped into level 2 through level 3 during the lifetime of level 2. The ruby laser[2] is a well-known example of such a three-level system.

[10] P. P. Sorokin and M. J. Stevenson, *Phys. Rev. Lett.* **5**, 557 (1960); *IBM J. Res. Dev.* **5**, 56 (1960).

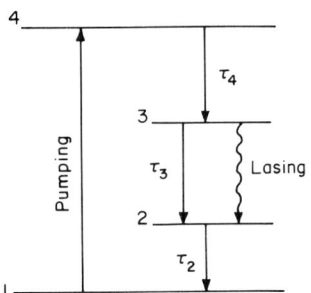

Fig. 3. Schematic of a four-level laser medium. Relaxation from the excited states is assumed to proceed through the next lower state.

The requirement that half of the atoms must be pumped into the upper level when the lower laser level is the ground state of the system is often difficult to meet. The four-level system[10] avoids this difficulty. This is shown schematically in Fig. 3. The Nd^{3+}-ion doped crystalline or glass laser is a well-known example of such a system. Here, because the lower laser level is far above the ground level, it is generally empty. If the relaxation rate out of this level is greater than that out of the third level, population inversion between the third and second levels can exist regardless of the pumping rate. The magnitude of the population inversion would be determined by the pumping rate and the relaxation rate between levels 4 and 3 and levels 2 and 1, assuming that pumping takes place between levels 1 and 4.

The rate equations give the requirements on the pumping and relaxation rates for a given population inversion. The magnitude of the population inversion required for laser oscillation in turn depends upon the electromagnetic feedback structure or the optical resonator and vice versa.

1.2.2. Optical Resonator

The primary function of the optical resonator is to provide the feedback mechanism needed to return the radiation emitted by the laser medium back to the medium for repeated amplification. The simplest and most commonly used form of optical resonator is basically a Fabry–Perot interferometer consisting of a pair of highly reflecting mirrors facing each other as shown schematically in Fig. 1.

Most laser mirrors are metallic or dielectric coated semitransparent mirrors or in the form of coatings deposited on the laser crystal. In the case of semiconductor lasers, because of the small dimensions and the intrinsically high refractive index of the materials, the cavity is often formed simply by cleaving the ends of the laser crystal and making use of the high Fresnel reflections at the cleaved facets; reflection-coated facets are also used sometimes. For far-infrared lasers, partially transparent mirrors are not always available; the radiation can simply be coupled out of the cavity through a small hole in the reflecting mirrors.

Plane mirrors or curved mirrors with properly chosen radius of curvature and spacing are used. Cavities with plane mirrors that are not exactly parallel will have large "walk-off" losses. With suitably chosen curved mirrors, it is easy to find some portion of the mirrors that are parallel and is therefore easier to align. They are particularly useful for low-gain lasers.

Because the fluorescence linewidth or the spectral width of the gain curve often encompasses several resonances of the cavity, the spectral characteristics of the laser output are determined by the resonance characteristics of the cavity. The resonances of Fabry–Perot cavities formed from infinite plane mirrors are determined by the condition that the total round-trip phase shift of the light wave is equal to the integral multiples of 2π:

$$(4\pi L/\lambda) + \phi_1 + \phi_2 = 2N\pi, \qquad (1.2.12)$$

where L is the optical length between the mirrors, and ϕ_1 and ϕ_2 the phase shifts at the mirrors. With no phase shifts at the mirrors, N corresponds to the number of half-wavelengths between the mirrors. Each value of N corresponds to a resonance of the cavity. These are the longitudinal modes of the cavity.

Numerically, L is typically a few tenths of a millimeter for semiconductor lasers, a few centimeters for insulating solid-state lasers, and tens of centimeters or meters in the case of dye lasers and gas lasers. In all cases, N is invariably a very large number. The resonances are therefore relatively closely spaced in the laser wavelength range. Because L is rarely known to the accuracy of a wavelength, the absolute resonance frequencies of the laser are not readily predictable. The separations between the resonance frequencies are, however, readily predictable:

$$\Delta \nu = \nu_N - \nu_{N-1} = c/2L \qquad (1.2.13)$$

in those cases where the dispersion in the medium can be neglected. In semiconductor lasers, this is often not negligible and the longitudinal mode spacings are

$$\Delta \nu = c/2L \left(1 - \frac{\lambda}{n}\frac{dn}{d\lambda}\right). \qquad (1.2.14)$$

$\Delta \nu$ ranges typically from on the order of 100 MHz in the case of dye lasers and gas lasers to 10^{11}–10^{12} Hz in the case of semiconductor lasers.

The spectral width $\delta \nu_c$ of the longitudinal modes of the passive cavity depends upon the total loss ΔI suffered by a wave of intensity I in each round trip through the passive cavity. It is approximately

$$\delta \nu_c \cong \frac{\Delta I}{2\pi I} \Delta \nu, \qquad (1.2.15)$$

which is typically a few percent of the mode spacing $\Delta \nu$.

The spectral width of each longitudinal mode of the laser emission is generally still much smaller then both the fluorescence line width or the cavity line width [Eq. (1.2.15)] and is often determined by such things as mirror vibrations or other fluctuations in the laser cavity.

The transverse dimensions of the mirrors for the Fabry–Perot cavity are determined on one hand by considerations of diffraction losses and beam divergence and on the other hand by the size limitations of the laser gain medium. Because the Fabry–Perot type of cavity is an open structure,[11] there are diffraction losses at the edges of the mirrors. The amount of such losses depends upon the Fresnel number F of the cavity, which is defined as the product of the mirror widths divided by the product of the mirror spacing and the wavelength. When F is smaller than 1, the diffraction loss can be substantial.[11] In addition, the laser beam divergence is inversely proportional to the transverse dimensions of the laser beam at the mirror. If the mirrors are too small, the laser beam becomes poorly collimated. In the case of semiconductor lasers, for example, the transverse dimension is limited in one direction by the width of the junction to typically on the order of a micron. Thus, the laser beam is highly divergent in that direction. For other lasers, beam divergence on the order of a milliradian or less can easily be obtained.

The finite transverse dimensions of the mirrors also lead to closely spaced resonances corresponding to the transverse modes of the laser cavity. The transverse mode spacings are typically much smaller than the longitudinal mode spacings. Because the intensity of the radiation in the cavity has to approach zero at the edges of the mirrors and the fields satisfy the wave equation, the intensity distributions of the tranverse modes[11] are essentially sinusoidal with integral numbers of half-waves between the edges. The higher order transverse modes have higher diffraction losses and in general are less likely to be excited unless there is spatially nonuniform gain or losses. Higher order transverse modes have

---

[11] A. G. Fox and T. Li, *Bell Syst. Tech. J.* **40**, 453 (1961).

larger beam divergence and lead to multiple maxima and nodal lines in the intensity across the beam. To improve the beam quality, it is common to use intracavity apertures to limit the laser action over a uniform region and to limit the beam size to increase the difference between the diffraction losses of the lowest and the higher transverse modes.

1.2.3. The Laser Oscillator

The gain medium and the optical resonator are the two basic structural components of the laser. Insofar as the operational characteristics of the laser are concerned, several questions are of particular importance: the threshold condition for oscillation, the saturation characteristics above the threshold, the stability of the oscillator, and the spectral characteristics of the laser emission.

1.2.3.1 Threshold Condition. The basic threshold condition for oscillation is that the wave after each round trip through the cavity and the laser medium should have the same amplitude and phase, or

$$(r_1 r_2)^{1/2} \exp\left[(g_{th} - \alpha)l + i(\phi_1 + \phi_2) + i\frac{4\pi L}{\lambda}\right] = 1, \quad (1.2.16)$$

where r_1 and r_2 are the intensity reflectivities of the mirrors, g_{th} refers to the intensity gain at the threshold, l is the length of the active medium, which could be shorter than the cavity length L, and α here is an equivalent distributed loss factor to take into account all the intracavity losses. Since the phase condition is automatically satisfied for the cavity resonances according to Eq. (1.2.12), the required gain at the threshold for osscillation is for nondegenerate laser levels:

$$g_{th} = (N_2 - N_1)_{th}\sigma_{st} = \alpha - \frac{1}{2l}\ln(r_1 r_2). \quad (1.2.17)$$

The distributed loss parameter α for homojunction semiconductor lasers can be substantial due to the fact that a substantial fraction of the laser light may travel in the regions of the laser where the loss may be high. For other lasers, it is often relatively small. The size of the laser medium, the minimum gain, and mirror reflectivities required for laser action are therefore all interdependent through this threshold condition.

1.2.3.2. Saturation Characteristics. Above the threshold, the gain in the system is greater than the loss. Any small signal or fluctuation will grow continually until the nonlinear saturation effect in the medium sets in to limit the amplitude of the oscillation. The output power, the spectral characteristics of the laser emission, and the stability of the laser oscillator all depend upon the nature of the saturation effect. The satura-

1.2. THE LASER

tion characteristics depend in turn upon the broadening mechanism for the fluorescence line—homogeneous or inhomogeneous broadening.

1.2.3.2.1. HOMOGENEOUS BROADENING. Laser transitions in semiconductors, dye molecules, color centers in solids, and rare-earth or transition-metal ion-doped solids at room temperature are typically predominantly homogeneously broadened. For homogeneously broadened lines, all the active atoms see essentially the same environment and have the same velocity. The fluorescence or absorption spectrum of any subgroup of active atoms or any macroscopic sample of the medium is in this case the same and typically Lorentzian:

$$g_f(\nu) = \frac{1}{\pi} \frac{\Delta\nu}{(\nu - \nu_{21})^2 + \Delta\nu^2} \qquad (1.2.18)$$

for a weak field of ν, with the half-width $\Delta\nu$ determined by the lifetimes of the upper and lower levels and the atomic coherence time. The finite lifetime could be due to the emission of photons or phonons or due to collisions in the cases of gases. Numerically, $\Delta\nu$ could vary from the order of 10^8 Hz, in the case of strongly allowed transitions in free atoms or ions, to over 10^{11} Hz for the red fluorescence lines of ruby at room temperature due to the two-phonon processes in solids.

In the case of absorption or stimulated emission of a very strong saturating field at ν_0, the normalized rate of absorption or stimulated emission $(1/I)\, dI/dt$ is reduced and no longer independent of I because the population difference is reduced due to the radiative transition process itself. It can be shown on the basis of the appropriate rate equations that the population of, for example, a nondegenerate two-level system under the influence of a strong field at ν_0 is

$$N_2 - N_1 = \frac{\overline{N}_2 - \overline{N}_1}{1 + I/I_s}, \qquad (1.2.19)$$

where $\overline{N}_2$ and $\overline{N}_1$ are the initial populations in the absence of the field, and I_s is a saturation parameter defined as

$$I_s = \frac{3\hbar^2[(\nu_{21} - \nu_0)^2 + \Delta\nu^2]c}{4p_{21}^2 \tau \Delta\nu} \qquad (1.2.20)$$

for a homogeneously broadened line with a corresponding transition dipole matrix element p_{21}. Note that the saturation parameter or the saturated population difference depends upon the frequency ν_0 relative to the transition frequency ν_{21}. Thus, the absorption line *shape* or the gain *profile* seen by this strong wave is different from that in the weak-field limit and depends upon the intensity of the wave as follows:

$$g_f(\nu_0, I) = \frac{1}{\pi} \frac{\Delta\nu}{(\nu_0 - \nu_{21})^2 + \Delta\nu^2 + (4p_{21}^2 I \Delta\nu\tau/3\hbar^2 c)}$$

$$= g_f(\nu_0, I \to 0) \frac{1}{1 + I/I_s}. \tag{1.2.21}$$

Note that, $g_f(\nu_0, I)$ as defined in (1.2.21) is not normalized to 1 when ν_0 is integrated from $-\infty$ to ∞.

In a laser above the threshold for oscillation, as the intensity grows, the gain of the medium changes from

$$g = (\overline{N}_2 - \overline{N}_1)\sigma_{st}$$

for low intensities near the threshold to the saturated value

$$g_{sat} = \frac{(\overline{N}_2 - \overline{N}_1)\sigma_{st}}{1 + I/I_s}. \tag{1.2.22}$$

The intensity will continue to grow until the saturated gain is reduced to the gain just enough to compensate for the losses or the gain at the threshold for oscillation. The actual calculation of the laser output intensity taking into account the saturation effect is, however, a rather complicated problem.

For a homogeneously broadened line, the line shape seen by a weak wave at ν in the presence of a strong wave at ν_0 remains Lorentzian [Eq. (1.2.18)] but the area under the line is reduced by the factor $(1 + I/I_s)$ due to the saturation effect. This fact has important implications in the spectral characteristics of the laser.

1.2.3.2.2. INHOMOGENEOUS BROADENING. Glass lasers or gas lasers at low pressures typically have inhomogeneously broadened laser transitions. In an inhomogeneously broadened line, the fluorescence or absorption spectra of various homogeneously broadened subgroups of active atoms are spectrally shifted from one another and the experimentally observed line profile results from the superposition of these homogeneously broadened lines. The frequency shifts are either due to differences in the environments of the subgroups of atoms or because they belong to different velocity classes. For example, the emission lines of gases at low pressures are usually inhomogeneously broadened due to the Doppler effect and those of rare earth ion-doped glasses are inhomogeneously broadened due to the differences in the crystal fields at different sites in the glass matrix. For example, the 1.06-μm line of Nd^{3+} ions has a predominantly inhomogeneous width of over 100 Å in glasses but a predominantly homogeneous width of a few angstroms in crystalline solids.

The actual line shape in each case will depend upon the convolution of the distribution function of the transition frequencies and the shape of the homogeneously broadened line shape of the subgroups of atoms. For weak fields, inhomogeneously broadened lines tend to be Gaussian:

$$g(\nu) = \frac{1}{\Delta\nu}\left(\frac{\ln 2}{\pi}\right)^{1/2} \exp[-(\ln 2)(\nu - \nu_{21})^2/\Delta\nu^2], \quad (1.2.23)$$

where $\Delta\nu$ is the half-width at half-maximum.

For a strong field at ν_0, the line profile of an inhomogeneously broadened line can be obtained from the convolution of the distribution function for the transition frequency with the saturated homogeneously broadened line shape function for the subgroups of atoms. The resulting absorption or stimulated emission line profile seen by the strong wave at ν_0 is

$$g_f(\nu_0, I) = g_f(\nu_0, I \to 0)(1 + I/I_s)^{-1/2} \quad (1.2.24)$$

for a predominantly inhomogeneously broadened Gaussian line, where

$$I_s = 3\hbar^2 c\, \Delta\nu/4p_{21}^2\tau. \quad (1.2.25)$$

Note that the dependence on the light intensity I is weaker in the case of an inhomogeneously rather than homogeneously broadened line. Also, since the saturation parameter [Eq. (1.2.25)] in this case does not depend upon the frequency of the strong wave, the line shape seen by the strong wave does not change with the intensity of the wave; this is very different from the homogeneously broadened case.

The line shape seen by a weak wave at ν is significantly influenced by the presence of a strong wave at ν_0, however. In the presence of a strong wave traveling in the same direction, the subgroups of atoms with transition frequencies closer to ν_0 will have larger transition rates leading to smaller population differences. Thus, the gain profile seen by a weak probing wave at ν will show a dip at ν_0 with a width corresponding roughly to the homogeneous width as shown schematically in Fig. 4. The strong wave is said to be able to "burn a hole"[12] at ν_0 in the line profile for the weak wave. On the other hand, if the strong wave is tuned, it will burn a hole everywhere by the same proportion; thus, the line shape seen by the strong wave itself is not altered in the inhomogeneous case.

---

[12] W. R. Bennett, Jr., *Phys. Rev.* **126**, 580 (1962); W. E. Lamb, Jr., *Phys. Rev.* **A134**, 1429 (1964).

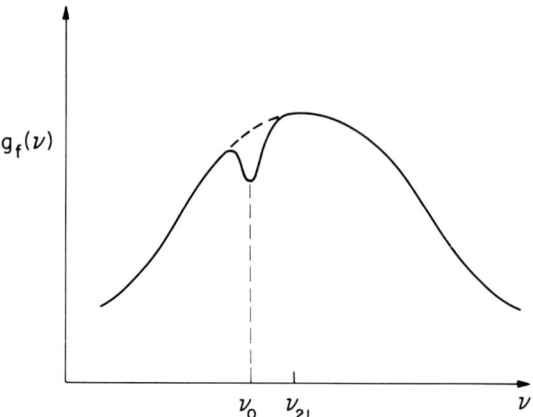

FIG. 4. Line profile of an inhomogeneously broadened line seen by a weak probing wave at ν in the presence of a strong saturating wave at ν_0 traveling in one direction.

These different saturation characteristics will have a profound influence on the spectral characteristics of the laser emission.

1.2.3.3. *Spectral Characteristics.* The total spectral width of the laser output depends on how many of the longitudinal modes of the laser can oscillate. After the mode with the lowest threshold has started to oscillate, whether another mode can go into oscillation or not depends upon the gain profile seen by the weak signal or noise at the other mode frequencies in the presence of the strong oscillating mode. From the discussion on the saturation characteristics of homogeneous and inhomogeneously broadened lines, it is clear that one would expect an inhomogeneously broadened laser to be able to oscillate in several modes with a resulting distorted gain profile and spectrum as shown schematically in Fig. 5, if the oscillating longitudinal modes are at least one homogeneous width apart. The intensity of each mode is such that the hole burned into the gain profile dips to the threshold value required for oscillation in that mode as given by Eq. (1.2.17). Without special efforts, such as the use of additional Fabry–Perot etalons inside the laser cavity, inhomogeneously broadened lasers are generally multimode lasers with a total spectral width not much less than that of the fluorescence line.

Crystalline solid-state lasers, including semiconductor and impurity ion-doped insulating crystalline solid lasers, and liquid lasers such as the dye lasers are typically homogeneously broadened at room temperature. Once one mode in such a laser has started to oscillate and the medium becomes illuminated by a strong wave, the gain profile seen by weak signals or noise at other mode frequencies becomes reduced in magnitude but has

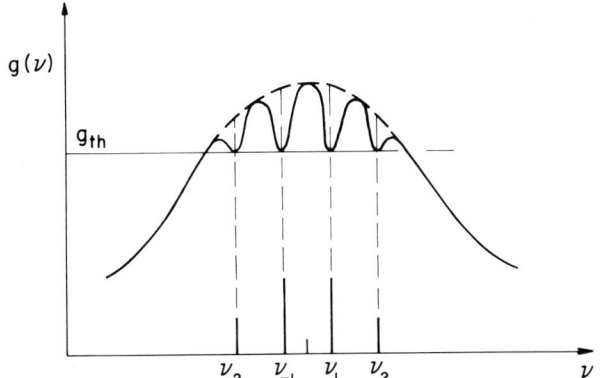

FIG. 5. Gain profile (solid curve) of an inhomogeneously broadened laser medium seen by a weak probing wave at ν in the presence of oscillating laser modes at $\nu_{\pm 1}$ and $\nu_{\pm 2}$. The holes burned into the gain profile dip down to the threshold level given by Eq. (1.2.17). The dashed curve shows the unsaturated gain profile.

the same shape as shown schematically in Fig. 6. Thus, in principle such a laser should oscillate in only one mode. But in reality even these lasers tend to oscillate in many modes. This has to do with the so-called spatial hole burning effect[13]: Different modes of the Fabry–Perot cavity have different spatial distributions. For cavities with highly reflecting mirrors,

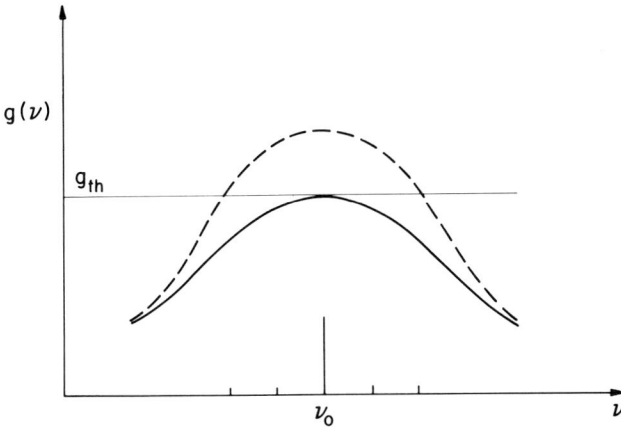

FIG. 6. Gain profile (solid curve) of a homogeneously broadened laser line seen by a weak probing wave at ν in the presence of an oscillating laser mode at ν_0. Only one mode can oscillate at a time if the spatial hole-burning effect is neglected. The dashed curve gives the unsaturated gain profile.

[13] C. L. Tang, H. Statz, and G. deMars, *J. Appl. Phys.* **34**, 2289 (1963); *Phys. Rev.* **128**, 1013 (1962).

the cavity modes are primarily standing waves. Near the nodal planes of each mode the atoms are not saturated by the oscillating mode and can therefore support oscillation in another mode, which would have nodal planes elsewhere. Thus, to obtain single-mode oscillation in all lasers is in general difficult and requires special techniques such as using additional intracavity etalons or, in the case of homogeneously broadened lines, avoiding standing waves by using cavities that can support traveling-wave resonant modes.[13] Examples of some techniques used with different lasers are discussed in the following chapters.

1.2.3.4. Laser Rate Equations and Stability Considerations. The question of the stability and transient behavior of the laser oscillator above the threshold is an important one, since some of the well-known lasers, such as the rare-earth ion or transition-metal ion-doped solid-state lasers, almost never have a stable output but have an output consisting mainly of random sharp spikes. Others such as most dye, semiconductor, and gas lasers have stable outputs above the threshold.

The basic reason for the instabilities in lasers is that energy can be stored in a laser either in the form of electromagnetic energy in the cavity or in the population inversion in the laser medium. It consists, therefore, of two coupled dynamic systems, and energy can oscillate back and forth between the two. Depending upon the coupling constant between the parts and the damping constant of each, the system may or may not be stable, resulting in very different dynamic behaviors of the lasers. In general, if the cavity lifetime is long compared to the lifetime of the population inversion, the laser tends to be stable and shows a smooth output. If the lifetime of the population inversion is long compared to the cavity lifetime, the laser tends to be less stable and show spiking. This can be demonstrated on the basis of the laser rate equations, which describe the dynamic or transient behavior of the laser above the threshold.

For a nondegenerate two-level system, the rate equations including stimulated emission for the population inversion n and the electromagnetic energy ϵ in the cavity are

$$\frac{dn}{dt} = -2Bg_f n\epsilon - n\left(p + \frac{1}{\tau}\right) + N_0\left(p - \frac{1}{\tau}\right), \qquad (1.2.26)$$

$$\frac{d\epsilon}{dt} = -\frac{\omega_0}{Q}\epsilon + B\hbar\omega_0 g_f n\epsilon, \qquad (1.2.27)$$

neglecting the spontaneous emission into the cavity mode, where B is the Einstein B coefficient, which is related to the corresponding Einstein A coefficient by

$$B = Ac^3/8\pi h\nu^3. \qquad (1.2.28)$$

N_0 in Eq. (1.2.26) is the total volume density of the atoms; the rate of pumping from the lower-level is assumed to be $pN_1 = \frac{1}{2}p(N_0 - n)$. A more complete description will have to be based upon the full density matrix equations for the medium and the wave equation for the field to be given in Chapter 1.3. However, when the time variations of all the relevant quantities are slow compared to the atomic coherence time or the "transverse" relaxation time T_2, these two coupled equations, which are sometimes also known in laser literature as the Statz–deMars equations,[14] for the population difference and the intensity of the field are sufficient to describe the dynamic characteristics of the laser above the threshold.

Because of the stimulated emission terms, Eqs. (1.2.26) and (1.2.27) are nonlinear and in general cannot be solved for the general time dependences of the population difference and the laser intensity for arbitrary initial conditions. The stability of the oscillator against small perturbations around the steady-state solution can be analyzed by linearizing the nonlinear equation around this point, however. It can thus be shown that a small perturbation around the steady state will have an exponential time dependence $\exp(\beta t)$, where

$$\beta \cong -\frac{2\eta}{\tau} \pm \left[\left(\frac{\eta}{\tau}\right)^2 - \frac{4\omega_0(\eta - 1)}{Q\tau}\right]^{1/2}, \qquad (1.2.29)$$

$$\eta = \tfrac{1}{2}N_0(p\tau - 1)B\hbar g_f Q, \qquad (1.2.30)$$

if the laser is not too far above the threshold or $p\tau \approx 1$; η also has the meaning of being a pumping parameter in the following sense:

$$\eta = \frac{(\text{pump power}) - (\text{pump power required for inversion})}{(\text{pump power at threshold for oscillation}) - (\text{pump power required for inversion})} \qquad (1.2.31)$$

Numerically, η is typically on the order of 1 to 10. For convenience let us say that it is 2; in which case, Eq. (1.2.29) shows that whether any small deviation from the steady state will damp back to the steady-state point in an oscillatory fashion or not will depend upon the relative magnitudes of the cavity lifetime Q/ω_0 and the lifetime of the population inversion τ. If

$$\omega_0/Q \gg 1/\tau,$$

Eq. (1.2.29) shows that such a perturbation will lead to weakly damped strong oscillatory variation in the laser intensity. Numerically solutions of the full nonlinear equations[14] (1.2.26) and (1.2.27) show that under such conditions, large deviations from the steady state could indeed lead to

---

[14] H. Statz and G. deMars, in "Quantum Electronics" (C. H. Townes, ed.). Columbia Univ. Press, New York, 1960.

sharp spikes in the laser output as often observed in solid-state lasers. It must be borne in mind, however, that even the full nonlinear equations (1.2.26) and (1.2.27) are still a relatively crude approximation of what really takes place in the laser far above the threshold. Indeed, although these equations predict that any deviations from the steady state will eventually always damp back to the steady-state solution, experimentally there are many cases where the sharp spikes in the laser output persist indefinitely. The precise reason for this is not yet clear.

If the lifetime of the population inversion is short compared to the cavity lifetime, Eq. (1.2.29) shows that any small deviation from the steady state will simply be smoothly damped, and the oscillator is stable. This is apparently the case for most gaseous and dye lasers. Semiconductor lasers are a borderline case where it can be either way, depending upon the actual numbers involved for each particular type of laser.

1.3. Density Matrix

1.3.1. Definition and Properties

For coherent optical effects and short-pulse phenomena, it is necessary to introduce the concept of density matrix and the corresponding density matrix equation.[9]

If the atoms in a macroscopic sample are all in one energy level or another, then the state of the sample can be specified in terms of the population distribution over these levels. If, however, the atoms are not all in such stationary states but some are in nonstationary mixed states in the sense that the wavefunctions are linear combinations of the eigenstates of the Hamiltonian of the atom, then the state of the macroscopic sample also depends upon the relative phases of the expansion coefficients of the wavefunctions, or the "atomic coherence" must also be specified in order to specify the state of the macroscopic sample. With the large number of atoms expected in a macroscopic sample, it is impossible to specify all the relative phases and the numbers of atoms in each level. On the other hand, because the number is large, it is possible to use a statistical description and specify only the average number of atoms in each state and the average of the relative phases for pairs of levels. The array of numbers characterizing such averages form the so-called density matrix. We now make these points more explicit.

If the atom is known to be in the state represented by the state function $|\Psi\rangle$, then the expectation value of any observable represented by the dynamic variable $\hat{A}$ is

$$\langle \hat{A} \rangle = \langle \Psi | \hat{A} | \Psi \rangle. \qquad (1.3.1)$$

1.3. DENSITY MATRIX

For a uniform sample containing a large number of atoms, it is not possible to know precisely what state each atom is in. At best, one may know how the atoms are distributed statistically over all possible states $|\Psi\rangle$. Suppose this probability distribution function is P_Ψ. The expectation value of the same observable averaged over the sample is then

$$\overline{\langle \hat{A} \rangle} = \sum_{|\Psi\rangle} P_\Psi \langle \Psi | \hat{A} | \Psi \rangle. \qquad (1.3.2)$$

With respect to some complete set of orthonormal basis states $|m\rangle$, this averaged expectation value can be written as

$$\begin{aligned} \overline{\langle \hat{A} \rangle} &= \sum_{|\Psi\rangle} \sum_{m,n} P_\Psi \langle \Psi | m \rangle \langle m | \hat{A} | n \rangle \langle n | \Psi \rangle \\ &= \sum_{m,n} \langle n | \left[\sum_{|\Psi\rangle} |\Psi\rangle P_\Psi \langle \Psi| \right] |m\rangle \langle m | \hat{A} | n \rangle \qquad (1.3.3) \\ &= \mathrm{Tr} \left[\left\{ \sum_{|\Psi\rangle} |\Psi\rangle P_\Psi \langle \Psi| \right\} \hat{A} \right] \end{aligned}$$

A density operator $\hat{\rho}$ can then be defined as

$$\hat{\rho} \equiv \sum_{|\Psi\rangle} |\Psi\rangle P_\Psi \langle \Psi| \qquad (1.3.4)$$

and the corresponding density matrix element is

$$\langle n | \hat{\rho} | m \rangle \equiv \sum_{|\Psi\rangle} \langle n | \Psi \rangle P_\Psi \langle \Psi | m \rangle. \qquad (1.3.5)$$

The averaged expectation value per atom of any dynamic variable $\hat{A}$ is therefore

$$\overline{\langle \hat{A} \rangle} = \mathrm{Tr}[\hat{\rho}\hat{A}]. \qquad (1.3.6)$$

Expanding the state function with respect to the same set of basis functions

$$|\Psi\rangle = \sum_m \langle m | \Psi \rangle | m \rangle = \sum_m C_m | m \rangle, \qquad (1.3.7)$$

it is clear that the density matrix element ρ_{nm} also has the meaning that it is the average value of the product of the expansion coefficients:

$$\rho_{nm} = \sum_{|\Psi\rangle} P_\Psi C_m^* C_n = \overline{C_m^* C_n}, \qquad (1.3.8)$$

which depends upon the relative phases of the wavefunctions.

The diagonal elements give the volume density N_n of atoms in each state $|n\rangle$:

$$N_n = N \overline{|C_n|^2} = N \rho_{nn}, \qquad (1.3.9)$$

where N is the volume density of the atoms. With the eigenstates of the Hamiltonian as the basis states, the diagonal elements of the density matrix give the population distribution over the energy levels.

The off-diagonal elements ρ_{nm} depend upon the relative phases of the expansion coefficients C_m and C_n; therefore, the average of the product $C_m^* C_n$ is a measure of the "coherence" of the wavefunctions of the atoms in the sample. Another way of seeing this is the following. Consider an ensemble of two-level atoms. The induced macroscopic polarization per volume is, according to Eq. (1.3.6),

$$\mathbf{P} = N[\rho_{12}\langle 2|\mathbf{p}|1\rangle + \rho_{21}\langle 1|\mathbf{p}|2\rangle] \qquad (1.3.10)$$

assuming the atoms have no permanent dipole moments. Since the macroscopic dipole moment is the sum of the induced dipole moments of the individual atoms, it clearly depends upon the phase coherence of these constituent dipoles. Equation (1.3.10) shows that the macroscopic dipole moment is proportional to the off-diagonal elements of the density matrix ρ_{mn}. These off-diagonal elements therefore characterize the phase coherence of the induced dipole moments of the individual atoms or the phase coherence of the atomic wavefunctions.

With the above interpretation of the meanings of the diagonal and off-diagonal elements of the density matrix, it follows that the density matrix of a sample at, for example, thermal equilibrium at a temperature T should be

$$\rho_{mn}^{(\text{th.eq.})} = \delta_{mn} \exp(-E_m/kT) \Big/ \sum_{n'} \exp(-E_{n'}/kT) \qquad (1.3.11a)$$

or

$$\hat{\rho}^{(\text{th.eq.})} = \exp(-\hat{H}/kT)/\text{Tr}[\exp(-\hat{H}/kT)], \qquad (1.3.11b)$$

because the population distribution over the energy levels follows the Boltzmann distribution and the relative phases must be completely random leading to zero off-diagonal elements.

It also follows from Eq. (1.3.5) or (1.3.8) that the density matrix must be Hermitian,

$$\hat{\rho} = \hat{\rho}^+, \qquad (1.3.12)$$

and from Eq. (1.3.9) its trace must be equal to 1:

$$\text{Tr}\,\hat{\rho} = 1. \qquad (1.3.13)$$

1.3.2. Density Matrix Equation

The time dependence of the density matrix is governed by the density matrix equation, which is the quantum-mechanical analog of the classical

1.3. DENSITY MATRIX

Boltzmann equation. Assume first that there is no relaxation process and the atoms are independent of one another but experience exactly the same forces. In this case, the probability distribution function for a given density matrix $P_{\Psi(t)}$ will not depend explicitly upon time; that is, the fraction of atoms associated with a given state $|\Psi(t)\rangle$ remains the same even though the state function itself may be changing with time according to Schrödinger's equation. This is so since the time evolution of the state of an atom is uniquely determined according to Schrödinger's equation and the initial conditions. As long as there is no relaxation process and all the atoms in the sample see the same forces, if two atoms are initially in the same state, they will remain in the same state; if they are initially in different states, they will always be in different states. Thus, $P_{\Psi(t)}$ associated with each state $|\Psi(t)\rangle$ can neither increase nor decrease with time; in other words, it does not depend explicitly upon time. It therefore follows from the definition of $\hat{\rho}$ and Schrödinger's equation that the time rate of change of the density operator is

$$\frac{\partial \hat{\rho}}{\partial t} = \sum_{\Psi} \left[\left(\frac{\partial |\Psi\rangle}{\partial t}\right) P_{\Psi} \langle \Psi| + |\Psi\rangle \left(\frac{\partial P_{\Psi}}{\partial t}\right)^0 \langle \Psi| + |\Psi\rangle P_{\Psi} \frac{\partial \langle \Psi|}{\partial t} \right]$$

$$= \frac{i}{\hbar} [\hat{\rho}, \hat{H}], \qquad (1.3.14)$$

where $\hat{H}$ is the Hamiltonian of the atoms. It contains only those terms corresponding to the forces that are common to all the atoms in the sample but excludes specifically all the forces that vary from atom to atom and fluctuate in time.

On the other hand, if there are relaxation processes present and different atoms experience different forces due to the relaxation mechanisms, then atoms in different states at a given time may end up in the same state at a later time and vice versa. As a result, the probability distribution function $P_{\Psi(t)}$ depends explicitly upon time, or the number of atoms in the sample associated with each state $|\Psi(t)\rangle$ changes with time as the state function $|\Psi(t)\rangle$ evolves with time according to Schrödinger's equation. Thus, the time rate of change of $\hat{\rho}$ has an additional term due to the fact that $\partial P_{\Psi}/\partial t \neq 0$:

$$\frac{\partial \hat{\rho}}{\partial t} = \frac{i}{\hbar} [\hat{\rho}, \hat{H}] + \left(\frac{\partial \hat{\rho}}{\partial t}\right)_{\text{relax}} \qquad (1.3.15)$$

Since this additional term is due to the random forces acting on the atoms associated with the relaxation processes present, it is designated as $(\partial \hat{\rho}/\partial t)_{\text{relax}}$ in Eq. (1.3.15). This relaxation term can only be given approximately. If the sample is not too far from thermal equilibrium, it is not unreasonable to assume that the elements of the density matrix relax exponentially back to the equilibrium values $\bar{\rho}_{mn}$, or

$$\frac{\partial \rho_{mn}}{\partial t} = \frac{i}{\hbar} \sum_l (\rho_{ml} H_{ln} - H_{ml} \rho_{ln}) - \frac{\rho_{mn} - \bar{\rho}_{mn}}{T_{mn}}. \quad (1.3.16)$$

If there is an external pumping mechanism that maintains the population inversion, $\bar{\rho}_{mn}$ is the equilibrium density matrix in the presence of this pumping mechanism, not the thermal equilibrium density matrix in the absence of pumping.

Note that the relaxation times T_{mn} for different density matrix elements may be different. The relaxation time T_{mn} for the diagonal elements are sometimes called the longitudinal relaxation time, a terminology borrowed from magnetic resonance work since the diagonal elements describe the longitudinal component of the magnetic moment of a macroscopic sample in that case. The relaxation times for the off-diagonal elements are called the transverse relaxation times or the atomic coherence times.

The dynamics of the medium are now predictable from some initial state using the density matrix equations once a convenient set of basis functions is chosen. The choice is often dictated by the nature of the problem.

Suppose there is no externally applied perturbation on the sample first and $\hat{H} = \hat{H}_0$ is the Hamiltonian of the atoms containing only the force terms that are common to all atoms. Assuming one can solve the eigenvalue equation

$$\hat{H}_0 |m\rangle = E_m |m\rangle \quad (1.3.17)$$

and using the eigenstates of $\hat{H}_0$ as the basis states, the density matrix equation can be solved immediately:

$$\rho_{mn}(t) = \bar{\rho}_{mn} + [\rho_{mn}(0) - \bar{\rho}_{mn}] \exp\left[-\left(i\omega_{mn} + \frac{1}{T_{mn}}\right) t\right]. \quad (1.3.18)$$

Suppose now the medium is subject to a perturbation $\hat{V}$:

$$\hat{H} = \hat{H}_0 + \hat{V}. \quad (1.3.19)$$

With respect to the same basis functions, the eigenfunctions of $\hat{H}_0$, the corresponding density matrix equation becomes

$$\frac{\partial \rho_{mn}}{\partial t} + i\omega_{mn} \rho_{mn} + \frac{\rho_{mn} - \bar{\rho}_{mn}}{T_{mn}} = \frac{i}{\hbar} \sum_l [\rho_{ml} V_{ln} - V_{ml} \rho_{ln}]. \quad (1.3.20)$$

This is the general equation describing the linear and nonlinear and the steady-state and dynamic behaviors of the sample under the influence of an external perturbation taking into account the effects of all relevant relaxation processes. As such it is fundamental to almost all laser-related phenomena and devices.

As an example of its use, let us consider the important case of near-resonant electric-dipole interaction of a linearly polarized monochromatic wave with a sample of nondegenerate two-level atoms. Equation (1.3.20), in this case for m and n each equal to 1 and 2, leads to

$$\frac{\partial}{\partial t}(\rho_{22} - \rho_{11}) + \frac{1}{T_1}[(\rho_{22} - \rho_{11}) - (\bar{\rho}_{22} - \bar{\rho}_{11})]$$

$$= -\frac{i}{\hbar}[\rho_{21}p_{z12}Ee^{+i\omega_0 t} - \rho_{12}p_{z21}Ee^{-i\omega_0 t}], \quad (1.3.21)$$

$$\frac{\partial}{\partial t}\rho_{21} + i\omega_{21}\rho_{21} + \frac{1}{T_2}\rho_{21} = -\frac{i}{2\hbar}(\rho_{22} - \rho_{11})p_{z21}Ee^{-i\omega_0 t}, \quad (1.3.22)$$

for $\omega_{21} \approx \omega_0$, so that the $e^{i\omega_0 t}$ term in the equation for the off-diagonal element and the $2\omega_0 t$ terms in the diagonal equation can be neglected. The important point here is that the dynamics of the medium must be described by one complex equation for the population difference and another for the off-diagonal element or the atomic coherence. If, however, the relaxation rate T_2^{-1} is large compared to the rates of change of $\rho_{21} = \rho_{12}^*$ and $\rho_{22} - \rho_{11}$, one can neglect the transient solution of Eq. (1.3.22); the off-diagonal element in this case is simply proportional to the population difference:

$$\rho_{21} = -\frac{ip_{z21}}{2\hbar}\frac{(\rho_{22} - \rho_{11})}{i(\omega_{21} - \omega_0) + (1/T_2)}Ee^{-i\omega_0 t} \quad (1.3.23)$$

There is then only one independent rate equation to describe the dynamics of the medium:

$$\frac{\partial}{\partial t}(\rho_{22} - \rho_{11}) + \frac{1}{T_1}[(\rho_{22} - \rho_{11}) - (\bar{\rho}_{22} - \bar{\rho}_{11})]$$

$$= -2B(\rho_{22} - \rho_{11})\epsilon g_f(\nu), \quad (1.3.24)$$

where

$$B = 2\pi|p_{z21}|^2/\hbar^2$$

is the Einstein B coefficient,

$$g_f(\nu_0) = \frac{1}{\pi}\frac{(1/2\pi T_2)}{(\nu_{21} - \nu_0)^2 + (1/2\pi T_2)^2}$$

is the fluorescence line shape function, and $\epsilon = |E|^2/8\pi$ is the electromagnetic energy density. Equation (1.3.24) is the same equation as that

[Equation (1.2.26)] in the rate equations approach, or the Statz–deMars equation approach, to describe the dynamics and the stability of the laser oscillator described in Chapter 1.2. It is now clear under what conditions the dynamics of the medium can be described by a single rate equation for the population difference.

If the coherence time T_2 is not short, which is often the case when picosecond pulses are involved, the time dependences of the population difference and the atomic coherence will in general be quite different; in this case, both density matrix equations (1.3.21) and (1.3.22) must be used to describe the dynamics of the medium and its interaction with the incident light pulse. This is the situation with mode-locked lasers and picosecond spectroscopy to be described below.

Finally, the steady-state solution of the density matrix equation can be obtained using the standard perturbation technique of expanding the matrix elements in successive orders of V:

$$\rho_{mn} = \rho_{mn}^{(0)} + \rho_{mn}^{(1)} + \rho_{mn}^{(2)} + \cdots, \qquad (1.3.25)$$

giving the result

$$\rho_{mn}^{(n)} = -\frac{i}{\hbar} \sum_{l} \int_{-\infty}^{t} (V_{ml}\rho_{ln}^{(n-1)} - \rho_{ml}^{(n-1)}V_{ln})$$

$$\exp\left(i\omega_{mn} + \frac{1}{T_{mn}}\right)(t' - t)\, dt' \qquad (1.3.26)$$

where $\rho_{mn}^{(0)} = \bar{\rho}_{mn}$. The nth-order density matrix element gives the nth-order macroscopic polarization $\mathbf{p}^{(n)}$ of the medium.

$$\mathbf{p}^{(n)} = N \sum_{l,m} [\rho_{ml}^{(n)}\mathbf{p}_{lm} + \mathbf{p}_{ml}\rho_{lm}^{(n)}].$$

Within the limits of the general applicability of the density matrix formalism and perturbation theory, this general result can be used for a very wide variety of optical problems in both the linear and nonlinear regimes (see, for example, Bloembergen[4]).

1.4. Special Techniques

It is often possible to modify the laser output characteristics rather drastically by simple modifications of the resonator characteristics. Two well-known examples are the Q-switching and mode-locking techniques. These are used to obtain from a given laser significantly increased intensity and shorter pulse duration than normally possible. In this chapter, we discuss briefly these two widely used experimental techniques. Ex-

1.4. SPECIAL TECHNIQUES

amples of others are discussed in connection with various types of lasers in the following chapters.

1.4.1. Q-Switching[15]

Q-switching is an important technique for obtaining short (on the order of nanoseconds) and intense pulses from some types of lasers, mainly those with a relatively long (much longer than nanoseconds) lifetime for the population inversion. The basic idea is the following. When the laser pumping mechanism is first turned on, once the population inversion being created reaches the threshold for oscillation [Eq. (1.2.17)] stimulated emission will immediately build up. Further pumping will lead to more stimulated emission lasting the duration of the pump pulse, but the population inversion will be locked to the threshold value once the transients die out. If the laser can be prevented from oscillating until a large population inversion is first created, then once the laser is allowed to oscillate, stimulated emission will build up much more rapidly leading to a more intense but shorter pulse for the same available pump pulse energy. This can be achieved by first making the Q of the resonator very low so that the laser will not reach oscillation. When a large population inversion is created, in a time shorter than the lifetime of the population inversion, the Q is suddenly switched to a high value. At this point, because the population inversion is much greater than the threshold value, stimulated emission will build up much more rapidly, leading to a more intense pulse than if the Q were initially set at the same high value. By holding off the laser oscillation, the energy stored in the form of population inversion will be dumped in a shorter period of time leading to a shorter but more intense laser pulse.

In practice, this can be done by using, for example, an electro-optic shutter (Kerr or Pockels cell), a rotating mirror, or retroreflecting prism (see Fig. 7) such that a low-loss Fabry–Perot cavity is formed only after some delay from the onset of the pumping of the laser medium. Another frequently used method is to insert inside the laser cavity a saturable ab-

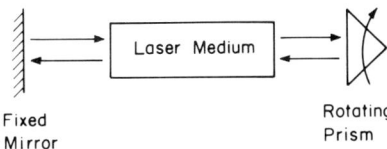

FIG. 7. Schematic of a Q-switched laser using a rotating prism.

---

[15] R. W. Hellwarth, *in* "Advances in Quantum Electronics" (J. R. Singer, ed.). Columbia Univ. Press, New York, 1961.

sorber, which has a high absorption coefficient at low light intensity levels but becomes essentially transparent at high intensity levels. Thus, initially the cavity loss is high until the intracavity light intensity reaches a certain level, at which point the saturable absorber starts to become more transparent. As this happens, the light intensity increases faster due to stimulated emission, which leads to further reduction of the intracavity loss. This is a highly nonlinear process leading to rapid switching of the cavity Q. In all these cases, the cavity Q must be switched in a time much shorter than the lifetime of the population inversion, so that after the Q-switching the population inversion is still much greater than the threshold value. Otherwise, the population inversion will adjust continuously during the switching process and there will be no excess population inversion afterwards. This is the reason why Q-switching can only be employed in systems with a long lifetime, such as rare-earth ion or transition-metal ion-doped solid-state lasers. For most gaseous and dye lasers, the lifetime is on the order of a few nanoseconds; it would be difficult to switch the cavity Q that rapidly.

The Q-switching effect can be analyzed readily using the laser equations outlined in Chapter 1.2.3.4. Experimentally, with Q-switching, one can routinely change the output of, for example, a ruby laser from a millisecond pulse consisting of randomly spaced microsecond spikes to a single pulse of 5 to 10 nsec long of approximately the same total energy content. This means an increase of four to five orders of magnitude in peak intensity.

1.4.2. Mode-Locking[16]

As pointed out earlier, most lasers tend to oscillate in many longitudinal modes. The phases of these modes are in general independent of each other. If the longitudinal modes can be forced to be exactly equally spaced and have the same phase at some instant of time, then the amplitudes of all the oscillating modes will add due to constructive interference giving rise to a high peak intensity at that instant of time. After a time interval equal to the inverse of the mode spacing times the number of oscillating modes the intensity will quickly go down due to destructive interference of the modes. Such peaking of the laser intensity will repeat with a period equal to the inverse of the mode spacing $C/2L$, which is also the round trip time of the light pulse inside the laser cavity. To achieve mode-locking of the laser one can for example introduce into the cavity an optical element that favors repetitive peaking of the laser intensity. This

---

[16] A. J. deMaria, D. H. Stetser, and H. Heynau, *Appl. Phys. Lett.* **8**, 22 (1966); P. W. Smith, *Proc. IEEE* **58**, 1342 (1970).

1.4. SPECIAL TECHNIQUES

can be either an active or passive element, leading to the so-called active or passive mode-locking schemes.

Active mode-locking generally involves the use of an intracavity element whose loss is actively modulated with a period equal to the cavity roundtrip time. This forces the phases of the modes to be locked such that the intensity peaks when the intracavity loss is low. It also tends to force the modes to be precisely equally spaced in spite of possible dispersion in the cavity so that the peaking of the laser intensity will correspond to the periodic loss-modulation. One of the most commonly used methods of loss modulation is the use of an acousto-optic deflector, which periodically deflects the light beam from the laser cavity through Bragg diffraction in the acousto-optic crystal.

In an otherwise ideal system, the pulse duration of the mode-locked laser output is often limited by the bandwidth of the laser. For gas lasers such as the Ar^+ lasers, the pulse duration is typically on the order of 100 psec, limited primarily by the Doppler linewidth of the argon laser line. For dye lasers with a tuner bandwidth of 10 Å or more, subpicosecond pulses can be obtained.

Not only the loss but also the gain of the medium can be actively modulated to achieve mode-locking. For example, mode-locking of an Ar^+ laser pumped dye laser can be achieved by mode-locking the argon laser and making the optical lengths of the two laser cavities equal. This is sometimes called mode-locking by synchronous pumping. Picosecond pulses from dye lasers synchronously pumped with 100-psec pulses from argon lasers are routinely achieved.

Passive mode-locking usually involves the use of some kind of saturable absorber inside the cavity. Saturable absorbers have lower losses at higher laser intensity levels. If the phases of the oscillating modes are otherwise degenerate, with the saturable absorber the phases will tend to adjust themselves to form mode-locked pulses so that the overall loss seen by all the oscillating modes tends to be lower. Another way of visualizing this is that as the laser intensity builds up from noise fluctuations, the highest noise spike in the laser cavity will see the lowest loss in the saturable absorber as it propagates back and forth inside the laser cavity. It will therefore grow the fastest at the expense of the other noise spikes and develop into a mode-locked laser pulse inside the cavity. Every time the intracavity laser pulse reaches the output coupling mirror, a fraction will be transmitted as the laser output. The laser emission therefore consists of a train of pulses separated by the round-trip time of the intracavity pulse.

Short pulses from mode-locked lasers are widely used to observe or generate ultrafast effects in physics, chemistry, and biology. This will be discussed later in the part on picosecond spectroscopy by Ippen and Shank.

Acknowledgment

I would like to thank S. Blit, C. H. Dugan, and D. Fröhlich for their helpful comments on the manuscript.

2. ATOMIC AND IONIC GAS LASERS*

Gaseous lasers employing excited atoms or their ions as the gain medium account for by far the largest number of lasers in operation today. The familiar red helium–neon laser is manufactured at a rate exceeding 50,000 units per year; the more powerful and more costly blue–green argon ion laser accounts for about 1000 units per year. Although these two laser types are the most popular today, many other atomic and ionic lasers have been demonstrated in the past 16 years. This period of intense research and development has yielded laser oscillation on over 1000 wavelengths in 51 elements as atoms or ions and even more in molecules.[1-3] In some cases the levels are excited by collisions with plasma electrons, and the inversion results simply from a favorable situation of upper and lower level lifetimes. In other cases the inversion occurs through selective excitation of the upper level by resonant collisions with another atomic or ionic species. The periodic chart shown in Fig. 1 indicates the elements in which oscillation has been obtained, the form of the active species (atom or ion), and the nature of the collision partner (electron or atom/ion). In many cases different methods of excitation produce quite different laser characteristics in the same element. The variety of combinations possible is evident in the noble gases; this variety arises not as much from inherent advantages in energy level structure but rather from the ease with which noble gases are employed in discharges. The more difficult to vaporize and chemically reactive materials are only now beginning to exhibit a comparable variety of laser wavelengths as techniques such as sputtering are developed to put them in convenient form for gas laser discharges.[4]

[1] C. S. Willett, *in* "Handbook of Lasers" (R. J. Pressley, ed.), Chapter 6. CRC Press, Cleveland, Ohio, 1971; W. B. Bridges and A. N. Chester, *ibid.*, Chapter 7; M. A. Pollack, *ibid.*, Chapter 8.
[2] C. S. Willett, "Introduction to Gas Lasers; Population Inversion Mechanisms," Appendix, Tables 1–85. Pergamon, Oxford, 1974.
[3] R. Beck, W. Englisch, and K. Gürs, "Table of Laser Lines in Gases and Vapors." Springer-Verlag, Berlin and New York, 1976.
[4] L. Csillag, M. Jánossy, K. Rózsa, and T. Salamon, *Phys. Lett.* A **50**, 13 (1974).

* Part 2 is by **William B. Bridges**.

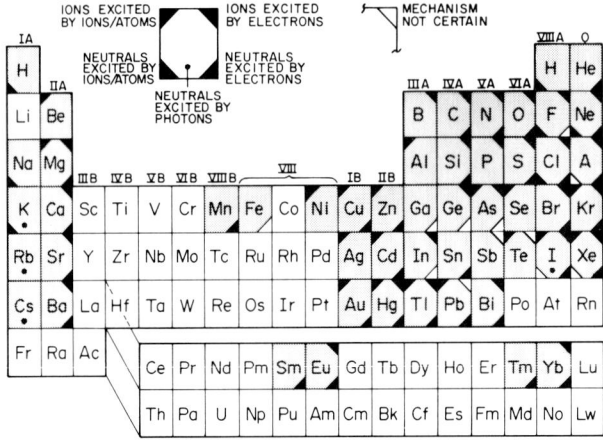

FIG. 1. Periodic table of the elements indicating which elements have exhibited laser oscillation as gaseous atoms or ions. The means of obtaining the population inversion is indicated in general categories.

The simple picture of a system of energy levels in thermal equilibrium results in a distribution of level populations that decreases exponentially with increasing energy according to the Boltzmann factor $\exp(-E/kT)$. In this expression E is the energy of the level with respect to that of the lowest level of the system and T is the absolute temperature of the equilibrium system. By definition, no population inversion can exist in such a system, and hence no laser oscillation. It is tempting to ascribe such thermal equilibrium properties to plasmas, and thus it would seen that "heroic" measures would be necessary to make a gas laser by disturbing the equilibrium sufficiently to obtain a population inversion in a gas discharge. Early attempts to realize a gas discharge laser included such measures as optical pumping with fortuitously resonant radiation in the He–Cs system,[5] binary gas mixtures employing resonant collisional excitation transfer in binary gas mixtures such as helium–neon,[6] mercury–sodium,[7] and mercury–zinc.[8] Some of these means were successful in producing laser oscillation. In the course of research it was soon discovered (often by accident) that heroic measures were *not* required. Inverted populations occur quite naturally in plasmas as simple as the positive column of a low-pressure noble-gas discharge. In retrospect it is al-

[5] P. Rabinowitz, S. Jacobs, and G. Gould, *Appl. Opt.* **1**, 513 (1962).
[6] A. Javan, *Phys. Rev. Lett.* **3**, 86 (1959).
[7] S. G. Rautian and I. I. Sobelman, *Sov. Phys.—Tech. Phys.* (*Engl. Transl.*) **12**, 156 (1960).
[8] V. K. Ablekov, M. S. Pesin, and I. L. Fabelinskii, *Sov. Phys.—JETP* (*Engl. Transl.*) **12**, 618 (1961).

most surprising that the gas laser was not discovered accidentally by prior generations of spectroscopists; certainly all the experimental techniques required were known. It remained only to ask the right question in the right context. In fact, the wide variety of gas lasers and the relative ease with which they can be realized suggests that it is difficult or perhaps impossible to make a gas discharge in which energy level populations *are* described by the simple Boltzmann distribution!

The discussion of gas laser mechanisms and characteristics that follows is organized along the four categories used in Fig. 1: (1) neutral atoms excited by electron collision, (2) ions excited by electron collision, (3) neutral atoms excited by collisions with atoms, and (4) ions excited by collisions with atoms or ions. Within each category the most widely used or best-understood laser (or two) is described in some detail in terms of internal excitation mechanisms and the resulting external characteristics. The more important variations are described by comparison. This organizational scheme is not the only one possible, nor is it necessarily the best in all cases; for example, the cw neutral xenon laser and the pulsed neutral copper vapor laser are both excited by electron collision (category 1), but they differ in their external characteristics far more than the He–Ne (category 3) differs from the He–Cd$^+$ (category 4). In some cases the actual internal mechanisms are not known well enough for definite categorization, and in other cases more than one process may contribute simultaneously. The four categories serve only as a starting point in understanding the details of each laser. The discussions given in the following sections are intended to serve as a brief outline of mechanisms and characteristics and as a guide to the literature. Further details on all gas lasers are found in the excellent book on mechanisms by Willett[9]; reviews of the status of specific laser types are in the discussions of those particular types.

2.1. Neutral Atom Lasers Excited by Electron Collision

Two distinct modes of electron-collision-excited laser oscillation have been demonstrated to date in the energy levels of neutral atoms: (1) an inherently transient oscillation exhibiting very high gain ("superradiant") and high peak output powers, but with a pulse duration substantially less than the upper level radiative lifetime; and (2) continuous oscillations with

---

[9] C. S. Willett, "Introduction to Gas Lasers; Population Inversion Mechanisms." Pergamon, Oxford 1974.

gains ranging from low to high and output powers on the order of milliwatts. These two modes differ greatly in their internal mechanisms even though they may be both observed in the same atom, or even on the same energy levels; for example, the xenon 3.508-μm transition exhibits both modes of operation. The transient lasers, sometimes referred to as *cyclic lasers* or *self-terminating lasers*, are the best understood in their internal mechanisms, and are treated first.

2.1.1. Cyclic Lasers

The specific requirements for an ideal cyclic laser were defined by Walter et al.[10] Their model is illustrated in Fig. 2. An upper laser level (2) is connected to a ground state (0) of opposite parity by a highly allowed transition with an A coefficient of the order of 10^8 sec^{-1}, and to a lower laser level (1), also of opposite parity, by an allowed or partially forbidden transition with a somewhat smaller A coefficient ($10^4 < A < 10^8$ sec^{-1}). Levels 1 and 0 are of the *same* parity so that the transition $1 \rightarrow 0$ is forbidden and occurs only nonradiatively by collisions with the tube walls, electrons, or other atoms. A population inversion between levels 1 and 2 is obtained by applying a short, high-intensity current pulse so that electron collisions quickly excite level 2 via the $0 \rightarrow 2$ transition. The population of level 1 remains low since the excitation rate via the "forbidden" $0 \rightarrow 1$ transition is low for electron collisions. The inversion persists

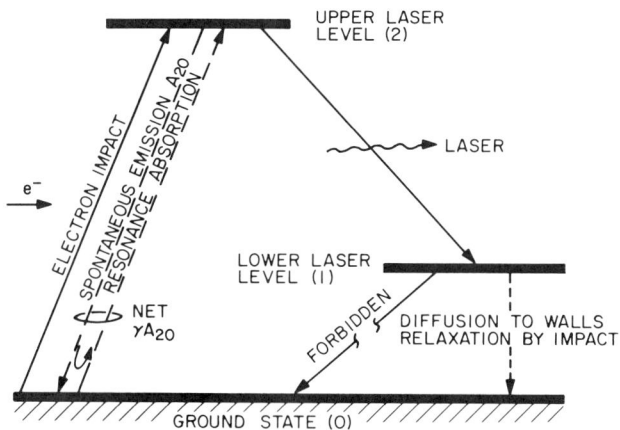

FIG. 2. Energy level model for an ideal cyclic laser as defined by Walter et al.[10]

[10] W. T. Walter, N. Solimene, M. Piltch, and G. Gould, *IEEE J. Quantum Electron.* **qe-2**, 474 (1966).

until the populations of levels 1 and 2 are equalized by spontaneous and stimulated radiation via transitions $2 \rightarrow 1$ or by electron collisional deexcitation. Loss of level 2 population via spontaneous emission back to the ground state is minimized by arranging the atom density and discharge tube diameter so that this radiation is highly "trapped;" that is, the $2 \rightarrow 0$ radiation is reabsorbed by ground state atoms ($0 \rightarrow 2$) before it can escape from the discharge. The radiative deexcitation of level 2 is thus governed by *diffusion* rather than direct radiation, according to the theory developed by Holstein.[11] The net spontaneous emission rate can be expressed as γA_{20}, with γ a factor that can be made much less than unity depending on the ground-state atom density and discharge dimensions. It is desirable to make $\gamma A_{20} \ll A_{21}$. For maximum efficiency of such a laser system it is also desirable that no other lower levels are radiatively connected to level 2 so that the branching ratio A_{21}/A_2 is thus effectively unity to the lower laser level. In addition, the lower laser level should be close to the ground state so that the minimum amount of energy is thrown away on deexcitation but not so close that the level is populated thermally from the ground state. The electrical exciting pulse must rise very steeply compared to A_{21}^{-1} so that no population is lost by spontaneous emission before the oscillation threshold is reached. The exciting pulse should, of course, last no longer than the duration of the self-terminating laser pulse. The pulse repetition rate is limited by the time taken for the population in level 1 to relax to the ground state by collisions.

An extensive review of cyclic laser investigations published through mid-1970 was made by Petrash,[12] covering ionic and molecular systems as well as neutral atomic systems. Several new laser systems predicted in that review (Au, Ba) have now been demonstrated, as well as some that were not predicted (Eu). The power performance levels, especially of some of the metal vapor lasers, have also increased greatly since Petrash's review.

2.1.1.1. Metal Vapor Cyclic Lasers. Examples of transient lasers that fit some or all of these requirements have been demonstrated in the vapors of several metals and in the noble gases. Table I lists the wavelengths for laser transitions in the metallic elements that best fit the model of the cyclic laser. A gas laser employing a metal vapor has the added complication that a sufficient pressure of the vapor, usually 0.1 to 1 torr, must be produced by evaporating the pure metal or dissociating a compound containing the metal. The latter process has the advantage of proceeding at a lower temperature. Table I indicates the operating tempera-

---

[11] T. Holstein, *Phys. Rev.* **72**, 1212 (1947); **83**, 1159 (1951).
[12] G. G. Petrash, *Sov. Phys.—Usp.* (*Engl. Transl.*) **14**, 747 (1972).

TABLE I. Cyclic Metal Vapor Lasers

Atom	λ (μm)	Ground	Upper	Lower	T (°C)	References
Ca	5.5457	$4s^2\ ^1S_0$	$4p\ ^1P_1^o$ →	$3d\ ^1D_2$	>550	a
Sr	6.4567	$5s^2\ ^1S_0$	$5p\ ^1P_1^o$ →	$4d\ ^1D_2$	>460	a
Ba	1.1303	$6s^2\ ^1S_0$	$6p\ ^1P_1^o$ →	$5d\ ^3D_2$	700–800	b,c
Ba	1.5000		$6p\ ^1P_1^o$ →	$5d\ ^1D_2$		b,c
Ba	4.33		$7p\ ^1P_1^o$ →	$6d\ ^1D_2$		c
Ba	5.0309		$8p\ ^1P_1^o$ →	$8s\ ^3S_1$	760	b
Mn	0.5341	$4s^2\ a\ ^6S_{5/2}$	$4p\ y\ ^6P_{7/2}^o$ →	$4s\ a\ ^6D_{9/2}$	1100	
⋮	⋮		(six lines total)		to	d,e
Mn	0.5538		$4p\ y\ ^6P_{3/2}^o$ →	$4s\ a\ ^6D_{1/2}$	1300	
Mn	1.28997		$4p\ z\ ^6P_{7/2}^o$ →	$4s\ a\ ^6D_{7/2}$		
⋮	⋮		(six lines total)			d
Mu	1.39975		$4p\ z\ ^6P_{3/2}^o$ →	$4s\ a\ ^6D_{7/2}$		
Cu	0.5105	$4s\ ^2S_{1/2}$	$4p\ ^2P_{3/2}^o$ →	$4s^2\ ^2D_{5/2}$	1500	f
Cu	0.5700		$4p\ ^2P_{3/2}^o$ →	$4s^2\ ^2D_{3/2}$	600[CuI]	g
Cu	0.5782		$4p\ ^2P_{1/2}^o$ →	$4s^2\ ^2D_{3/2}$	1500	f
Au	0.3122	$6s^2\ S_{1/2}$	$6p\ ^2P_{3/2}^o$ →	$6s^2\ ^2D_{5/2}$?	1500	h
Au	0.6278		$6p\ ^2P_{1/2}^o$ →	$6s^2\ ^2D_{3/2}$		i
Cd	0.7132	$5s^2\ ^1S_0$	$7p\ ^1P_1^o$ →	$6s\ ^3S_1$	600	j
Cd	1.1654		$8p\ ^1P_1^o$ →	$5d\ ^3D_1$		j
Cd	1.1869		$6p\ ^1P_1^o$ →	$6s\ ^3S_1$		j
Hg	1.1177	$6s^2\ ^1S_0$	$7p\ ^1P_1^o$ →	$7s\ ^3S_1$	85	k,l
Tl	0.5350	$6p\ ^2P_{1/2}^o$	$7s\ ^2S_{1/2}$ →	$6p\ ^2P_{3/2}^o$	800	m
Pb	0.3639	$6p^2\ ^3P_0$	$7s\ ^3P_1^o$ →	$6p^2\ ^3P_1$	800–1000	n
Pb	0.4058		$7s\ ^3P_1^o$ →	$6p^2\ ^3P_2$		n
Pb	0.4062		$6d\ ^3D_1^o$ →	$6p^2\ ^1D_2$		n
Pb	0.7229		$7s\ ^3P_1^o$ →	$6p^2\ ^1D_2$		n,o
Eu	1.7596	$6s^2\ ^8S_{7/2}^o$	$6p\ ^8P_{9/2}$ →	$5d\ ^8D_{11/2}^o$	>610°C	p,q
Eu	5.0646		$6p\ ^8P_{5/2}$ →	$5d\ ^8D_{9/2}^o$		p,q
Eu	5.4292		$6p\ ^8P_{5/2}$ →	$5d\ ^8D_{5/2}^o$		p,q

[a] J. S. Deech and J. H. Sanders, *IEEE J. Quantum Electron.* **qe-4**, 474 (1968).

[b] Ph. Cahuzac, *Phys. Lett. A* **32**, 150 (1970).

[c] A. A. Isaev, M. A. Kazaryan, S. V. Markova, and G. G. Petrash, *Sov. J. Quantum Electron. (Engl. Transl.)* **5**, 285 (1975).

[d] M. Piltch, W. T. Walter, N. Solimene, G. Gould, and W. R. Bennett, Jr., *Appl. Phys. Lett.* **7**, 309 (1965).

[e] W. T. Silfvast and G. R. Fowles, *J. Opt. Soc. Am.* **56**, 832 (1966).

[f] W. T. Walter, N. Solimene, M. Piltch, and G. Gould, *IEEE. J. Quantum Electron* **qe-2**, 474 (1966).

[g] L. A. Weaver, C. S. Liu, and E. W. Sucov, *IEEE J. Quantum Electon.* **qe-10**, 140 (1974).

[h] A. A. Isaev, M. A. Kazaryan, and G. G. Petrash, *J. Appl. Spectrosc. (Engl. Transl.)* **18**, 357 (1973).

2.1. NEUTRALS EXCITED BY ELECTRON COLLISION

ture for the pure metal lasers as determined in the references cited. Usually a noble gas at a pressure of several torr to several tens of torr is used as a buffer to keep the metal atoms from depositing on the optical surfaces of the discharge, which are often outside the hot zone. Because of the chemical reactivity of most metals at high temperature, the materials used in the discharge tube envelope are necessarily limited. Fused silica can be used below 1000°C, but alumina or other ceramic materials are usually employed at 1000°C or above. Only simple, pure-metal electrodes are usually used for the same reason, and these are often located outside the hot zone of the oven. The laser plasma is usually formed by simply switching a charged capacitor or pulse delay line across the tube electrodes. The magnitude of the pulse current is determined by the initial charging voltage, typically 3 to 30 kV, and the discharge impedance, which is a function of the tube dimensions and the gas pressures employed. (The ratio of voltage to current in such discharges typically varies with current, so that the concept of a "discharge impedance" is somewhat misleading; however, the term is used by many workers.) The rate of current rise in the discharge can be limited by the stray inductance in the discharge circuit if care is not taken in arranging the circuit elements. The ultimate limitation, however, is the propagation time of the ionization wave from one electrode to the other. Often a transverse discharge is used to minimize this propagation time. The current pulse length is determined by the value of the capacitor or the length of the delay line. High pulse repetition rates (1–100 kHz) are possible because the higher buffer gas pressures relax the lower laser levels quickly. At sufficiently high repetition rates no oven is required, since the average input power can become high enough to evaporate the metal, provided the discharge tube has sufficient thermal insulation ("self-heating" operation).

Stimulated-emission pulses of 3 to 30 nsec are typical of the transitions listed in Table I. Since light travels only 1–10 m in this time, high-Q op-

---

[i] W. T. Walter, *IEEE J. Quantum Electron.* **qe-4**, 355 (1968).

[j] A. N. Dubrovin, A. S. Tibilov, and M. K. Shevtsov, *Opt. Spectrosc. (Engl. Transl.)* **32**, 685 (1972).

[k] A. L. Bloom, W. E. Bell, and F. O. Lopez, *Phys. Rev.* **135**, A578 (1964).

[l] K. B. Bockasten, M. Garavaglia, B. A. Lengyel, and T. Lundholm, *J. Opt. Soc. Am.* **55**, 1051 (1965).

[m] A. A. Isaev and G. G. Petrash, *JETP Lett. (Engl. Transl.)* **7**, 156 (1968).

[n] A. A. Isaev and G. G. Petrash, *JETP Lett. (Engl. Transl.)* **10**, 119 (1969).

[o] G. R. Fowles and W. T. Silfvast, *Appl. Phys. Lett.* **6**, 236 (1965).

[p] Ph. Cahuzac, *Phys. Lett. A* **31**, 541 (1970).

[q] P. A. Bokhan, V. M. Klimkin, V. E. Prokop'ev, and V. I. Solomonov, *Sov. J. Quantum Electon. (Engl. Transl.)* **7**, 81 (1977).

tical resonators are not particularly effective and are not usually used. Output mirror couplings of the order of 50% or more are typical. The observed lasers are all of the "superradiant" variety, possessing sufficient gain to saturate the transitions with a single- or double-pass amplification of spontaneous emission. Thus, when operated without an optical resonator the laser output exhibits spatial narrowing, with an angular divergence typical of the geometrical diameter-to-length ratio of the active medium. Spectral narrowing is also observed, with the linewidth of the laser output being narrower than the Doppler-broadened linewidth but, of course, no smaller than the reciprocal of the stimulated emission pulse width.

The first of these cyclic metal vapor lasers to oscillate was the 0.7229-μm line of Pb vapor, obtained by Fowles and Silfvast.[13] Later measurements of Silfvast and Deech[14] showed that the small-signal gain on this transition was 600 dB/m. They obtained a peak output power of 2 kW from a tube 10 cm long by 10 mm in diameter. The remaining Pb transitions listed in Table I were obtained by Isaev and Petrash by using faster-rising excitation pulses.[15] The A coefficients for all the other Pb laser lines are larger than that for the 0.7229-μm line, and so their upper levels must be populated at a faster rate to be inverted. Accordingly, the gain on the 0.4057-μm line should be even larger than the value for the 0.7229-μm line. Using a discharge tube 4.5 mm in diameter by 40 cm long and pulses at a 2.5-kHz repetition rate, Isaev *et al.*[16] produced 10-mW average power at 0.4062 μm in 2.5-nsec, 1.6-kW pulses and 25-mW average power at 0.7229 μm in 5-nsec, 2.0-kW pulses. In a somewhat larger[17] tube, 15 mm diameter by 60 cm long, they increased the 0.7229-μm output over an order of magnitude to 0.3 W average, 34 kW peak in 3.5-nsec pulses. Neon was used as a buffer gas at 5 torr. Anderson *et al.*[18] have recently examined the hypothesis that the visible lines in Pb require fast-rising current pulses with the rather surprising result that the 0.4057-μm line prefers an order of magnitude *slower* current pulse than 0.4062. In a tube 3.9 mm in diameter and 20 cm long, oscillation at 0.4062 μm was obtained with rate of rise of 5×10^{11}A/cm^2 sec (300 A in 6 nsec) with buffer gas pressures of 4 torr or less of He, Ne, Ar or air.

[13] G. R. Fowles and W. T. Silfvast, *Appl. Phys. Lett.* **6**, 236 (1965).
[14] W. T. Silfvast and J. S. Deech, *Appl. Phys. Lett.* **11**, 97 (1967).
[15] A. A. Isaev and G. G. Petrash, *JETP Lett.* (*Engl. Transl.*) **10**, 119 (1969).
[16] A. A. Isaev, M. A. Kazaryan, and G. G. Petrash, *J. Appl. Spectrosc.* (*Engl. Transl.*) **18**, 357 (1973).
[17] A. A. Isaev, M. A. Kazaryan, and G. G. Petrash, *Sov. J. Quantum Electron.* (*Engl. Transl.*) **2**, 470 (1973).
[18] R. S. Anderson, B. G. Bricks, T. W. Karras, and L. W. Springer, *IEEE J. Quantum Electron* **qe-12**, 313 (1976).

The 0.4057 μm line would not oscillate under these excitation conditions, but it would oscillate at 8×10^{10} A/cm² sec with 8 torr of Ne. Under these latter conditions the 0.7228-μm line would also oscillate, but it was delayed 12–14 nsec relative to the 0.4057-μm emission. An average power of 3.5 mW was obtained at 3 kHz repetition rate; the measured gain at 0.4057 μm was 165 dB/m.

The first of the group IIA elements to exhibit cyclic laser oscillation were Ca and Sr, each with one infrared line on transitions that fit the cyclic laser description, as listed in Table I. Deech and Sanders[19] measured small-signal gains of 300 dB/m. Cahuzac later obtained oscillation on the analogous transition in Ba at 1.5000 μm, as well as many others, including some transitions that do not fit the cyclic laser model.[20] Further work by Isaev, et al[21,22] on Ba has produced a somewhat different but overlapping set of oscillating wavelengths. Table I lists only those best fitting the cyclic model. The strongest lasers are the 1.5000- and 1.1303-μm lines that originate from the lowest resonance level, $6p^1P_1^o$. Cahuzac measured 40 to 65 dB/m gain for these transitions.[20] Isaev et al., obtained a multiline peak output power of 10 kW from a tube 15 mm in diameter by 65 cm long, but noted that the output was primarily at 1.5000 μm. The average output was 1.6 W at 6.25 kHz pulse repetition rate with an efficiency of 0.28%. Bokhan and Solomonov[22a] have made a study of the 1.5000-μm line as a function of discharge diameter, pulse capacitance and voltage, and operating temperature. They obtained a maximum specific energy output of 8 μJ/cm³ and an efficiency of 0.16% from a tube only 13 cm long. The higher lying atomic barium resonance levels $7p^1P_1^o$ and $8p^1P_1^o$ are also pumped strongly enough to yield laser oscillation as listed in Table I. Other laser transitions in Ba originate from levels that have partially forbidden transitions to the ground state ($6p'\ ^1P_1^o$, $6p\ ^3P_1^o$, $6p'\ ^3P_1^o$) and probably operate as cyclic lasers even though they have been omitted from Table I. Other laser lines listed in references (20)–(22) originate from levels with ΔJ-forbidden transitions to the ground state; these lasers are probably pumped by cascade from higher-lying states or collisions with buffer-gas metastables. The lightest member of the group IIA elements, Mg, does not exhibit laser oscillation on the transition with upper

---

[19] J. S. Deech and J. H. Sanders, *IEEE J. Quantum Electron* **qe-4**, 474 (1968).

[20] Ph. Cahuzac, *Phys. Lett. A* **32**, 150 (1970).

[21] A. A. Isaev, M. A. Kazaryan, and G. G. Petrash, *Sov. J. Quantum Electron. (Engl. Transl.)* **3**, 358 (1974).

[22] A. A. Isaev, M. A. Kazaryan, S. V. Markova, and G. G. Petrash, *Sov. J. Quantum Electron. (Engl. Transl.)* **5**, 285 (1975).

[22a] P. A. Bokhan and V. I. Solomonov, *Sov. J. Quantum Electron. (Engl. Transl.)* **8**, 184 (1978).

level $3p\,{}^1P_1{}^o$, analogous to the strong lines observed in Ca, Sr, and Ba because the analogous lower level does not exist. The only laser line in Mg originating from a higher-lying resonance level, $7p\ {}^1P_1{}^o$, is evidently pumped by collisions with buffer gas atoms, since it oscillates in the discharge afterglow.[23]

Many pulsed laser transitions have been observed in the group IIB elements Zn, Cd, and Hg, but only a few fit the cyclic model, three in Cd[23] and one in Hg.[24,25] Another lasing transition has been found in Zn, $6p\ {}^1P_1 \to 4d\ {}^1D_2$, with levels fitting the model, but the output is observed only in the afterglow of the noble-gas-buffered discharge, not during the fast-rising current pulse.[23]

Oscillation on 11 of the 12 Mn lines listed in Table I was obtained by Piltch et al.[26] with gains up to 37 dB/m. Peak power outputs on the order of 300 W were observed on the green lines and slightly less on the infrared lines in a discharge tube 100 cm long and 10 mm in diameter. An additional 0.5481-μm line was observed by Silfvast and Fowles.[27] Manganese provides an example of a medium that is cyclic but does not fulfill all of the requirements set down by Walter et al.[10] for the *ideal* cyclic laser; the excitation is split among six upper levels, each communicating with at least two lower levels. Parametric studies of Mn laser output characteristics have been made by Chen,[28] Bokhan et al.,[29] and Isaev et al.[30] Bokhan et al. find that 70–80% of the output can be obtained on the 0.5341-μm line if a helium buffer gas pressure of 30–35 torr is used to promote collisional coupling between the several upper levels. They obtained a total average power of 3.5 W at 5 kHz pulse repetition rate (50 kW peak power, 15-nsec pulses) at an efficiency of 0.17% from a tube 20 mm in diameter by 46 cm long. Isaev et al.[30] obtained 2.1 W average power at 5 kHz pulse repetition rate (24 kW peak power, 20-nsec pulses) at an efficiency of 0.2% from a tube 20 mm in diameter by 70 cm operated at 1150°C and 12 torr He pressure. The best average power output with neon as the buffer gas was 1.0 W.

---

[23] A. N. Dubrovin, A. S. Tibilov, and M. K. Shevtsov, *Opt. Spectrosc.* (*Engl. Transl.*) **32**, 685 (1972).

[24] A. L. Bloom, W. E. Bell, and F. O. Lopez, *Phys. Rev.* **135**, A578 (1964).

[25] K. B. Bockasten, M. Garavaglia, B. A. Lengyel, and T. Lundholm, *J. Opt. Soc. Am.* **55**, 1051 (1965).

[26] M. Piltch, W. T. Walter, N. Solimene, and G. Gould, *Appl. Phys. Lett.* **7**, 309 (1965).

[27] W. T. Silfvast and G. R. Fowles, *J. Opt. Soc. Am.* **56**, 832 (1966).

[28] C. J. Chen, *J. Appl. Phys.* **44**, 4246 (1973).

[29] P. A. Bokhan, V. D. Burlakov, V. A. Gerasimov, and V. I. Solomonov, *Sov. J. Quantum Electron.* (*Engl. Transl.*) **6**, 672 (1976).

[30] A. A. Isaev, M. A. Kazaryan, G. G. Petrash, and V. M. Cherezov, *Sov. J. Quantum Electron.* (*Engl. Transl.*) **6**, 978 (1976).

Oscillation on the single thallium line at 5350 Å was first demonstrated by Isaev et al.[31] with thallium metal at 600–900°C and later by decomposition[32] of thallium iodide (TlI) in the range 370–440°C. Subsequent experiments[33] with TlI indicate that the inversion results in this case because the 7s $^2S_{1/2}$ upper level is preferentially formed when the molecule dissociates, rather than by electron collision excitation of the dissociated Tl atoms. For example, TlBr fails to produce oscillations. Petrash reports 300 W peak power output from a tube 3 mm in diameter by 20 cm long in 1–3-nsec pulses.[12]

Copper vapor has received by far the greatest attention and development of the cyclic lasers. The energy level arrangement shown in Fig. 3 almost perfectly matches the specifications for an ideal laser except that two transitions are possible instead of just one. (A third line at

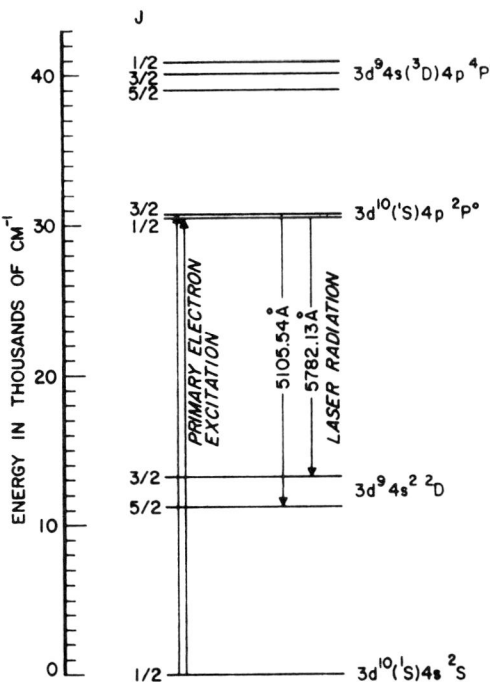

FIG. 3. Energy levels in neutral atomic copper showing the two principal laser lines. (From Walter et al.[10] by courtesy of the IEEE.)

[31] A. A. Isaev, P. I. Ishchenko, and G. G. Petrash, *JETP Lett.* (Engl. Transl.) **6**, 118 (1967).
[32] A. A. Isaev and G. G. Petrash, *JETP Lett.* (Engl. Transl.) **7**, 156 (1968).
[33] A. A. Isaev, M. A. Kazaryan, and G. G. Petrash, *Opt. Spectrosc.* (Engl. Transl.) **31**, 180 (1971).

0.5700 μm connecting the 3/2 upper level with the $J = 3/2$ lower level has also been observed to lase by Weaver et al.,[34] but its A coefficient is sufficiently small that it does not affect the performance on the green and yellow lines.) Laser oscillation on the green and yellow lines was first obtained by Walter et al.[10] in 1965, shortly after the first demonstration of Pb vapor.[13] Peak powers of 200 kW and average powers of 15 W at 15 kHz repetition rate was reported by Isaev et al.[35] in 1972. Table II lists the performance obtained by these and other groups using lasers of various sizes and excited by various techniques. In some experiments copper vapor was obtained by evaporating bits of copper inside the discharge tube with an external oven. In other experiments the electrical energy supplied to the discharge is used to self-heat the tube and evaporate the copper; thermal insulation around the discharge tube envelope and high pulse repetition frequencies are required. The optimum temperature for either heating method is about 1500° to 1700°C, corresponding to a Cu vapor pressure of 0.3 to 2.5 torr.[36] Laser oscillation has been obtained with He, Ne, Ar, Kr, Xe, and N_2 as buffer gases[36a], but the optimum buffer gas seems to be neon at pressures of 10–20 torr. Recently, Bokhan and Shcheglov[36b] have found a second optimum neon pressure at 470 torr in a transverse discharge. Copper atoms have also been obtained by dissociating the copper halides CuCl,[37,38,39] CuBr[38,39] and CuI[40,34] in the discharge at much lower temperatures, 400–500°C, where sufficient vapor pressure of the halide is obtained. Double pulses have been used, with the first dissociating the compound and the second exciting the Cu atoms.[37] Gabay et al.[41] have recently compared the three halides in the same laser tube and observed a ratio of output powers of 6:3:2 for CuBr:CuCl:CuI, each optimized for temperature, buffer gas, and delay between the double exciting pulses.

[34] L. A. Weaver, C. S. Liu, and E. W. Sucov, *IEEE J. Quantum Electron.* **qe-10,** 140 (1974).

[35] A. A. Isaev, M. A. Kazaryan, and G. G. Petrash, *JETP Lett.* (*Engl. Transl.*) **16,** 27 (1972).

[36] R. E. Honig, *RCA Rev.* **23,** 567 (1962).

[36a] C. M. Ferrar, *IEEE J. Quantum Electron.,* **qe-10,** 655 (1974).

[36b] P. A. Bokhan and V. B. Shcheglov, *Sov. J. Quantum Electron.* (*Engl. Transl.*) **8,** 219 (1978).

[37] C. J. Chen, N. M. Nerheim, and G. R. Russell, *Appl. Phys. Lett.* **23,** 514 (1973).

[38] O. S. Akirtava, V. L. Dzhikiya, and Yu. M. Oleinik, *Sov. J. Quantum Electron.* (*Engl. Transl.*) **5,** 1001 (1975).

[39] A. A. Isaev, M. A. Kazaryan, G. Yu. Lemmerman, G. G. Petrash, and A. N. Trofimov, *Sov. J. Quantum Electron.* (*Engl. Transl.*) **6,** 976 (1976).

[40] C. S. Liu, E. W. Sucov, and L. A. Weaver, *Appl. Phys. Lett.* **23,** 92 (1973).

[41] S. Gabay, I. Smilanski, L. A. Levin, and G. Erez, *IEEE J. Quantum Electron.* **qe-13,** 364 (1977).

Table II shows that the pulse lengths obtained in pure Cu lasers vary from 5 to 20 nsec, with the shortest pulses occurring at the highest laser power levels. Pulse lengths in the dissociated halides are typically the same length as pure Cu vapor lasers since the ground state atoms are created by dissociation during previous pulses. Longer pulses have been observed in rapidly-flowing vapors. Asmus and Moncur[42] obtained 65-nsec pulses with a plasma-gun (exploding-wire) Cu vapor source. Russell et al.[43] observed pulse lengths up to 185 nsec in supersonic flowing vapor. Fast-flowing vapor should also allow shorter interpulse periods than that required by simple relaxation of the lower laser level to the ground state. Ferrar[44] has demonstrated undegraded laser outputs with interpulse periods of 20 μsec, equivalent to a 50-kHz repetition rate, at transverse flow velocities corresponding to thermal evaporation at 1700°C. Alaev et al.[45] studied repetition rates up to 100 kHz in a non-flowing longitudinal discharge, obtaining undegraded pulse outputs up to 15–18 kHz, followed by a decrease in peak power above 20 kHz, faster than the inverse of the repetition rate. Bokhan and Shcheglov[36b] have also found maximum repetition rates of 10 kHz (less than 5% decrease) to 18 kHz (63% decrease) in a transverse discharge at the usual buffer gas conditions (neon at 10 torr) and correspondingly faster maximum rates of 20 to 36 kHz at the higher optimum buffer gas pressure of 470 torr. Gordon et al.[45a] report an average output power proportional to repetition rate up to 50 kHz for bursts of pulses in a longitudinal laser using CuCl. Fahlen[46] has observed that laser oscillation switches from almost all green at repetition rates below 100 kHz to almost all yellow at a repetition rate of 150 kHz.

A split-cylinder hollow-cathode laser configuration has been successfully demonstrated by Fahlen[47]; the Cu atoms are produced by both ion bombardment sputtering of the copper-coated cathode and thermal evaporation by discharge self-heating. A hollow-cathode configuration using CuCl has also recently been demonstrated by Smilanski et al.[48] A com-

---

[42] J. F. Asmus and N. K. Moncur, *Appl. Phys. Lett.* **13**, 384 (1968).

[43] G. R. Russell, N. M. Nerheim, and T. J. Pivirotto, *Appl. Phys. Lett.* **21**, 565 (1972).

[44] C. M. Ferrar, *IEEE J. Quantum Electron.* **qe-9**, 856 (1973).

[45] M. A. Alaev, A. I. Baranov, N. M. Vereshchagin, I. N. Gnedin, Yu. P. Zherebtsov, V. F. Moskalenko, and Yu. M. Tsukanov, *Sov. J. Quantum Electron. (Engl. Transl.)* **6**, 610 (1976).

[45a] E. B. Gordon, V. G. Egorov, and V. S. Pavlenko, *Sov. J. Quantum Electron. (Engl. Transl.)* **8**, 266 (1978).

[46] T. S. Fahlen, *IEEE J. Quantum Electron.* **qe-13**, 546 (1977).

[47] T. S. Fahlen, *J. Appl. Phys.* **45**, 4132 (1974).

[48] I. Smilanski, A. Kerman, L. A. Levin, and G. Erez, *IEEE J. Quantum Electron.* **qe-13**, 24 (1977).

TABLE II. Copper Vapor Laser Performance

Ref.	Date	Tube dimensions[a]			Optical output parameters[a]											
		Diam. (mm)	Length (cm)	Volume (cm³)	Peak P (kW)	Avg. P (W)	Pulse τ (nsec)	Rep rate (kHz)	E/V (μJ/cm³)	Effic. (%)	T (°C)	Heat	Source	Buffer gas	Press (torr)
[d]	1966	10	80	(63)	2	0.02	20	0.66	0.6	0.1	1500	Oven	Cu	He	1–3
[e]	1968	50	80	(1570)	>40	>1	16	>1	(0.6)	>1	1520	Oven	Cu	—	—
[f]	1972	8	70	35	—	6	—	18	(9.5)	—	1500	Self	Cu	"Inert"	50?
		15	70	125	200	15	5	15	(8)	0.8					
[g]	1973	12	65	75	—	—	40	0.005	17	—	400	Oven	CuCl	He	2
[h]	1974	7.6	91	(41)	—	0.27	—	12	(0.5)	0.25	~1500?	Self	Cu[b]	Ar	1–6
[i]	1975	6	7	(2)	—	0.27	—	4.2	33	—	<1850	Self	Cu	He,Ne,Ar	10–30
		8.5	8.5	(4.8)	—	1.3	—	6.8	39	—					
[j]	1975	7	40	(15)	25	2	10	8	16	0.2	1500	Self	Cu	He	20
[k]	1975	12	80	(90)	—	1	40	12.5	(0.9)	0.1	400	Self	CuCl	Ne	8
		18	80	(203)	—	3	40	12.5	(1.2)	0.3	400	Self	CuCl	Ne	10
		30	66	(466)	—	6	—	12.5	(1)	0.8	—	Self	CuBr	Ne	10
[l]	1975	c	25	12	—	0.1	20	1	8	(0.004)	—	Self	Cu[c]	He,Ne	40
[m]	1976	7	50	19	24	3.5	7	20	(9)	—	1500	Self	Cu	Ne	15
					6	1.5	5	50	(1.6)	—					10
					1	1	5	100	(0.5)	—					10
[n]	1976	11	40	(38)	—	3	—	15	(5.2)	0.2	540	Self	CuCl	Ne	25
		18	90	(230)	—	6	—	12	(2.2)	0.17	540	Self	CuCl	Ne	10
		30	50	(353)	—	6	—	14.4	30	0.29	540	Self	CuBr	Ne	4
[o]	1977	30	(50)	(353)	10–50	0.4	20	0.925	(1.2)	0.06	430	Oven	CuCl[b]	Ne	5
[p]	1977	c	40	16	80	(0.008)	10	0.001	50	—	450	Oven	CuCl	Ne	5–60
					64	(0.0064)	10	0.001	40	—	650	Oven	CuI	Ne	

q	1977	28	80	(492)	100 200	43.5 36	— —	16.7 9	5 8	1 0.9	—	Self	Cu	Ne	30
			6 15	(2.9) 15	10 —	(0.075) (0.025)	7 —	1 —	26 55	0.4 —				He	14
r	1978	c													
									25	0.64		1700	Oven	Cu	Ne	10,470

[a] Numbers in parentheses are calculated from authors' dimensions and do not take into account mode filling factors, etc.
[b] Hollow cathode discharge; copper vapor produced by both evaporation and sputtering.
[c] Transverse discharge.
[d] W. T. Walter, N. Solimene, M. Piltch, and G. Gould, *IEEE J. Quantum Electron.* **qe-2**, 474 (1966).
[e] William T. Walter, *IEEE J. Quantum Electron.* **qe-4**, 355 (1968).
[f] A. A. Isaev, M. A. Kazaryan, and G. G. Petrash, *JETP Lett. (Engl. Transl.)* **16**, 27 (1972).
[g] C. J. Chen, N. M. Nerheim, and G. R. Russell, *Appl. Phys. Lett.* **23**, 514 (1973).
[h] Theodore S. Fahlen, *J. Appl. Phys.* **45**, 4132 (1974).
[i] R. S. Anderson, L. Springer, B. G. Bricks, and T. W. Karras, *IEEE J. Quantum Electron.* **qe-11**, 172 (1975).
[j] P. A. Bokhan, V. N. Nikolaev, and V. I. Solomonov, *Sov. J. Quantum Electron. (Engl. Transl.)* **5**, 96 (1975).
[k] O. S. Akirtava, V. L. Dzhikiya, and Yu. M. Oleinik, *Sov. J. Quantum Electron. (Engl. Transl.)* **5**, 1001 (1975).
[l] I. S. Aleksandrov, Yu. A. Babeiko, A. A. Babaev, O. I. Buzhinskii, L. A. Vasil'ev, A. V. Efimov, S. I. Krysanov, G. N. Nikolaev, A. A. Slivitskii, A. V. Sokolov, L. V. Tatarintsev, and V. S. Tereshchenkov, *Sov. J. Quantum Electron. (Engl. Transl.)* **5**, 1132 (1975).
[m] M. A. Alaev, A. I. Baranov, N. M. Vereshchagin, I. N. Gnedin, Yu. P. Zherebtsov, V. F. Moskalenko, and Yu. M. Tsukanov, *Sov. J. Quantum Electron. (Engl. Transl.)* **6**, 610 (1976).
[n] A. A. Isaev, M. A. Kazaryan, G. Yu. Lemmerman, G. G. Petrash, and A. N. Trofimov, *Sov. J. Quantum Electron. (Engl. Transl.)* **6**, 966 (1976).
[o] I. Smilanski, A. Kerman, L. A. Levin, and G. Erez, *IEEE J. Quantum Electron.* **qe-13**, 24 (1977).
[p] G. V. Abrosimov and V. V. Vasil'tsov, *Sov. J. Quantum Electron. (Engl. Transl.)* **7**, 512 (1977).
[q] A. A. Isaev and G. Yu. Lemmerman, *Sov. J. Quantum Electron. (Engl. Transl.)* **7**, 799 (1977).
[r] P. A. Bokhan and V. B. Shcheglov, *Sov. J. Quantum Electron. (Engl. Transl.)* **8**, 219 (1978).

parison of CuCl, CuBr, and CuI in the same laser has been made by Gabay et al.[48a] showing that CuBr yields the highest output power by a factor of about 2. Transverse discharges between planar electrodes[49,45a] and coaxial electrodes[50] have also been used with Cu vapor to obtain lower-inductance configurations that allow faster-rising current pulses. An annular output beam of 0.5 W average at a 1 kHz repetition rate and 0.3% efficiency was obtained with the radial discharge version. The highest reported specific output energy of 55 μJ/cm^3 has been obtained by Bokhan and Shcheglov[45a] in a small transverse discharge (6 cm length). Poorer results (25 μJ/cm^3) were obtained with a 15-cm version which the authors attribute to a poorer impedance match to their pulse generator.

The high gains, short-pulsed operation, and larger-diameter discharges favored by the copper-vapor laser make diffraction-limited operation difficult to obtain with a conventional low-loss ("stable") optical resonator. Zemskov et al.[51] have employed an unstable resonator[52] with magnification factor $M = 250$ to obtain a nearly diffraction-limited output of 5 W average power from a 20-mm diameter by 60-cm-long laser. With plane mirrors the same laser produced 10 W average output power but with about 60 times larger beam divergence.

Leonard[53] has given a simple theoretical description of the Cu vapor laser employing a model analogous to that used by Gerry[54] to describe the UV N_2 laser. Theoretically estimated electron excitation cross sections from the ground state to upper and lower laser levels and the ionization limit are integrated over an assumed Maxwellian velocity distribution with electron temperature as a parameter. Differential equations describing the time evolution of the external circuit voltage and current, the plasma resistivity, the electron density and temperature, the excitation rates of upper and lower laser levels, and the saturated laser flux are then integrated numerically. Leonard obtained good agreement with the experimental results of Walter et al.,[10] the only results then available.

[48a] S. Gabay, I. Smilanski, L. A. Levin, and G. Erez, *IEEE J. Quantum Electron.* **qe-13**, 364 (1977).

[49] I. S. Aleksandrov, Yu. A. Babeĭko, A. A. Babaev, O. I. Buzhinskii, L. A. Vasil'ev, A. V. Efimov, S. I. Krysanov, G. N. Nikolaev, A. A. Slivitskii, A. V. Sokolov, L. V. Tatarintsev, and V. S. Tereshchenkov, *Sov. J. Quantum Electron.* (*Engl. Transl.*) **5**, 1132 (1975).

[50] Yu. A. Babeĭko, L. A. Vasil'ev, V. K. Orlov, A. V. Sokolov, and L. V. Tatarintsev, *Sov. J. Quantum Electron.* (*Engl. Transl.*) **6**, 1258 (1976).

[51] K. I. Zemskov, A. A. Isaev, M. A. Kazaryan, G. G. Petrash, and S. G. Rautian, *Sov. J. Quantum Electron.* (*Engl. Transl.*) **4**, 474 (1974).

[52] A. E. Siegman, *Proc. IEEE* **53**, 277 (1965).

[53] D. A. Leonard, *IEEE J. Quantum Electron.* **qe-3**, 380 (1967).

[54] E. T. Gerry, *Appl. Phys. Lett.* **7**, 6 (1965).

More recently, Eletskii et al.[55] and Isaev and Petrash[55a] have made theoretical estimates of laser performance and the conditions under which they should be obtained. Littlewood and Webb[55b] have made a rate equation analysis of the Cu vapor system using the latest values of excitation cross sections and including the electron collisional deexcitation of the upper laser levels. They obtain excellent agreement with experimentally observed pulse magnitudes and shapes. Their analysis shows that the performance of the Cu vapor laser (and likely all electron-excited cyclic lasers) will ultimately be limited in output and efficiency by electron collisional deexcitation. Batenin et al.[55c] have recently made a time-resolved interferometric measurement of the electron density in a Cu laser plasma, obtaining values of 10^{13}–10^{14} cm^{-3} during the laser pulse period.

Gold has an energy level structure analogous to that of copper, and until recently the performances of gold-vapor lasers have been substantially inferior to that of copper. Oscillation was first reported on the red 0.6278-μm line by Walter[56] and later demonstrated in an exploding-wire arrangement by Asmus and Moncur.[42] Isaev et al.[16] obtained oscillation on the 0.3122-μm line. Fahlen[57] has obtained simultaneous oscillation on the Au red line and the Cu green and yellow lines in a self-heated discharge, although the Au output was more than two orders of magnitude weaker. Subsequently Markova and Cherezov[58] have obtained much better performance from gold vapor. Using a discharge tube 16 mm in diameter and 65 cm long in a self-heated mode they obtained 2.1 W average output power on the 0.6278-μm line at a pulse repetition rate of 11 kHz and 1.4 kW average input power. Neon at 25 torr was the optimum buffer gas, with helium nearly as good, and argon and xenon a factor of 3 worse. Power output increased linearly with pulse repetition rate, indicating that the optimum gold vapor pressure had probably not been reached in the self-heated discharge. Using an 8-mm-diameter tube, an average output of 0.2 W was obtained on the 0.3122-μm line at a 10-kHz repetition rate with neon at 7.5 torr as a buffer gas.

[55] A. V. Eletskii, Yu. K. Zemtsov, A. V. Rodin, and A. N. Starostin, Sov. Phys—Dokl. (Engl. Transl.) **20**, 42 (1975).

[55a] A. A. Isaev and G. G. Petrash, Proc. P. N. Lebedev Phys. Inst. **81**, (1975) (in Russian); translated Consultants Bureau, New York (1976).

[55b] I. M. Littlewood and C. E. Webb, IEEE J. Quantum Electron., to be published.

[55c] V. M. Batenin, V. A. Burmakin, P. A. Vokhmin, A. I. Evtyunin, I. I. Klimovskii, M. A. Lesnoi, and L. A. Selezneva, Sov. J. Quantum Electron. (Engl. Transl.) **7**, 891 (1977).

[56] W. T. Walter, IEEE J. Quantum Electron. **qe-4**, 355 (1968).

[57] T. S. Fahlen, IEEE J. Quantum Electron. **qe-12**, 200 (1976).

[58] S. V. Markova and V. M. Cherezov, Sov. J. Quantum Electron. (Engl. Transl.) **7**, 339 (1977).

Laser oscillation has been obtained on over 40 wavelengths in the infrared in the rare earths samarium, europium, thulium, and ytterbium by Cahuzac,[59-61] who has also made level assignments for most of these lines. The laser lines in Eu fit the cyclic model quite well except for the multiplicity of lines allowed, as in Mn. The observed lines in Tm and Yb do not fit the cyclic model; their upper levels have the same parity as the ground state and are thus probably not excited directly from the ground state by electron collision. The lines in Sm are largely unidentified. Klimkin[62] has made a further study of the Yb laser and observed additional wavelengths, but otherwise disappointing output power performance. Bokhan et al.[63] have obtained excellent performance with Eu, producing 1.65 W average power from an 11-mm by 50-cm discharge tube and 1.95 W from a 20-mm by 46-cm tube operating at a 10-kHz repetition rate. The optimum buffer gas pressure was 15–25 torr helium and the operating temperature was greater than 600°C. Pulse lengths of about 20 nsec were observed for the 1.7596-μm line, which comprised about half of the output power. The spectral composition of the remaining half of the output power was not determined, but occurred at wavelengths greater than 4 μm, which is consistent with Cahuzac's observations[60] of relative line strengths.

Chou and Cool[63a] have obtained many additional infrared laser transitions in As, Bi, Ga, Ge, Hg, In, Pb, Sb, and Tl atoms by dissociation of metal halides, hydrides, and methyl complexes. The exact excitation mechanisms are not known, but electron collision excitation is likely.

2.1.1.2. Noble Gas Cyclic Lasers. The noble gases exhibit several transient laser lines with characteristics similar to those of the metal-vapor lasers. These lasers require fast-rising excitation pulses, possess metastable or quasi-metastable lower levels and are thus self-terminating, and exhibit very high (superradiant) gains and output powers. The line observed in helium by Isaev et al.[31]

$$\text{He. } 2.0581 \; \mu\text{m:} \quad 2p \; ^1P_1^o \longrightarrow 2s \; ^1S_0 \quad (1s^2 \; ^1S_0 \text{ ground state})$$

fits the cyclic laser definition fairly well. However, the situation in the heavier noble gases differs significantly from the cyclic laser definitions in that some of the observed upper laser levels are *not* connected to the

---

[59] Ph. Cahuzac, *Phys. Lett. A* **27**, 473 (1968).
[60] Ph. Cahuzac, *Phys. Lett. A* **31**, 541 (1970).
[61] Ph. Cahuzac, *J. Phys. (Paris)* **32**, 499 (1971).
[62] V. M. Klimkin, *Sov. J. Quantum Electron. (Engl. Transl.)* **5**, 326 (1975).
[63] P. A. Bokhan, V. M. Klimkin, V. E. Prokop'ev, and V. I. Solomonov, *Sov. J. Quantum Electron. (Engl. Transl.)* **7**, 81 (1977).
[63a] M. S. Chou and T. A. Cool, *J. Appl. Phys.* **47**, 1055 (1976).

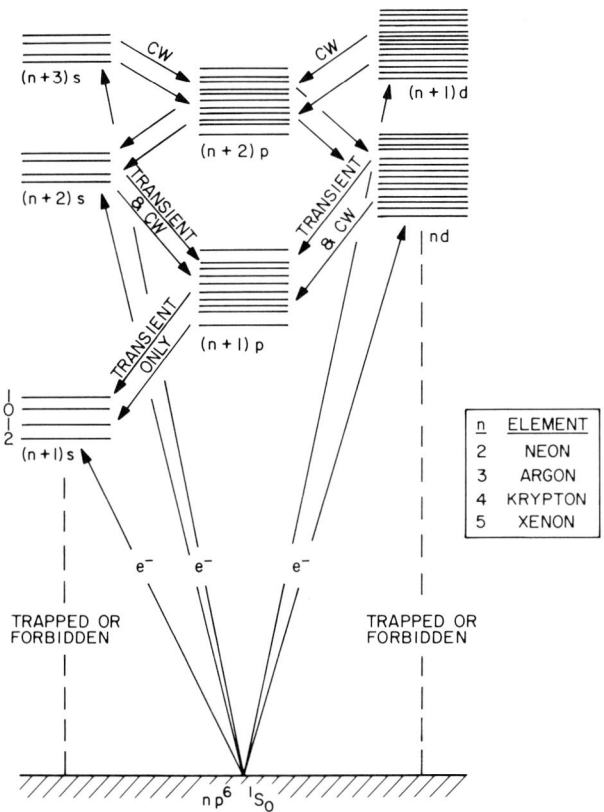

FIG. 4. General energy level structure of the neutral noble gases, indicating the cyclic (transient) and cw laser transitions. The lowest d level in neon is 3d.

ground states by dipole-allowed transitions. They also differ from the ideal (as does the helium line) in that *both* upper and lower laser levels are high-lying with respect to the ground state, thus limiting the ultimate (quantum) efficiency. Typically, several strong transitions branch from the upper laser levels, further reducing the efficiency. Nevertheless, respectable output powers have been generated; for example, the Ne 0.5401-μm line has produced 85 kW peak power in 5-nsec pulses, with a specific energy per pulse of the order of 1.5 μJ/cm^3 in a transverse discharge configuration.[64]

Figure 4 illustrates the general energy level structure of the heavier noble gases. The ground state consists of six np electrons in a $^1S_0$ configuration ($n = 2,3,4,5$ for Ne, Ar, Kr, Xe). The lowest excited states con-

[64] D. A. Leonard, *IEEE J. Quantum Electron.* **qe-3**, 133 (1967).

sists of four $(n + 1)$s levels, two of which are truly metastable with ΔJ-forbidden transitions to the ground state. The remaining two $(n + 1)$s levels have $J = 1$, with allowed transitions to the ground state and relatively large electron excitation cross sections. However, at the gas pressures typical of laser operation the radiation from these transitions is highly trapped,[11] rendering the $J = 1$ levels effectively metastable as well. The next-highest excited states are the ten $(n + 1)$p levels, which have the same parity as the ground state, with both radiation and electron excitation thus dipole-forbidden. As Fig. 4 indicates, one set of transient laser lines occurs between the $(n + 1)$s and $(n + 1)$p levels. Other laser lines have been observed between $(n + 2)$s or nd upper levels and $(n + 1)$p lower levels; some of these higher lying transitions are capable of continuous oscillation as well, in contrast to the transient-only nature of the $(n + 1)$p $\rightarrow$ $(n + 1)$s lines, and may or may not be cyclic lasers according to the definition given in Walter et al.[10] Still other transitions between higher-lying states have been observed in pulsed and continuous oscillation.

Table III lists the laser lines that have been observed between the $(n + 1)$s and $(n + 1)$p levels shown in Fig. 4. The levels are given in the pair-coupling or J-l notation suggested by Racah.[65] Note that while the Paschen designation of the levels has a fixed correspondence to the J-l notation for the $(n + 1)$s levels in all the noble gases, the correspondence changes for the $(n + 1)$p levels from gas to gas. The lines listed in italics in Table III are observed only in gas *mixtures* at high pressure and are assumed to be pumped by processes other than electron excitation[66]; they are not considered further here. Three different hypotheses have been proposed for the preferential excitation of the $(n + 1)$p states over the $(n + 1)$s states for the remaining lines:

(1) Leonard et al.[67] propose that the $(n + 1)$p levels are produced directly from the ground state by electron collision:

$$n\text{p}^6 + e^- + \text{K.E.} \longrightarrow (n + 1)\text{p} + \text{K.E.},$$

citing Druyvesteyn's data for excitation of the 3s and 3p states of neon by electrons with energies near the 3p excitation threshold.[68] The excitation function for these forbidden transitions is sharply peaked and of significant magnitude only within 1–2 eV of threshold, then falls rapidly at higher energies. (see Fig. 21 in Section 2.3 for cross sections of similar

---

[65] G. Racah, *Phys. Rev.* **61**, 537 (1942).
[66] P. L. Chapovsky, V. N. Lisitsyn, and A. R. Sorokin, *Opt. Commun.* **16**, 33 (1976).
[67] D. A. Leonard, R. A. Neal, and E. T. Gerry, *Appl. Phys. Lett.* **7**, 175 (1965).
[68] M. J. Druyvesteyn and F. M. Penning, *Rev. Mod. Phys.* **12**, 87 (1940).

TABLE III. Transient Laser Lines Observed Between Lowest Neutral Noble Gas Energy Levels[a]

	$np'[1/2]_0$	$np'[1/2]_1$	$np'[3/2]_2$	$np'[3/2]_1$	$np[1/2]_0$	$np[3/2]_2$	$np[3/2]_1$	$np[5/2]_2$	$np[5/2]_3$	$np[1/2]_1$
$ns'[1/2]_1$ ($1s_2$)	Ne 5852[e] Ar 7503[e]									Ar 11488[k]
$ns'[1/2]_0$ ($1s_3$)	x	Ar 7724?[h]	x	Ne 6266[h] Ar 7948[h]	x	x		x	x	Ar 10473[k]
$ns[3/2]_1$ ($1s_4$)	Ne 5401[d]					Ne 6305[l]		Ne 6506[l]	x	Ar 9658[k]
$ns[3/2]_2$ ($1s_5$)	x	Ar 6965[i]	Ne 5944[b] Ar 7067[i]		x	Ne 6143[b] Ar 7635[h] Kr 7603[j] Xe 8231[k]	Ar 7724?[h] Xe 8409[g]	Kr 8104[c] Xe 9045[c]		Ar 9122[k] Kr 8928[k] Xe 9799[c]

[a] Wavelengths in angstroms; x = ΔJ – forbidden
[b] D. Rosenberger, *Phys. Lett.* **13**, 228 (1964).
[c] D. Rosenberger, *Phys. Lett.* **14**, 32 (1965).
[d] M. Clunie, R. S. A. Thorn, and K. E. Trezise, *Phys. Lett.* **14**, 28 (1965).
[e] W. B. Bridges and A. N. Chester, *Appl. Opt.* **4**, 573 (1965).
[f] D. A. Leonard, R. A. Neal, and E. T. Gerry, *Appl. Phys. Lett.* **7**, 175 (1965).
[g] O. Andrade, M. Gallardo, and K. Bockasten, *Appl. Phys. Lett.* **11**, 99 (1967).
[h] A. A. Isaev and G. G. Petrash, "Electronic Engineering, Series 3: Gas-Discharge Devices," No. 3, p. 17 (1967). (Translation available from National Technical Information Services, Springfield, Virginia 22161; Document No. AD696877.)
[i] V. A. Tolkachev, *J. Appl. Spectrosc.* (*Engl. Transl.*) **8**, 449 (1968).
[j] G. J. Linford, *IEEE J. Quantum Electron.* **qe-8**, 477 (1972).
[k] P. L. Chapovsky, V. N. Lisitsyn, and A. R. Sorokin, *Opt. Commun.* **16**, 33 (1976).
[l] V. M. Kaslin and G. G. Petrash, *J. Appl. Spectrosc.* (*Engl. Transl.*) **12**, 414 (1970).

transitions in helium). Leonard *et al.* estimate that the energies in their discharge are of the right order of magnitude to match this peaked cross section.

(2) Clunie *et al.*,[69] Rosenberger,[70] and Andrade *et al.*[71] propose that the $(n + 1)$p levels are excited by radiative cascade from higher lying s and d states that are connected to the ground state by allowed transitions. Andrade *et al.* cite examples in Kr and Xe of strong superradiant laser lines that terminate in the appropriate $(n + 1)$p upper levels (but which do not necessarily optimize at the same gas pressure). Unfortunately, they did not investigate the time behavior of these line pairs.

(3) Bennett and Kindlmann[72] and Knyazev and Petrash[73] propose that the $(n + 1)$p levels are populated by a two-step excitation process via the $(n + 1)$s levels with $J = 1$:

$$n\text{p}^6 + \text{e}^- + \text{K.E.} \longrightarrow (n + 1)\text{s}\Big|_{J=1} + \text{e}^-, \qquad (2.1.1)$$

followed by

$$(n + 1)\text{s}\Big|_{J=1} + \text{e}^- + \text{K.E.} \longrightarrow (n + 1)\text{p} + \text{e}^-. \qquad (2.1.2)$$

They point out that the observed delay of only a few nanoseconds between current and laser pulses would require that the cascading lines be superradiant as well, and that this is *not* observed in several important cases, e.g., Ne 0.6143 μm. Leonard[74] has calculated the magnitude of the two-step contribution to be of the order of 10% of the single-step process and considers it sufficiently large to warrant further consideration.

These three hypotheses are discussed in further detail in the review by Isaev and Petrash.[75] It is probably not possible to decide finally among the three hypotheses with only the experimental evidence now at hand, but an additional observation can be made: Laser transitions are seen from the $(n + 1)$p upper levels with $J = 0, 1, 2$ but not $J = 3$. The transition $(n + 1)\text{p}[5/2]_3 \rightarrow (n + 1)\text{s}[3/2]_2$ should be a very strong laser line if its upper level is properly pumped. In hypothesis (1) there is no a priori reason for excluding collisional pumping to the $J = 3$ level, since *all* tran-

---

[69] D. M. Clunie, R. S. A. Thorn, and K. E. Trezise, *Phys. Lett.* **14**, 28 (1965).
[70] D. Rosenberger, *Phys. Lett.* **13**, 228 (1964).
[71] O. Andrade, M. Gallardo, and K. Bockasten, *Appl. Phys. Lett.* **11**, 99 (1967).
[72] W. R. Bennett, Jr. and P. J. Kindlmann, *Phys. Rev.* **149**, 38 (1966).
[73] I. N. Knyazev and G. G. Petrash, *J. Appl. Spectrosc.* (*Engl. Transl.*) **4**, 401 (1966).
[74] D. A. Leonard, *IEEE J. Quantum Electron.* **qe-3**, 133 (1967).
[75] A. A. Isaev and G. G. Petrash, *Sov. Phys.—JETP* (*Engl. Transl.*) **29**, 607 (1969).

sitions are parity-forbidden. Under hypotheses (2) and (3) the lack of oscillation from the $J = 3$ upper level is understandable, since an intermediate level with $J = 1$ is required in both cases, and only $J = 0, 1, 2$ levels would be pumped from this level. Hypothesis (3) has difficulty explaining the five lines in Table III with $J = 1$ lower levels. However, these five lines are somewhat different in their characteristics than the others. The 0.5401-μm line requires very short excitation pulses and optimizes at one or two orders of magnitude higher gas pressure (2-20 torr) than the other neon lasers. The Ne 0.5852- and Ar 0.7503-μm lines were seen only in low-pressure mixtures of He, Ne, and Ar in a very long (3 m) discharge tube.[76] The Ne 0.6305- and 0.6506-μm lines require a mixture of neon and SF_6.[77] It is quite likely that different mechanisms are responsible for oscillation on these five lines.

2.1.1.3. Other Transient Noble Gas Lasers. Another general class of transient laser lines in the neutral noble gases are those with $(n + 2)$s or nd upper levels and $(n + 1)$p lower levels, as indicated in Fig. 4. Over 50 wavelengths have been observed to oscillate on these multiplets in the 1-5-μm range by Clark,[78] Bockasten et al.,[79] Andrade et al.,[71] and Linford.[80-82] Table IV lists the strongest lines, most of which possess gains high enough to be superradiant. The upper levels of many of these lines are connected to the $np^6\ ^1S_0$ ground state by strongly allowed transitions and would fit the model of direct electron excitation nicely. However, there are a few lines originating from upper levels with $J = 2$, for which transitions from the ground state are ΔJ-forbidden and which should thus have low electron excitation cross sections; among these are the strongly oscillating lines Xe 3.508 and 5.575 μm. The route by which these $J \geq 2$ levels are populated is not known; it could occur by cascade from higher levels or by one-step electron impact excitation on the forbidden transition. Clark[78] and Linford[80] both distinguished two distinct groups of lines in xenon according to their time behavior: those which oscillate promptly at the beginning of the current pulse and self-terminate in a microsecond or less and those that build up strength over a several-microsecond pulse and terminate when the current pulse terminates. It would be satisfying if the short-pulse lines were those with $J = 1$ upper levels and the long-pulse lines with $J \geq 2$ upper levels so that one could attribute direct elec-

---

[76] W. B. Bridges and A. N. Chester, *Appl. Opt.* **4**, 573 (1965).
[77] V. M. Kaslin and G. G. Petrash, *J. Appl. Spectrosc. (Engl. Transl.)* **12**, 414 (1970).
[78] P. O. Clark, *Phys. Lett.* **17**, 190 (1965).
[79] K. Bockasten, T. Lundholm, and O. Andrade, *Phys. Lett.* **22**, 145 (1966).
[80] G. J. Linford, *IEEE J. Quantum Electron.* **qe-8**, 477 (1972).
[81] G. J. Linford, *IEEE J. Quantum Electron.* **qe-9**, 610 (1973).
[82] G. J. Linford, *IEEE J. Quantum Electron.* **qe-9**, 611 (1973).

TABLE IV. Other Strong, Transient, Neutral Noble-Gas Lasers

Atom	Wavelength (μm)	Transition	Ref.
Ar	1.21396	$3d'[3/2]_1^o \to 4p'[3/2]_1$	a
Ar	1.24028	$3d'[3/2]_1^o \to 4p[3/2]_1$	a,b
Ar	1.27022	$3d'[3/2]_1^o \to 4p'[1/2]_1$	a,b,c
Ar	1.69406^f	$3d[3/2]_1^o \to 4p[3/2]_2$	b,c,d
Ar	2.31332^f	$3d[1/2]_1^o \to 4p'[1/2]_1$	b,c,d
Ar	3.708	$4d[3/2]_1^o \to 5p[3/2]_1$	b
Kr	1.44268	$6s[3/2]_1^o \to 5p[3/2]_1$	b,d
Kr	1.47654	$6s[3/2]_1^o \to 5p[5/2]_2$	b,d
Kr	1.68535	$4d[7/2]_3^o \to 5p[5/2]_3$	b,d
Kr	2.19025^f	$4d[3/2]_2^o \to 5p[3/2]_2$	b,d
Kr	2.52338^f	$4d[1/2]_1^o \to 5p[3/2]_2$	b,d
Kr	3.956	$5d[3/2]_1^o \to 6p[5/2]_2$	b
Xe	1.60533	$7s[3/2]_1^o \to 6p[3/2]_2$	d
Xe	1.73258^f	$5d[3/2]_1^o \to 6p[5/2]_2$	b,d,e
Xe	2.02623^f	$5d[3/2]_1^o \to 6p[3/2]_1$	b,d,e
Xe	3.5080^f	$5d[7/2]_3^o \to 6p[5/2]_2$	b,e
Xe	5.5754^f	$5d[7/2]_4^o \to 6p[5/2]_3$	b,e

[a] K. Bockasten, T. Lundholm, and O. Andrade, *Phys. Lett.* **22**, 145 (1966).
[b] G. J. Linford, *IEEE J. Quantum Electron.* **qe-8**, 477 (1972).
[c] D. G. Sutton, L. Galvan, P. R. Valenzuela, and S. N. Suchard, *IEEE J. Quantum Electron* **qe-11**, 54 (1975).
[d] O. Andrade, M. Gallardo, and K. Bockasten, *Appl. Phys. Lett.* **11**, 99 (1967).
[e] P. O. Clark, *Phys. Lett.* **17**, 190 (1965).
[f] CW oscillation also observed.

tron excitation to $J = 1$ and cascade excitation to $J \geq 2$. While this happens to be the case for the last four Xe lines listed in Table IV, it is not the case for *all* the Xe lines observed by Clark and Linford. The exact excitation route remains unknown.

Which lines actually oscillate within a given multiplet and their relative strengths also depends on the transition probabilities of the individual lines as well as their population inversions. The theoretical calculation of A coefficients for the neutral noble gases is generally complicated[83–85] but some guidelines can be given. With levels designated by their Racah notation $nl\,[k]_J$, the strongest transitions within a given multiplet are those with $\Delta k = \Delta J$ ($\Delta \equiv$ upper–lower). Of these, the strongest are $\Delta k =$

[83] G. F. Koster and H. Statz, *J. Appl. Phys.* **32**, 2054 (1961).
[84] H. Statz, C. L. Tang, and G. F. Koster, *J. Appl. Phys.* **34**, 2625 (1963).
[85] W. L. Faust and R. A. McFarlane, *J. Appl. Phys.* **35**, 2010 (1964).

$\Delta J = \Delta l$, the next strongest $\Delta k = \Delta J = 0$, and the next strongest $\Delta k = \Delta J = -\Delta l$. For any of these, the larger J is, the larger the transition probability. (A similar set of rules exists for levels described by $L-S$ coupling[86].) These rules are only approximate, but are reasonably well obeyed in the numerical calculations[85] and in experimental transition probabilities. The strong laser lines of Table IV also show reasonable agreement with these rules; of the 17 lines listed, 14 have $\Delta J = \Delta k$, 5 have $\Delta J = \Delta k = +\Delta l$ and 7 have $\Delta J = \Delta K = 0$. Of course, exact agreement with these rules is not expected for lasers, since the population inversions are not the same for all lines in the multiplet.

Only a little has been written on the external operating characteristics of these transient lines. Typically, currents of tens to hundreds of amperes are used in tubes of a few millimeters diameter. Andrade et al.[71] give optimum gas pressures from 0.05 to over 2 torr for several lines in a 7-mm diameter by 50-cm discharge excited by a 3-nf capacitor charged to 19–26 kV. Sutton et al.[87] have observed that the particular lines that oscillate in the 3d → 4p multiplet in argon are affected by the addition of SF_6 to the discharge. With pure Ar at 5 torr they obtained oscillation only on the 1.27022-μm line; with 3 torr of SF_6 added, oscillation at 1.27022 μm ceases and six other lines originating from the $3d[3/2]_1^0$ and $3d[1/2]_1^0$ levels (including the two listed in Table IV) oscillate with two orders of magnitude increase in output. The explanation for this switching and enhancement is not known. The laser used by Sutton et al.[87] was a 25-mm diameter, 3-m transverse discharge with a helical pin arrangement.

2.1.2. Continuous Neutral Noble Gas Lasers

From the preceding examples of transient laser oscillation it is evident that a continuous population inversion can also be maintained by electron collisions, provided the lower laser level can relax rapidly enough to a still lower-lying level, faster than the stimulated emission rate. However, the decay rate from this lower-lying level to the ground state need not be so rapid; for low-power lasers it is usually sufficient that it be relaxed to the ground state by collisions with the discharge tube walls. The population in the lower-lying level may build up to values much larger than that in the lower laser level; it becomes a bottleneck only when its population is sufficiently large that repopulation of the lower laser level can occur from it by radiation trapping or electron collision.

---

[86] E. V. Condon and G. H. Shortley, "The Theory of Atomic Spectra." Cambridge Univ. Press 1963.

[87] D. G. Sutton, L. Galvan, P. R. Valenzuela, and S. N. Suchard, *IEEE J. Quantum Electron.* **qe-11**, 54 (1975).

Figure 4 indicates the cw laser lines fitting this description in the noble gases. The lowest-lying cw lasers are the nd → $(n + 1)$p and $(n + 2)$s → $(n + 1)$p transitions. The $(n + 1)$p levels are depopulated by radiation to the $(n + 1)$s levels, which in turn are relaxed primarily by wall collisions. Continuous oscillation is observed to originate from nd and $(n + 2)$s levels with $J = 1$; these levels are strongly connected to the ground state and should have large electron excitation cross sections. However, oscillation is *also* observed from levels with $J = 0$ and $J > 2$, with ΔJ-forbidden transitions to the ground state, just as in the case of the transient lasers. In fact, some of the stronger cw lines are also the stronger transient lines, as indicated in Table IV. Again, the exact route by which the upper laser levels are populated is not known. Continuous oscillation is also observed in the higher-lying multiplets, for example, $(n + 1)$d → $(n + 2)$p, and $(n + 2)$p → nd and $(n + 2)$s, although not all multiplets are represented in each noble gas. The existence of sufficient excitation to oscillate on these high-lying transitions suggests that spontaneous emission cascades from these levels may also contribute to the excitation of the lower-lying states.

Over 300 continuously oscillating laser lines are known in the neutral noble gases,[1–3] spanning the wavelength range of about 1 to over 100 μm with about half of the total occurring in neon. The majority of these laser lines were first obtained by the group at the Bell Telephone Laboratories including Bennett, Faust, Garrett, McFarlane, and Patel,[88–93] and have received little study since. Gains of a few to several percent per meter were measured, with optimum gas pressures of 0.01 to 0.1 torr and discharge currents less than 0.1 A/cm^2.

Among the few lines that have received significant study and optimization since their initial discovery are the last three xenon lines listed in Table IV, primarily because of the high optical gains exhibited by these lines. The 3.508-μm line received particular attention because its wavelength falls in a region of low atmospheric attenuation[94] and (prior to the

[88] C. K. N. Patel, W. R. Bennett, Jr., W. L. Faust, and R. A. McFarlane, *Phys. Rev. Lett.* **9**, 102 (1962).

[89] W. L. Faust, R. A. McFarlane, C. K. N. Patel, and C. G. B. Garrett, *Appl. Phys. Lett.* **4**, 85 (1962).

[90] W. L. Faust, R. A. McFarlane, C. K. N. Patel, and C. G. B. Garrett, *Phys. Rev.* **133**, A1476 (1964).

[91] C. K. N. Patel, R. A. McFarlane, and C. G. B. Garrett, *Appl. Phys. Lett.* **4**, 18 (1964).

[92] R. A. McFarlane, W. L. Faust, C. K. N. Patel, and C. G. B. Garrett, *Proc. IEEE* **52**, 318 (1964).

[93] C. K. N. Patel, W. L. Faust, R. A. McFarlane, and C. G. B. Garrett *Proc. IEEE* **52**, 713 (1964).

[94] W. Eppers, in "Handbook of Lasers" (R. J. Pressley, ed.), p. 53. CRC Press, Cleveland, Ohio, 1971.

discovery of the CO_2 laser) it was considered a good candidate for optical communication or radar systems operating in the atmosphere. This line was first observed as a strongly oscillating line in a helium–xenon discharge by Faust et al.[89] and later determined to have an optical of gain of over 50 dB/m in a helium–xenon discharge by Paananen and Bobroff.[95] The role of helium in the internal mechanism of this laser was at first puzzling, since no resonant transfer can take place from the helium metastables to the xenon upper laser levels as it does in the helium–neon laser (see Section 2.3); in fact, the ionization potential of xenon lies below the lowest helium metastable level. Gains of 50 dB/m in a pure xenon discharge and 60 dB/m in a helium–xenon mixture were measured by Bridges,[96] demonstrating that helium is not essential to high-gain operation, but that it does modify the discharge characteristics in a favorable manner. With a plasma probe measurement, Aisenberg determined that a typical helium–xenon laser discharge exhibited a threefold increase in electron density compared to a pure xenon discharge at the same input power.[97] It may be that the additional electron density results from Penning ionization of xenon atoms by collision with helium metastables (see Section 2.4 for a description of the Penning process). One practical difficulty encountered with dc-excited helium–xenon mixtures is that the gas separates by cataphoresis[98] at modest currents. To alleviate this problem, rf excitation is often used.

A detailed study of the Xe 3.508-μm gain dependence on pressure, tube diameter, and signal level was made by Clark,[99] who determined that the small-signal gain and saturated output power continue to increase as the pressure is lowered, down to the lowest value of pressure he was able to measure accurately, 10 mtorr. At such low pressures the optimum discharge current becomes very large and the cleanup of gas by the tube walls and electrodes becomes so rapid that stable operation at lower pressures is difficult. At 10-mtorr pressure Clark observed an approximate D^{-1} dependence of gain on the diameter for diameters larger than about 5 mm, obtaining gain values of approximately $300D^{-1}$ dB/m (D in mm). He observed less than this with tubes of smaller diameter. Armstrong[100] has developed a servocontrolled, liquid-nitrogen-cooled xenon reservoir system to stabilize low xenon pressures; with this system he measured the 3.508-μm output powers shown in Fig. 5 for a 6-mm-diameter discharge,

---

[95] R. A. Paananen and D. L. Bobroff, *Appl. Phys. Lett.* **2**, 99 (1963).
[96] W. B. Bridges, *Appl. Phys. Lett.* **3**, 45 (1963).
[97] S. Aisenberg, *Appl. Phys. Lett.* **2**, 187 (1963).
[98] See, for example, F. H. Shair and D. S. Remer, *J. Appl. Phys.* **39**, 5762 (1968).
[99] P. O. Clark, *IEEE J. Quantum Electron.* **qe-1**, 109 (1965).
[100] D. R. Armstrong, *IEEE J. Quantum Electron.* **qe-4**, 968 (1968).

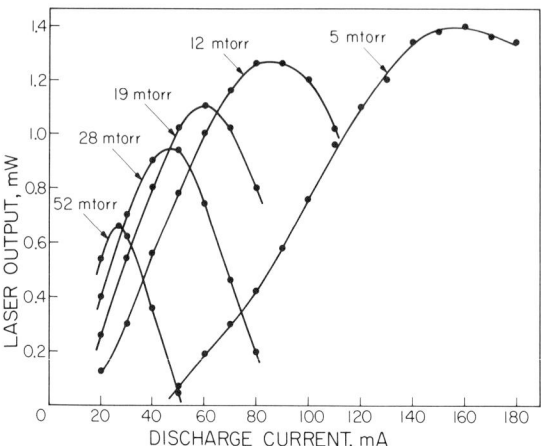

FIG. 5. Output power of a He–Xe 3.508-μm laser vs. discharge current and gas pressure (Armstrong,[100] by courtesy of the IEEE).

100 cm long. Later measurements of gain were made by Aleksandrov *et al.*[101] with a single even xenon isotope, ^{136}Xe, which substantially narrows the gain linewidth and increases the magnitude of the gain. They also observed a D^{-1} dependence, with gains of $350D^{-1}$ dB/m for the 3.508-μm line and $650D^{-1}$ dB/m for the 5.575-μm line, for a range of tube diameters of 5 to 26 mm, and report even higher gains at their lowest operating pressure, 2 mtorr. Aleksandrov *et al.* observed that the gain decreased substantially if the glass discharge tube was surrounded with a reflecting screen of paper or foil, demonstrating that the lower laser level can be repopulated by radiation trapping on the 0.8- and 0.9-μm lines of the 6p → 6s multiplet. Smith and Maloney[102] have extended the measured gain values to still smaller diameter tubes by using an optical waveguide for the discharge tube bore (see Section 2.5). They measured a net gain of 1000 dB/m on the 3.508-μm line in a 250-μm diameter waveguide laser using a 3:1 mixture of ^{3}He:^{136}Xe at a total pressure of 5.9 torr and a combination of rf and dc excitation. With dc excitation alone, 350 μA at 1000 V in a 2.5-cm-long discharge, the gain value was 840 dB/m; with natural xenon it was 30% less. These values agree reasonably with those calculated from the expression given by Aleksandrov *et al.*[101] derived from tubes of much larger diameter. The value of 1000 dB/m obtained by Smith and Maloney is an all-time record gain coefficient for any gas laser.

[101] E. B. Aleksandrov, V. N. Kulyasov, and A. B. Mamyrin, *Opt. Spectrosc.* (*Engl. Transl.*) **31**, 170 (1971).
[102] P. W. Smith and P. J. Maloney, *Appl. Phys. Lett.* **22**, 667 (1973).

The use of a laser discharge as a single-pass, high-gain, low-noise amplifier was first demonstrated by Paananen et al.[103] with a helium–neon amplifier at 3.39 μm. Bridges and Picus[104] used a xenon amplifier at 3.508 μm with 17 dB gain to improve the ultimate sensitivity of a room-temperature InAs detector by 16 dB. Insufficient optical spatial filtering was employed in that experiment to compare the results with the theoretical predictions of laser amplifier noise performance derived by Kogelnik and Yariv[105] and Steinberg,[106] except to place a lower bound on the fractional inversion $[N_2 - (g_2/g_1)N_1]/N_2$ of 0.1. (The fractional inversion of a laser amplifier plays a role similar to the quantum efficiency of a photon detector. If the lower-level population $N_1 = 0$, the laser amplifier is essentially ideal, as derived by Steinberg.) Careful amplifier noise measurements by Klüver[107] later showed that $[N_2 - (g_2/g_1)N_1]/N_2 = 0.62$ for the xenon 3.508-μm line in an optimized helium–xenon discharge operated as a small-signal amplifier, very close to being an ideal quantum amplifier. Klüver also measured the noise performance for partially saturated xenon amplifiers, obtaining excellent agreement with theory, and a value of 70 μW/mm² for the saturation flux. Thus, despite its very high gain characteristics, the xenon 3.508-μm line is essentially a low-power laser when excited by electron collision in a continuous low-pressure glow-discharge mode, although as much as 80 mW has been obtained from a He–Xe discharge 200 cm in length.[108] Much higher power operation has been observed in high-pressure helium–xenon mixtures, which are discussed in Section 2.3.

2.2. Ions Excited by Electron Collision

Ions as well as neutral atoms have exhibited electron-collision-excited laser oscillation. Although the exact routes of the excitation are not known in all cases, 400 or so laser lines fall within the *electron-excited* category. Oscillation is observed between the energy levels of multiply ionized as well as singly ionized atoms. Figure 6 indicates this schematically for argon and also illustrates the nomenclature commonly employed. Lines that originate from energy levels in neutral argon, for example,

---

[103] R. A. Paananen, H. Statz, D. L. Bobroff, and A. Adams, Jr., *Appl. Phys. Lett.* **4**, 149 (1964).
[104] W. B. Bridges and G. S. Picus, *Appl. Opt.* **3**, 1189 (1964).
[105] H. Kogelnik and A. Yariv, *Proc. IEEE* **52**, 165 (1964).
[106] H. Steinberg, *Proc. IEEE* **51**, 943 (1963).
[107] J. W. Klüver, *J. Appl. Phys.* **37**, 2987 (1966).
[108] J. T. LaTourrette, unpublished.

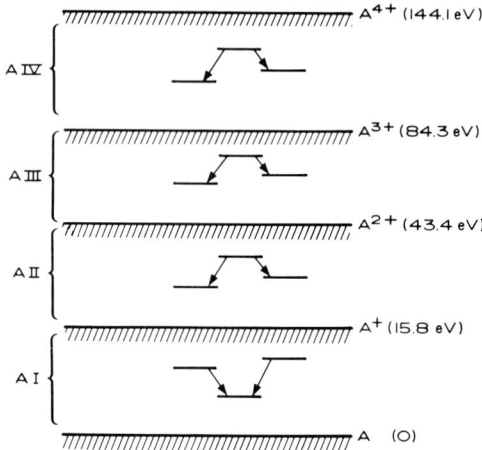

FIG. 6. Schematic representation of ionization states, indicating energies for argon. The collection of emission lines from energy levels in each ionization state is known as a *spectrum* and is designated by a Roman numeral as listed at the left.

make up the *first spectrum of argon,* designated Ar I; those originating from levels of the singly ionized atom Ar^+ make up the *second spectrum* Ar II, and so forth. Oscillation has been obtained between the excited states of as high as four-times-ionized atoms, or the *fifth spectrum* of some elements (Ne V, O V, S V, and possibly Xe V.)

Mercury was the first ion to exhibit laser oscillation in gaseous form.[109] (Recall that ruby, the very first laser, was actually an "ion" laser, with an active medium consisting of Cr^{3+} ions in an Al_2O_3 crystalline lattice.) However, the excitation of the Hg^+ upper laser levels occurs primarily by collision with noble gas atoms, and so it is discussed in Section 2.4. The first electron-collision-pumped ion laser was demonstrated almost simultaneously in 1964 by Bridges,[110] Convert *et al.*,[111,112] and Bennett *et al.*,[113] only a few months after the He–Hg^+ laser was first reported. Oscillation was obtained on several blue and green lines in the Ar II spectrum. Operation was pulsed at first, but the laser oscillation was observed to begin promptly with the leading edge of the current pulse and to last the entire duration of microseconds[110] to milliseconds.[113] There was no indi-

[109] W. E. Bell, *Appl. Phys. Lett.* **4**, 34 (1964).
[110] W. B. Bridges, *Appl. Phys. Lett.* **4**, 128 (1964).
[111] G. Convert, M. Armand, and P. Martinot-Lagarde, *C. R. Hebd. Seances Acad. Sci.* **258**, 3259 (1964).
[112] G. Convert, M. Armand, and P. Martinot-Lagarde, *C. R. Hebd. Seances Acad. Sci.* **258**, 4467 (1964).
[113] W. R. Bennett, Jr., J. W. Knutson, Jr., G. N. Mercer, and J. L. Detch, *Appl. Phys. Lett.* **4**, 180 (1964).

2.2. IONS EXCITED BY ELECTRON COLLISION

cation of self-termination by lower-level filling as in Cu vapor laser or operation only in the afterglow of a noble gas buffer as in the He–Hg$^+$ laser. Continuous oscillation was obtained by Gordon et al.[114] on the blue–green Ar II lines and on all of the Kr II and Xe II lines listed by Bridges.[115] These first results were followed rapidly by many others, obtaining oscillation on higher ionization states,[116-118] operation in the ultraviolet[116,117,119] and infrared,[120] and in the ions of many elements in addition to the noble gases.[116-122] Progress in ion laser spectroscopy during the first year was quite rapid and was summarized by Bridges and Chester.[123] More recent listings of ion laser lines including both electron- and atom-collision-excited lines, were given by Bridges and Chester,[1] Willett,[1,2] Davis and King,[124] and Beck et al.[3] New results continue to appear, even in the noble gases. For example, Marling has reinvestigated the ultraviolet region for the noble gases[125,126]; he has found several lines with pulses outputs greater than 1 kW, and wavelengths less than 0.2 μm.

Development of the argon ion laser as a practical, commercially available source was equally rapid. Units with 1-W cw output were available[127] only a few months after the publication of the initial discovery in April 1964. With efficiencies of less than 0.1% typical for this laser, new discharge tube techniques had to be developed to dissipate several kilowatts of power and withstand the severe ion bombardment on the confining walls before further advances in output power could be realized. Some of these techniques are described in Section 2.5. Many research efforts were undertaken very early to understand the basic mechanisms by which the population inversion was produced and to discover the ultimate limitations on power output and efficiency as well as the scaling laws for ion lasers. This research effort continues today, with improvements in high-power discharge technology and new mechanism insights proceeding hand in hand. Brief summaries of inversion mechanisms and

---

[114] E. I. Gordon, E. F. Labuda, and W. B. Bridges, *Appl. Phys. Lett.* **4**, 178 (1964).
[115] W. B. Bridges, *Proc. IEEE* **52**, 843 (1964).
[116] W. B. Bridges and A. N. Chester, *Appl. Opt.* **4**, 573 (1965).
[117] R. A. McFarlane, *Appl. Phys. Lett.* **5**, 91 (1964).
[118] H. J. Gerritsen and P. V. Goedertier, *J. Appl. Phys.* **35**, 3060 (1964).
[119] P. K. Cheo and H. G. Cooper, *J. Appl. Phys.* **36**, 1862 (1965).
[120] D. C. Sinclair, *J. Opt. Soc. Am.* **55**, 571 (1965).
[121] A. L. Bloom, W. E. Bell, and F. O. Lopez, *Phys. Rev.* **135**, A578 (1964).
[122] R. A. McFarlane, *Appl. Opt.* **3**, 1196 (1964).
[123] W. B. Bridges and A. N. Chester, *IEEE J. Quantum Electron.* **qe-1**, 66 (1965).
[124] C. C. Davis and T. A. King, *in* "Advances in Quantum Electronics" (D. W. Goodwin, ed.), pp. 169–473. Academic Press, New York, 1975.
[125] J. B. Marling, *IEEE J. Quantum Electron.* **qe-11**, 822 (1975).
[126] J. B. Marling and D. B. Lang, *Appl. Phys. Lett.* **31**, 181 (1977).
[127] Anonymous brochure, "1 Watt cw Argon Ion Laser Model LG 12." Raytheon Co., Waltham, Massachusetts, dated 1/65.

typical laser performance are given in the sections following the discussion of ion laser spectroscopy given in Section 2.2.1.

2.2.1. Spectroscopy of Noble-Gas Ion Lasers

The outer shell of electrons in a noble-gas atom consists of two s electrons and six p electrons, each with n quanta of energy, grouped together in an arrangement with zero net spin, zero net orbital angular momentum, and zero total momentum. The usual designation for this is ns^2np^6 $^1S_0$, where n = 2,3,4,5 for neon through xenon, respectively. Sometimes the designation is expanded to include the inner electron shells as well, e.g., the krypton ground state may be described completely as $1s^22s^22p^63s^23p^63d^{10}4s^24p^6$ $^1S_0$, which accounts for all 36 electrons, but more often it is abbreviated to $4s^24p^6$ $^1S_0$. The last portion of the designation gives the angular momenta in the form $^{(2S+1)}L_J$, where S is the net spin of all the electrons, L their net orbital angular momentum, and J the total angular momentum resulting from coupling L and S. By convention, L is given "alphabetical" values: 1 = S, 2 = P, 3 = D, 4 = F, 5 = G, etc. When one electron is removed to form an ion, the inner shells are unchanged and the remaining electrons in the outermost shell assume the configuration ns^2np^5 $^3P^o_{3/2}$ for the ion ground state. The superscript "o" indicates an odd-parity state; even-parity states omit the superscript. The excited states of the ion result when one of the five p electrons is further excited. The resulting energy level is usually best described by the Russell–Saunders, or LS, coupling scheme[86]; and the designation of the level contains the configuration of the remaining four p electrons (the "core"), the n and l quantum numbers of the excited ("running") electron, and the overall configuration resulting from the coupling of the electron to the core. For example, one of the lower excited states in singly ionized argon is $3s^23p^4(^3P)4s$ $^2P_{3/2}$, where (^3P) describes the configuration of the two 3s and four 3p electrons remaining in the core, and $^2P_{3/2}$ describes the overall coupling when the 4s electron is added to the core. Three different core configurations are known for the singly ionized noble gases (^3P), (^1D), and (^1S). A shorthand method of denoting the core is to use a prime or double prime on the running electron designation for the (^1D) or (^1S) cores, respectively. Thus, the argon level just mentioned would often be written simply as 4s $^2P_{3/2}$, while the level $3s^23p^4(^1S)4s$ $^2S_{1/2}$ would be written 4s″ $^2S_{1/2}$. Similar designations are used for more highly ionized states.

The laser lines observed in the singly and doubly ionized noble gases may be grouped together as shown schematically in Figs. 7 and 8. It is clear that by far the largest number of laser transitions occurs from the lowest-lying p states associated with each core to the lowest-lying s and d

2.2. IONS EXCITED BY ELECTRON COLLISION

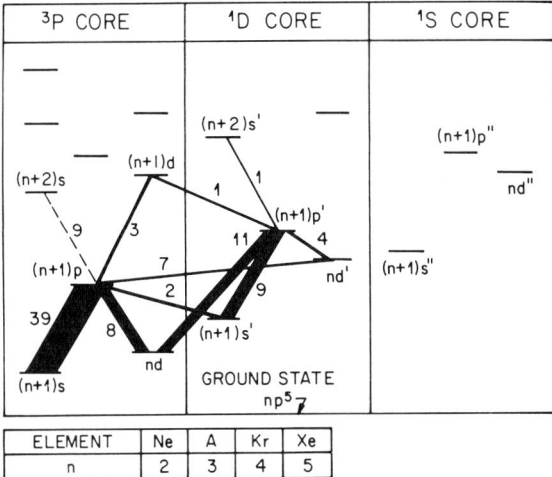

FIG. 7. Generalized energy level diagram for singly ionized noble gas atoms. The number shown beside each supermultiplet group is the number of laser transitions observed in that group. Note that the nd, nd', nd'' levels do not exist in neon (Bridges and Chester,[116] with revisions; by courtesy of American Institute of Physics.)

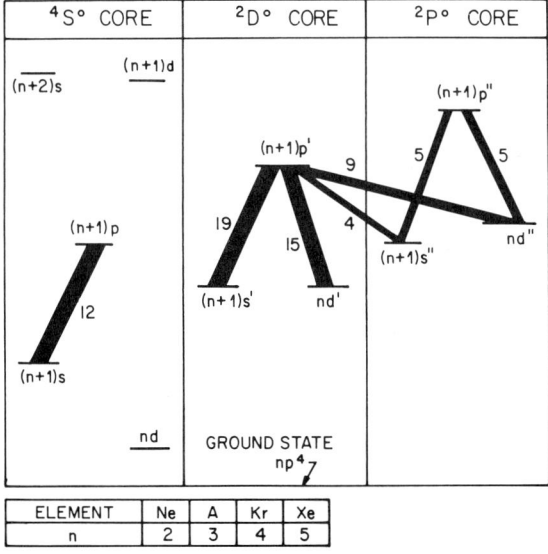

FIG. 8. Generalized energy level diagram for doubly ionized noble gas atoms. The number shown beside each supermultiplet group is the number of laser transitions observed in that group. Note that the nd, nd', and nd'' levels do not exist in neon (Bridges and Chester,[116] with revisions; by courtesy of American Institute of Physics).

states with the same core. There are also a significant number of transitions from the lowest p states of one core to the lowest s or d states of a different core; these are "forbidden" by the LS coupling model, since the core must reconfigure as well as the running electron. The remaining exceptions are few in number except for the nine $(n + 2)s \rightarrow (n + 1)p$ transitions shown dotted in Fig. 7. These laser lines are known to be excited by atom collision rather than electron collision, and they are discussed in Section 2.4. Within the supermultiplet groups of lines shown in the figures, the strongest laser lines are generally the strongest lines observed in spontaneous emission, and these in turn are generally the lines for which the "preferred" LS selection rules are obeyed[86]: Δcore = 0, $\Delta S = 0$, $\Delta J = \Delta L = +1$ ($\Delta \equiv$ upper–lower), and J large. The "strong" selection rules $\Delta J = 0, \pm 1$ but $J = 0 \not\rightarrow J = 0$ are strictly obeyed for all observed laser transitions. Some exceptions to the "preferred" rules occur; for example, the strong Ar II 0.5145-μm line has $\Delta S = -1$. However, the line is observed to be strong in spontaneous emission, so the discrepancy is in the ability of the LS model to describe the situation. Another note of caution: the spectroscopy of the ionized noble gases is not so well worked out that the energy level descriptions are immutable. Some of the level designations have been changed even since the observation of laser oscillation, e.g., Kr II.[128] For some strong laser lines spectroscopic designations are not known at all, e.g., the highly ionized xenon lines.

The particular laser transitions that oscillate in the 4p $\rightarrow$ 4s supermultiplet of singly ionized argon are shown in Fig. 9. All but one of the nine possible transitions from the 4p doublet levels allowed by the J selection rules are observed to oscillate, both pulsed and cw; only the 0.4376-μm 4p $^2S^o_{1/2} \rightarrow$ 4s $^2P_{3/2}$ line has not been reported. Only two of the lines from the quartet upper levels oscillate. Table V lists these ten lines, plus one other from a different supermultiplet that happens to fall in the same wavelength range, and compares their theoretical transition probabilities, experimental spontaneous emission intensities, laser oscillation threshold currents, and output powers in four different commercial lasers. The transition probabilities are those calculated by Statz et al.[129] for the pure LS-coupled model and also for an "exact" model employing linear combinations of LS-coupled wavefunctions. Calculations of A_{ij} by Marantz et al.,[130,131] similar to those of Statz et al. but including more levels in the

---

[128] L. Minnhagen, H. Strihed, and B. Petersson, *Ark. Fys.* **39**, 471 (1969).

[129] H. Statz, F. A. Horrigan, S. H. Koozekanani, C. L. Tang, and G. F. Koster, *J. Appl. Phys.* **36**, 2278 (1965).

[130] H. Marantz, Ph.D. Thesis, Cornell University, Ithaca, New York (1968).

[131] R. I. Rudko and C. L. Tang, *J. Appl. Phys.* **38**, 4731 (1967).

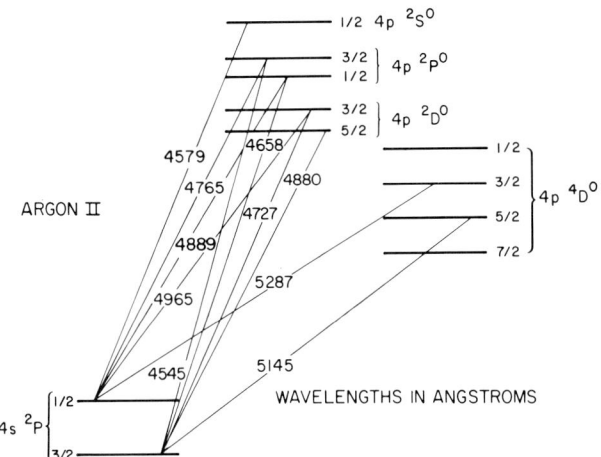

FIG. 9. Partial energy level diagram for singly ionized argon showing 10 of the laser transitions in the blue–green portion of the spectrum (Bridges,[110] with revisions; by courtesy of American Institute of Physics).

linear combinations, are also given. Comparison of the three columns of A_{ij} illustrates the degree of departure from pure *LS* coupling for this ion and also the sensitivity of the "exact" method to the number of terms included. The agreement between two sets of relative spontaneous emission measurements taken at different laboratories[131,132] is excellent; the agreement between the two sets of laser oscillation thresholds at two different laboratories[133,134] is not quite as good, as might be expected, since optical losses and mirror reflectivity variations are involved as well; the comparison among the power outputs of the four different lasers by two different manufacturers is excellent.

In the absence of collisional deexcitation the *total* spontaneous emission from a level will just equal the excitation rate for that level. The spontaneous emission and threshold data suggest that the $^2D^o_{5/2}$ upper level of the 0.4880-μm line must be excited more strongly than the $^2S^o_{1/2}$ and $^2P^o_{1/2}$ levels, since the 0.4579- and 0.4658-μm lines from these levels have similar A_{ij} but are weaker by three or four times in spontaneous emission and have higher oscillation thresholds. This higher pump rate is also re-

---

[132] W. B. Bridges and A. S. Halsted, "Gaseous Ion Laser Research," Final Tech. Rep., AFAL-TR-67-89. Hughes Res. Lab., Malibu, California, 1967 (unpublished). (Available from National Technical Information Service, Accession Number AD-814897.)

[133] R. A. Paananen, as reported in Statz *et al.*[129]

[134] E. F. Labuda and A. M. Johnson, *IEEE. J. Quantum Electron.* **qe-2**, 700 (1966).

TABLE V. Comparison of Theoretical Transition Probabilities, Spontaneous Emission Intensities, Oscillation Thresholds, and Output Powers for the Blue–Green Laser Lines in Ar II

λ (μm)	Transition	A_{ij} 10^7 sec^{-1}			Spontaneous int.		Oscillation threshold (A)		Commercial laser output power (W)			
		Pure LS[a]	Exact[a]	Exact[b]	mW/cm^3[c]	Arb.u.[d]	[e]	[f]	CR-2[g]	164-00[h]	CR-18[g]	170-03[h]
0.4545	4p $^2P^o_{3/2}$ → 4s $^2P_{3/2}$	10.2	2.77	5.19	46	53	10.5	3.2	0.05	—	0.7	0.5
0.4579	4p $^2S^o_{1/2}$ → 4s $^2P_{1/2}$	3.97	8.42	8.99	54	85	5.2	4.4	0.15	0.15	1.3	1.2
0.4658	4p $^2P^o_{1/2}$ → 4s $^2P_{3/2}$	3.78	7.55	8.55	57.3	53	6.9	3.0	0.05	0.05	0.7	0.5
0.4727	4p $^2D^o_{3/2}$ → 4s $^2P_{3/2}$	1.81	7.27	4.56	49	62	6.7	3.3	0.06	0.06	1.0	0.8
0.4765	4p $^2P^o_{3/2}$ → 4s $^2P_{1/2}$	1.76	7.15	4.47	63	63	3.8	2.2	0.3	0.3	2.5	1.5
0.4880	4p $^2D^o_{5/2}$ → 4s $^2P_{3/2}$	9.85	8.96	8.45	140	211	1.45	1.0	0.7	0.7	6.0	5.0
0.4889	4p $^2P^o_{1/2}$ → 4s $^2P_{1/2}$	6.53	1.98	1.25	9.7	9	[i]	[i]	—	—	—	—
0.4965	4p $^2D^o_{3/2}$ → 4s $^2P_{1/2}$	7.79	2.63	4.61	34	37	4.0	3.5	0.3	0.3	2.5	1.5
0.5017	4p' $^2F^o_{5/2}$ → 3d $^2D_{3/2}$	0	—	—	—	<25	—	5.0	0.14	0.14	1.5	0.7
0.5145	4p $^4D^o_{5/2}$ → 4s $^2P_{3/2}$	0	0.707	0.917	23	29	3.6	4.5	0.8	0.8	7.5	6.0
0.5287	4p $^4D^o_{3/2}$ → 4s $^2P_{1/2}$	0	0.14	0.196	2.2	3	10.3	17.2	0.1	—	1.5	1.3

[a] H. Statz, F. A. Horrigan, S. H. Koozekanani, C. L. Tang, and G. F. Koster, *J. Appl. Phys.* **36**, 2278 (1965).
[b] H. Marantz, Cornell University Thesis, as quoted in Rudko and Tang.
[c] R. I. Rudko and C. L. Tang. *J. Appl. Phys.* **38**, 4731 (1967); data for 2-mm-diam. tube at 5 A and 300 mtorr.
[d] W. B. Bridges and A. S. Halsted, "Gaseous Ion Laser Research," Final Tech. Rep. AFAL-TR-67-89 Hughes Res. Lab., Malibu, California, 1967 (unpublished). (Available from National Technical Information Service, Accession Number AD-814897.) Data for 3 mm tube at 15 A and 170 mtorr.
[e] Measurements by R. A. Paananen, as reported in Statz et al.[a]
[f] E. F. Labuda and A. M. Johnson, *IEEE J. Quantum Electron.* **qe-2**, 700 (1966).
[g] Power output specifications taken from Coherent Inc. ion laser brochure, dated 12/76.
[h] Power output specifications taken from Spectra-Physics data sheets, undated.
[i] This line oscillates both pulsed and cw, but is often overlooked because it is difficult to separate from 0.4880 μm with intracavity or external prisms.

flected in the five times higher saturated output power of the lasers given in the last four columns. By contrast the 0.5145-μm line has an order of magnitude smaller spontaneous emission intensity than 0.4880 μm, but a saturated laser output that is as high or higher than 0.4880 μm, which would indicate a comparable or higher pump rate. The reason for this situation is that when the 0.5145-μm line is not oscillating, approximately 80% of the $^4D^o_{5/2}$ population is radiated via the stronger 0.4426-μm $4p^4D^o_{5/2} \rightarrow 4s\ ^4P_{3/2}$ line, as shown by Rudko and Tang[131] and Bridges and Halsted.[132] These photons become available to 0.5145 μm when it oscillates. A similar situation occurs for the 0.5287-μm line. Other comparisons are less clear; for example, 0.4727 and 0.4965 μm originate from the same upper level and have about the same A_{ij} according to Marantz, but with a 3:1 ratio of A_{ij} according to Statz *et al.* The spontaneous emission measurements fall somewhere between, but the saturated output power results are 2:1 in the opposite direction. In the case of 0.4545 and 0.4765 μm, which also originate from a common upper level, the A_{ij} again disagree and the spontaneous emission measurements fall between, but the saturated power measurements agree in both sense and magnitude with the calculation of A_{ij} by Statz *et al.*

The laser lines observed in doubly ionized argon are shown in Fig. 10, along with all the known energy levels.[135] Here the strongest laser lines occur in the ultraviolet at 0.3511 and 0.3638 μm. The latter line was identified by McFarlane[117] as $4p'\ ^1F_3 \rightarrow 4s'\ ^1D^o_2$ although the Ar III singlet system is not known in the spectroscopic literature otherwise.[135] These levels have been drawn in Fig. 10 approximately where they should appear according to isoelectronic calculations.[136] The next strongest lines in Ar III are the 0.4183-μm line and the triplet of lines near 0.33 μm from the $4p'\ ^3F_{2,3,4}$ levels. The 0.4183-μm line does not originate from any levels shown in Fig. 10, and is probably $4p'\ ^1P_1 \rightarrow 4s'\ ^1D^o_2$ of the singlet system as determined from stimulated emission perturbation measurements.[136,137]

The energy level structure of singly ionized krypton is shown in Fig. 11. The strongest visible lines are 0.5682 and 0.6471 μm, followed by 0.4762, 0.5208, and 0.5309 μm in good agreement with the *LS* selection rules and the A_{ij} values calculated by Marantz *et al.*[138] using the method of linear combinations of *LS*-coupled wavefunctions. When oscillating simulta-

---

[135] C. E. Moore, *Natl. Bur. Stand. (U.S.), Circ.* **467**.
[136] W. B. Bridges and G. N. Mercer, "Ultraviolet Ion Lasers," Tech. Rep., ECOM-0299-F. Hughes Res. Lab., Malibu, California, 1969 (unpublished). (Available from National Technical Information Service, Accession Number AD-861927.)
[137] W. B. Bridges, unpublished.
[138] H. Marantz, R. I. Rudko, and C. L. Tang, *IEEE J. Quantum Electron.* **qe-5**, 38 (1965).

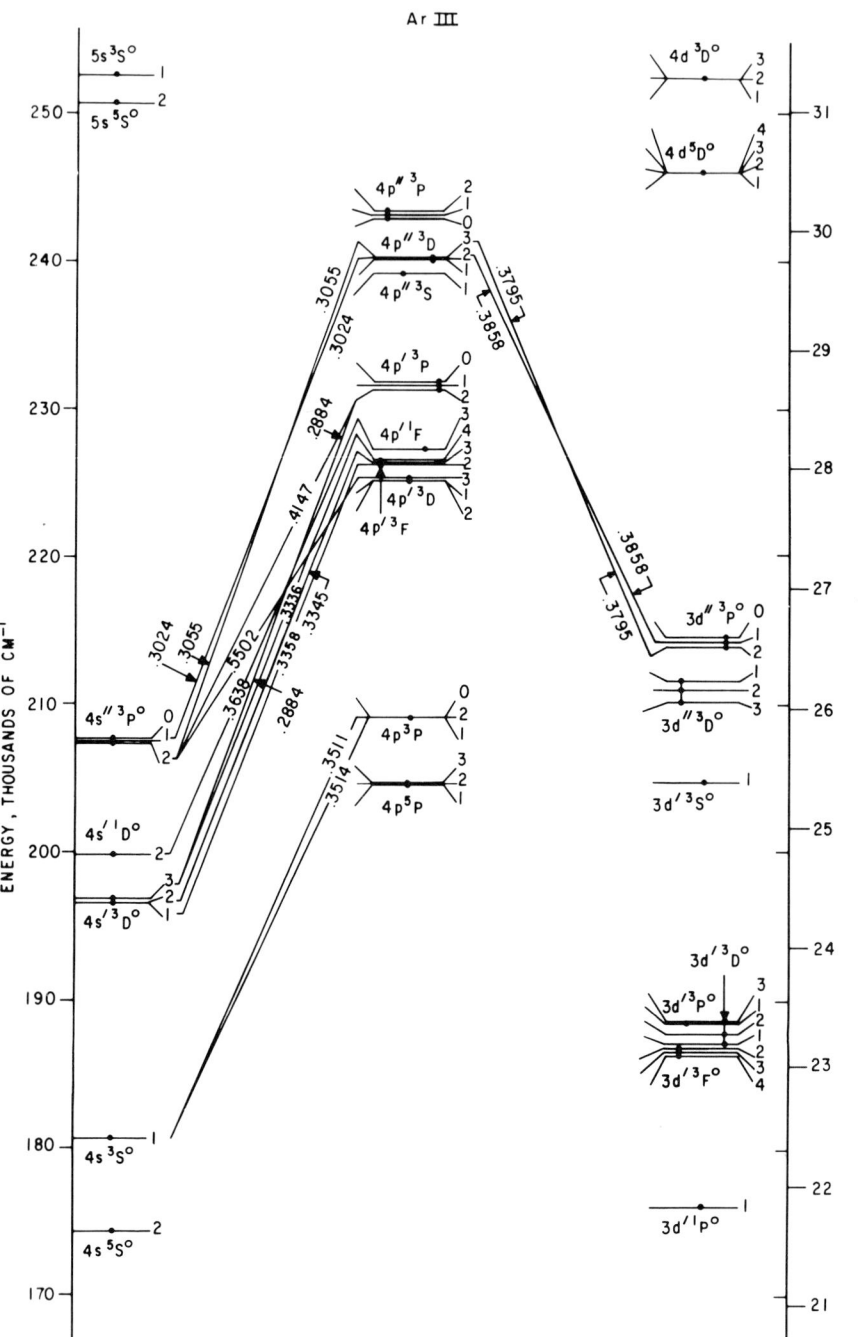

FIG. 10. Energy level diagram of double ionized argon showing all energy levels listed in Moore[135] (except the ground state) to approximate scale, and the observed laser transitions (Bridges and Chester,[1] by courtesy of CRC Press).

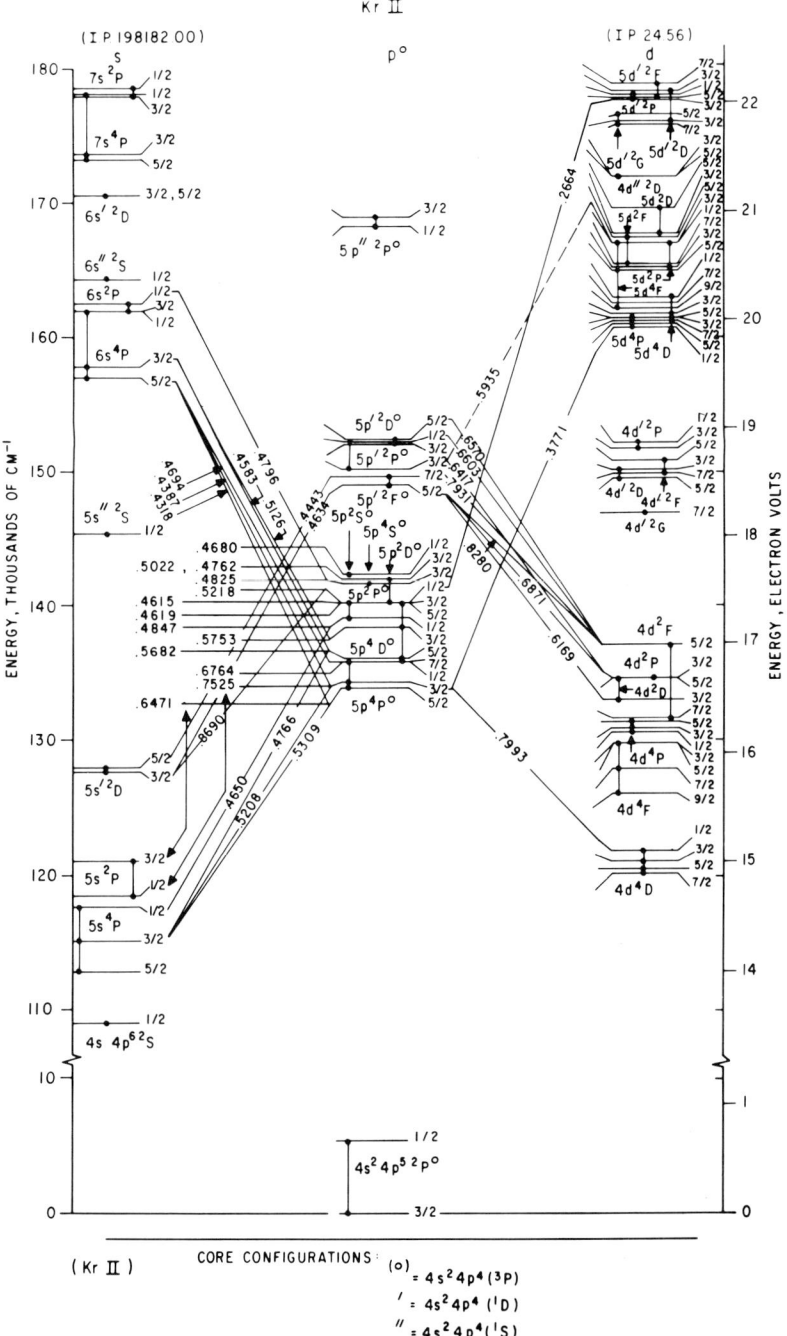

FIG. 11. Energy level diagram of singly ionized krypton, showing all energy levels listed in Moore[135] to approximate scale and the observed laser transitions. The level designations have not been corrected according to Minnhagen et al.[128] (Bridges and Chester,[1] by courtesy of CRC Press).

neously, this combination of wavelengths in krypton makes an almost "white" output beam. The total power available on the stronger krypton lines is half or less of the strongest lines in argon with the same discharge input power. However, the red line has found considerable application as a pump for the longer-wavelength cw dye lasers, so that the Kr II laser has received substantial commercial development as well as the Ar II laser. Strong ultraviolet output is also available from Kr III at 0.3507 μm on the 5p $^3P_2 \rightarrow$ 5s $^3S_1^0$ line, analogous to the 0.3511-μm line in Ar III, and in the violet at 0.4067 μm on the 5p' $^1F_3 \rightarrow$ 5s' $^1D_2^0$ line analogous to 0.3638 μm in Ar III.

Energy level diagrams and lists of laser wavelengths for neon and xenon are given by Bridges and Chester[1] and Willett.[1,2] These lasers have received comparatively little development, since argon and krypton are superior in output power and already cover the range of wavelengths offered by neon and xenon. The exception to this is the powerful blue, green, and uv lines exhibited by xenon at high pulsed-discharge currents. These lines are thought to originate in the Xe IV spectrum,[116,139–144] but their energy levels are not known—very few energy levels *are* known in Xe IV.[135] Many of these lines also oscillate continuously at much lower power levels.[144] Performance of the xenon laser is described later in this section.

Electron-excited laser oscillation has also been observed in other elements, with many lines observed in fluorine,[145,146] chlorine,[122,145] bromine,[147,148] oxygen,[116,117,119,149] sulfur,[150] nitrogen[116,117,119,151] phosphorus,[145,152] bismuth,[153] carbon,[117,154] silicon,[116,145,146] and a few each in

---

[139] J. A. Dahlquist, *Appl. Phys. Lett.* **6**, 193 (1965).
[140] M. Gallardo, M. Garavaglia, A. A. Tagliaferri, and E. Gallego Lluesma, *IEEE J. Quantum Electron.* **qe-6**, 745 (1970).
[141] C. C. Davis and T. A. King, *IEEE J. Quantum Electron.* **qe-8**, 755 (1972).
[142] E. Gallego Lluesma, A. A. Tagliaferri, C. A. Massone, M. Garavaglia, and M. Gallardo, *J. Opt. Soc. Am.* **63**, 362 (1973).
[143] V. Hoffmann and P. E. Toschek, *J. Opt. Soc. Am.* **66**, 152 (1976).
[144] W. B. Bridges and G. N. Mercer, *IEEE J. Quantum Electron* **qe-5**, 476 (1969).
[145] P. K. Cheo and H. G. Cooper, *Appl. Phys. Lett.* **7**, 202 (1965).
[146] H. P. Palenius, *Appl. Phys. Lett.* **8**, 82 (1966).
[147] W. M. Keefe and W. J. Graham, *Appl. Phys. Lett.* **7**, 263 (1965).
[148] W. M. Keefe and W. J. Graham, *Phys. Lett.* **20**, 643 (1966).
[149] Y. Hashino, Y. Katsuyama, and K. Fukuda, *Jpn. J. Appl. Phys.* **12**, 470 (1973).
[150] H. G. Cooper and P. K. Cheo, *Phys. Quantum Electron., Conf. Proc., 1965* p. 690 (1966).
[151] R. H. Neusel, *IEEE J. Quantum Electron.* **qe-3**, 207 (1967).
[152] G. R. Fowles, W. T. Silfvast, and R. C. Jensen, *IEEE J. Quantum Electron.* **qe-1**, 183 (1965).
[153] V. F. Keidan and V. S. Mikhalevskii, *J. Appl. Spectrosc. (Engl. Transl.)* **9**, 1154 (1968).
[154] R. W. Waynant, *Appl. Phys. Lett.* **8**, 419 (1973).

2.2. IONS EXCITED BY ELECTRON COLLISION

still more elements (see Fig. 1). Some of these laser lines have operated continuously as well as pulsed, depending on their energy level structure. Lines in ionized calcium,[10,155] strontium,[19,155] and barium[20,21,155,156] are essentially transient or cyclic lasers analogous to the Cu I laser described in the section on neutral atom lasers. The rare earths europium[63,157] and ytterbium[60-62] also exhibited ion laser oscillation, but little is known about their excitation mechanisms. To date, none of these ions have exhibited laser characteristics superior to argon or krypton, and as a consequence they have received little or no commercial development and little study beyond basic spectroscopy. Chemical reactivity and, in some cases, high-temperature operation, work against the development of gas lasers employing gases other than the noble gases.

2.2.2. Noble-Gas Ion Laser Characteristics and Mechanisms

Over 200 papers have appeared dealing with the external characteristics, internal excitation mechanisms, and plasma processes in the argon ion laser. Reviews incorporating the results of investigations up to 1970 were written by Kitaeva et al.,[158] Lin and Chen,[159] and Bridges et al.[160] More recently a review of Dunn and Ross[161] and a comprehensive monograph by Davis and King[124] have appeared and should be consulted for more extensive discussion and references to the literature than can be given in this short section. Only the basic characteristics and mechanisms are presented here, omitting many references to the interesting measurements that have led to these results.

The early work on cw ion lasers centered on parametric studies and investigations of the internal mechanisms of smaller diameter discharges (1–5 mm) with output powers in the 1–10-W range. Different methods of exciting the discharges were explored for their efficacy, and substantial developments were made in discharge tube materials and techniques. These investigations were carried out at many laboratories worldwide, with principal groups at Bell Laboratories,[114,162-171] Cor-

---

[155] E. L. Latush and M. F. Sém, *Sov. J. Quantum Electron.* (*Engl. Transl.*) **3**, 216 (1973).

[156] K. U. Baron and B. Stadler, *IEEE J. Quantum Electron.* **qe-11**, 852 (1975).

[157] P. A. Bokhan, V. M. Klimkin, and V. E. Prokop'ev, *JETP Lett.* (*Engl. Transl.*) **18**, 44 (1973).

[158] V. F. Kitaeva, A. I. Odintsov, and N. N. Sobolev, *Sov. Phys.—Usp.* (*Engl. Transl.*) **12**, 699 (1970).

[159] S.-C. Lin and C. C. Chen, *AIAA Pap.* **70-82** (1970) (unpublished).

[160] W. B. Bridges, A. N. Chester, A. S. Halsted, and J. V. Parker, *Proc. IEEE* **59**, 724 (1971).

[161] M. H. Dunn and J. N. Ross, *Prog. Quantum Electron.* **4**, Part 3, 233 (1976).

[162] E. I. Gordon and E. F. Labuda, *Bell Syst. Tech. J.* **43**, 4 (1964).

nell University,[129-131,138,171-173] the Hughes Research Laboratories,[110,114,132,136,144,160,174-176] the P. N. Lebedev Physics Institute,[158,177-188]

---

[163] E. F. Labuda, E. I. Gordon, and R. C. Miller, *IEEE J. Quantum Electron.* **qe-1**, 273 (1965).

[164] E. F. Labuda, C. E. Webb, R. C. Miller, and E. I. Gordon, *18th Gaseous Electron. Conf.*, 1965 unpublished (1965).

[165] C. E. Webb, *Proc. Symp. Mod. Opt.*, 1967, p. 366 (1967).

[166] E. I. Gordon, E. F. Labuda, R. C. Miller, and C. E. Webb, in "Physics of Quantum Electronics" (P. L. Kelly, B. Lax, and P. E. Tannenwald, eds.), p. 664. McGraw-Hill, New York, 1966.

[167] R. C. Miller, E. F. Labuda, and C. E. Webb, *Bell Syst. Tech. J.* **46**, 281 (1967).

[168] C. E. Webb, *J. Appl. Phys.* **39**, 5441 (1968).

[169] A. N. Chester, *Phys. Rev.* **169**, 172 (1968).

[170] A. N. Chester, *Phys. Rev.* **169**, 184 (1968).

[171] E. F. Labuda and A. M. Johnson, *IEEE J. Quantum Electron.* **qe-2**, 700 (1966).

[172] R. I. Rudko and C. L. Tang, *Appl. Phys. Lett.* **9**, 41 (1966).

[173] H. Marantz, R. I. Rudko, and C. L. Tang, *Appl. Phys. Lett.* **9**, 409 (1966).

[174] W. B. Bridges, P. O. Clark, and A. S. Halsted, *IEEE J. Quantum Electron.* **qe-2**, xix (1966).

[175] A. S. Halsted, W. B. Bridges, and G. N. Mercer, "Gaseous Ion Laser Research," Tech. Rep. AFAL-TR-68-227. Hughes Res. Lab., Malibu, California, 1968 (unpublished). (Available from National Technical Information Service, accession number AD-841834.)

[176] W. B. Bridges, P. O. Clark, and A. S. Halsted, "High Power Gas Laser Research," Tech. Rep., AFAL-TR-66-369. Hughes Res. Lab., Malibu, California, 1967 (unpublished). (Available from National Technical Information Service, Accession Number AD-807363.)

[177] V. F. Kitaeva, Yu. I. Osipov, and N. N. Sobolev, *JETP Lett. (Engl. Transl.)* **4**, 146 (1966).

[178] V. F. Kitaeva, Yu. I. Osipov, and N. N. Sobolev, *IEEE J. Quantum Electron.* **qe-2**, 635 (1966).

[179] V. F. Kitaeva, Yu. I. Osipov, N. N. Sobolev, and P. L. Rubin, *Sov. Phys.—Tech. Phys. (Engl. Transl.)* **12**, 850 (1967).

[180] P. L. Rubin, *Sov. Phys.—Tech. Phys. (Engl. Transl.)* **13**, 361 (1968).

[181] V. F. Kitaeva, Yu, I. Osipov, P. L. Rubin, and N. N. Sobolev, *IEEE J. Quantum Electron.* **qe-4**, 357 (1968).

[182] V. F. Kitaeva and Yu. I. Osipov, *Sov. Phys.—Tech. Phys. (Engl. Transl.)* **13**, 282 (1968).

[183] V. F. Kitaeva, Yu. I. Osipov, P. L. Rubin, and N. N. Sobolev, *IEEE J. Quantum Electron.* **qe-5**, 72 (1969).

[184] V. F. Kitaeva, L. F. Kocherga, L. Ya. Ostrovskaya, and N. N. Sobolev, *Proc. Int. Conf. Phenom. Ioniz. Gases, 9th 1969*, p. 255 (1969).

[185] V. F. Kitaeva, L. Ya. Ostrovskaya, and N. N. Sobolev, *Sov. J. Quantum Electron. (Engl. Transl.)* **1**, 341 (1972).

[186] V. F. Kitaeva, Yu. I. Osipov, L. S. Pavlova, V. M. Polyakov, N. N. Sobolev, and L. S. Fedorov, *Sov. Phys.—Tech. Phys. (Engl. Transl.)* **16**, 1509 (1972).

[187] V. F. Kitaeva, Yu. I. Osipov, and N. N. Sobolev, *IEEE J. Quantum Electron.* **qe-7**, 391 (1971).

[188] V. F. Kitaeva, Yu. I. Osipov, N. N. Sobolev, A. L. Shelekhov, V. P. Agheev, *IEEE J. Quantum Electron.* **qe-10**, 803 (1974).

RCA,[189-198] the Raytheon Research Division,[129,199-203] Spectra-Physics,[204-207] and Yale University.[113,208-215] More recent work on cw lasers has centered on the extension to larger-diameter discharges with output powers exceeding 100 W cw. This work is being carried out by investigators at the Technical University of Berlin,[216-226] the University of

---

[189] J. M. Hammer and C. P. Wen, *J. Chem. Phys.* **46**, 1225 (1967).
[190] I. Gorog and F. W. Spong, *Appl. Phys. Lett.* **9**, 61 (1966).
[191] I. Gorog and F. W. Spong, *RCA Rev.* **28**, 38 (1967).
[192] I. Gorog and F. W. Spong, *RCA Rev.* **30**, 277 (1969).
[193] I. Gorog, *RCA Rev.* **32**, 88 (1971).
[194] K. G. Hernqvist and J. R. Fendley, Jr., *Proc. Symp. Mod. Opt., 1967* p. 383 (1967).
[195] K. G. Hernqvist and J. R. Fendley, Jr., *IEEE J. Quantum Electron.* **qe-3**, 66 (1967).
[196] K. G. Hernqvist, *Appl. Opt.* **9**, 2249 (1970).
[197] J. R. Fendley, Jr., *IEEE J. Quantum Electron.* **qe-4**, 627 (1968).
[198] J. R. Fendley, Jr., and J. J. O'Grady, "Development, Construction and Demonstration of a 100 W cw Argon Ion Laser," Tech. Rep. ECOM-0246-F. RCA, Lancaster, Pennsylvania, 1970, (unpublished). (Available from National Technical Information Service.)
[199] R. A. Paananen, A. Adams, Jr., and D. T. Wilson, *NEREM Rec.* **7**, 238 (1965).
[200] R. A. Paananen, *IEEE Spectrum* **3**, 88 (1966).
[201] R. A. Paananen, *Appl. Phys. Lett.* **8**, 34 (1966).
[202] G. deMars, M. Seiden, and F. A. Horrigan, *IEEE J. Quantum Electron.* **qe-4**, 631 (1968).
[203] I. D. Latimer, *Appl. Phys. Lett.* **13**, 333 (1968).
[204] W. E. Bell, *Appl. Phys. Lett.* **7**, 190 (1965).
[205] A. L. Bloom, R. L. Byer, and W. E. Bell, *Phys. Quantum Electron., Proc. Conf., 1965* p. 688 (1966).
[206] J. P. Goldsborough, E. B. Hodges, and W. E. Bell, *Appl. Phys. Lett.* **8**, 137 (1966).
[207] J. P. Goldsborough, *Appl. Phys. Lett.* **8**, 218 (1966).
[208] W. R. Bennett, Jr., P. J. Kindlmann, G. N. Mercer, and J. Sunderland, *Appl. Phys. Lett.* **5**, 158 (1964).
[209] W. R. Bennett, Jr., E. A. Ballik, and G. N. Mercer, *Phys. Rev. Lett.* **16**, 603 (1966).
[210] W. R. Bennett, Jr., G. N. Mercer, P. J. Kindlmann, B. Wexler, and H. Hyman, *Phys. Rev. Lett.* **17**, 987 (1966).
[211] E. A. Ballik, W. R. Bennett, Jr., G. N. Mercer, and C. J. Elliot, *IEEE J. Quantum Electron.* **qe-2**, xiv (1966).
[212] E. A. Ballik, W. R. Bennett, Jr., and G. N. Mercer, *Appl. Phys. Lett.* **8**, 214 (1966).
[213] E. A. Ballik and C. J. Elliot, *Appl. Opt.* **5**, 1858 (1966).
[214] R. C. Sze and W. R. Bennett, Jr., *Phys. Rev. A* **5**, 837 (1972).
[215] G. N. Mercer, V. P. Chebotayev, and W. R. Bennett, Jr., *Appl. Phys. Lett.* **10**, 177 (1967).
[216] H. Boersch, G. Herziger, W. Seelig, and I. Volland, *Phys. Lett., A* **24**, 695 (1967).
[217] K. Banse, G. Herziger, G. Schäfer, and W. Seelig, *Phys. Lett. A* **27**, 682 (1968).
[218] K. Banse, H. Boersch, G. Herziger, G. Schäfer, and W. Seelig, *Phys. Lett. A* **28**, 6 (1968).
[219] G. Herziger and W. Seelig, *Z. Phys.* **215**, 437 (1968).
[220] G. Herziger and W. Seelig, *Z. Phys.* **219**, 5 (1969).
[221] H. Boersch, J. Boscher, D. Hoder, and G. Schäfer, *Phys. Lett. A* **31**, 188 (1970).
[222] G. Schäfer and W. Seelig, *Z. Angew. Phys.* **29**, 246 (1970).

Berne,[227-233] the University of California at San Diego,[159,234-240] and the Institute of Semiconductor Physics of the USSR Academy of Sciences.[241-247] In the sections that follow, brief summaries are given of the important characteristics of pulsed and cw noble gas ion lasers, the plasma parameters, excitation mechanism models proposed, processes important in producing the population inversion, and the discharge excitation techniques that have proven successful.

2.2.2.1. Characteristics of Pulsed Noble Gas Ion Lasers. Pulsed noble-gas ion lasers with discharge diameters of 1–20 mm have been studied up to current densities greater than 10 kA/cm^2; they typically operate at relatively low gas pressures, a few tens of mtorr for small diameter discharges. While no exact constant-electron-temperature scaling law, pD = const, has been observed for pulsed lasers, it is clear that the op-

---

[223] J. Boscher, T. Kindt, and G. Schäfer, Z. Phys. **241**, 280 (1971).
[224] J. Boscher, R. Finzel, J. Salk, and G. Schäfer, Appl. Phys. **5**, 203 (1974).
[225] K. Barthel, J. Boscher, J. Salk, and G. Schäfer, Appl. Phys. **8**, 79 (1975).
[226] J. Eichler and H. J. Eichler, Appl. Phys. **9**, 53 (1976).
[227] G. Herziger, H. R. Lüthi, and W. H. Seelig, Z. Phy. **264**, 61 (1973).
[228] K. Banse, H. R. Lüthi, and W. H. Seelig, Appl. Phys. **4**, 141 (1974).
[229] H. R. Lüthi and W. H. Seelig, Appl. Phys. **6**, 261 (1975).
[230] H. R. Lüthi, R. Mertenat, W. H. Seelig, and J. H. Steinger, Z. Angew. Math. Phys. **27**, 101 (1976).
[231] H. R. Lüthi and W. H. Seelig, J. Opt. Soc. Am. **66**, 1098 (1976).
[232] H. R. Lüthi, J. Appl. Phys. **48**, 664 (1977).
[233] H. R. Lüthi, W. H. Seelig, J. Steinger, and W. Lobsiger, IEEE J. Quantum Electron. **qe-13**, 404 (1977).
[234] S.-C. Lin and C. C. Chen, Bull. Am. Phys. Soc. [2] **14**, 839 (1969).
[235] C. P. Wang and S.-C. Lin, J. Appl. Phys. **43**, 5068 (1972).
[236] C. P. Wang and S.-C. Lin, J. Appl. Phys. **44**, 4681 (1973).
[237] C. C. Chen, "A Quantitative Theory of Noble Gas Ion Laser Discharge: Formulation," Paper P-5320. Rand Corp., Santa Monica, Calif., 1974 (unpublished).
[238] C. P. Wang and S.-C. Lin, J. Appl. Phys. **45**, 350 (1974).
[239] C. C. Chen, "A Quantitative Theory of Noble Gas Ion Laser Discharge: Argon Ion Laser," Paper P-5350. Rand Corp., Santa Monica, Calif., 1975 (unpublished).
[240] T. K. Tio, H. H. Luo, and S.-C. Lin, Appl. Phys. Lett. **29**, 795 (1976).
[241] V. I. Donin, V. M. Klement'ev, and V. P. Chebotayev, J. Appl. Spectrosc. **5**, 290 (1966).
[242] V. I. Donin, V. M. Klement'ev, and V. P. Chebotayev, Instrum. Exp. Tech. (Engl. Transl.) p. 932 (1967).
[243] V. I. Donin, J. Appl. Spectrosc. (Engl. Transl.) **11**, 889 (1969).
[244] V. I. Donin, Opt. Spectrosc. (Engl. Transl.) **26**, 160 (1969).
[245] V. I. Donin, Opt. Spectrosc. (Engl. Transl.) **26**, 128 (1970).
[246] V. I. Donin, Sov. Phys.—JETP. (Engl. Transl.) **35**, 858 (1972).
[247] G. N. Alferov, V. I. Donin, and B. Ya. Yurshin, JETP Lett. (Engl. Transl.) **18**, 369 (1973).

2.2. IONS EXCITED BY ELECTRON COLLISION

timum gas pressure and the discharge diameter are inversely related, at least over the range of diameters 1–20 mm. The maintenance of pressures below about 5 mtorr places a practical upper limit to the discharge diameter. Above the oscillation threshold current, the strength of the singly ionized laser lines increases rapidly up to current densities J of the order of 500 A/cm^2 for argon and then saturates and decreases with further increase in current. The doubly ionized laser lines usually reach threshold around the saturation point of the singly ionized lines or slightly before and increase rapidly, while the triply ionized laser lines reach threshold, and so on. This behavior was first noted for pulsed lasers by Cheo and Cooper,[119,248] although the effect was previously well known for spontaneous emission lines and was often used by spectroscopists to identify approximately the ionization states of various spectral lines. This technique has been used recently by Hoffmann and Toschek[143] to identify the ionization state of various xenon lines.

While the oscillation of pulsed noble gas ion lasers generally coincides with the current pulse, indicating direct electron collision as the excitation mechanism, the output waveform typically exhibits some temporal structure as shown in Fig. 12. As the discharge current is increased above threshold, a dip in the output waveform begins to develop as seen in Fig. 12 (right); eventually the dip deepens and widens sufficiently to separate the pulse into two portions: a short, self-terminating pulse and a second pulse that begins some time during the current pulse and then lasts until the current pulse decays. A similar behavior is exhibited if the pressure is increased at constant current as shown in Fig. 12 (left). Since the spontaneous emission from the upper level remains constant during the current pulse, the dip indicates a temporary increase in the population of the lower laser level; this was shown experimentally by Gordon *et al.*[166] They measured negative gain in a laser amplifier under discharge conditions that produced a "dead" interval in the same discharge when used as an oscillator. Application of a dc longitudinal magnetic field at constant current and pressure will also produce the same development of two pulses as shown by Toyoda and Yamanaka.[249] The application of magnetic fields smaller than those which cause the waveform to develop a dead region were shown to be beneficial by Birnbaum[250] and Toyoda and Yamanaka, with increases in peak output of 2–6 times for the same discharge current.

The same pressures, currents, and magnetic fields that promote the "dead" interval in the output waveform also produce a *spatial* develop-

[248] P. K. Cheo and H. G. Cooper, *Appl. Phys. Lett.* **6**, 177 (1965).
[249] K. Toyoda and C. Yamanaka, *Technol. Rep. Osaka Univ.* **17**, 407 (1967).
[250] M. Birnbaum, *Appl. Phys. Lett.* **12**, 86 (1968).

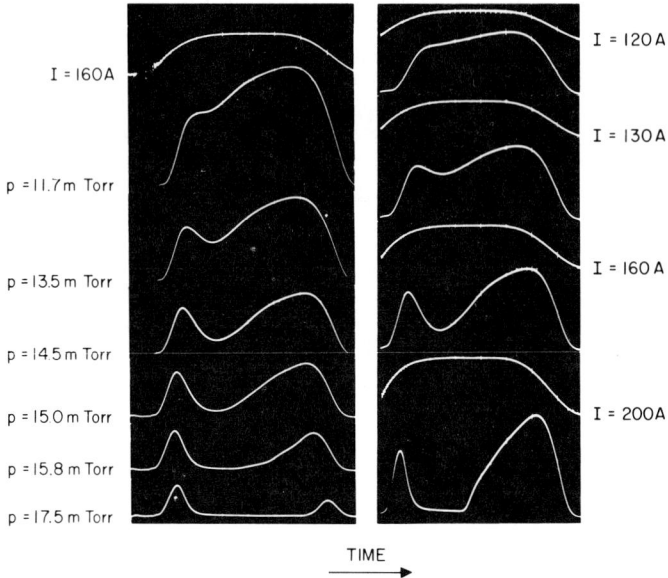

FIG. 12. Variation of the output waveform of a pulsed ion laser as the gas pressure is changed at constant current (left) or the discharge current is changed at constant gas pressure (right). The current pulse is about 50 μsec long in this 5-mm diameter laser.

ment in the laser output as well. At lower values of these parameters the gain is highest on the discharge tube axis, and optical cavity modes with peak intensities on the axis generally oscillate. As the current, pressure, or magnetic field is increased, the on-axis gain suffers a relative decrease, and modes with an annular distribution ("ring" modes) are encouraged, as shown by Cheo and Copper.[248] At very high values of the parameters, the laser output may consist of only a thin annulus near the discharge walls. The temporal and spatial behavior of the heavier noble gases krypton and xenon are similar to argon, except the onset of ring modes and temporal dead regions occurs at lower values of the pressure, current, and magnetic field. Both temporal and spatial variations described above are consistent with the proposition that the lower laser levels are depopulated radiatively, but that this radiation can be trapped by reabsorption[11] before it can escape the central portion of the discharge, as discussed later.

2.2.2.2. Characteristics of Continuous Noble Gas Ion Lasers. Continuous noble-gas ion lasers have been studied over a range of discharge diameters of about 1 to 16 mm, and with current densities up to 1000 A/cm². The current dependence of the multiline blue–green output of a

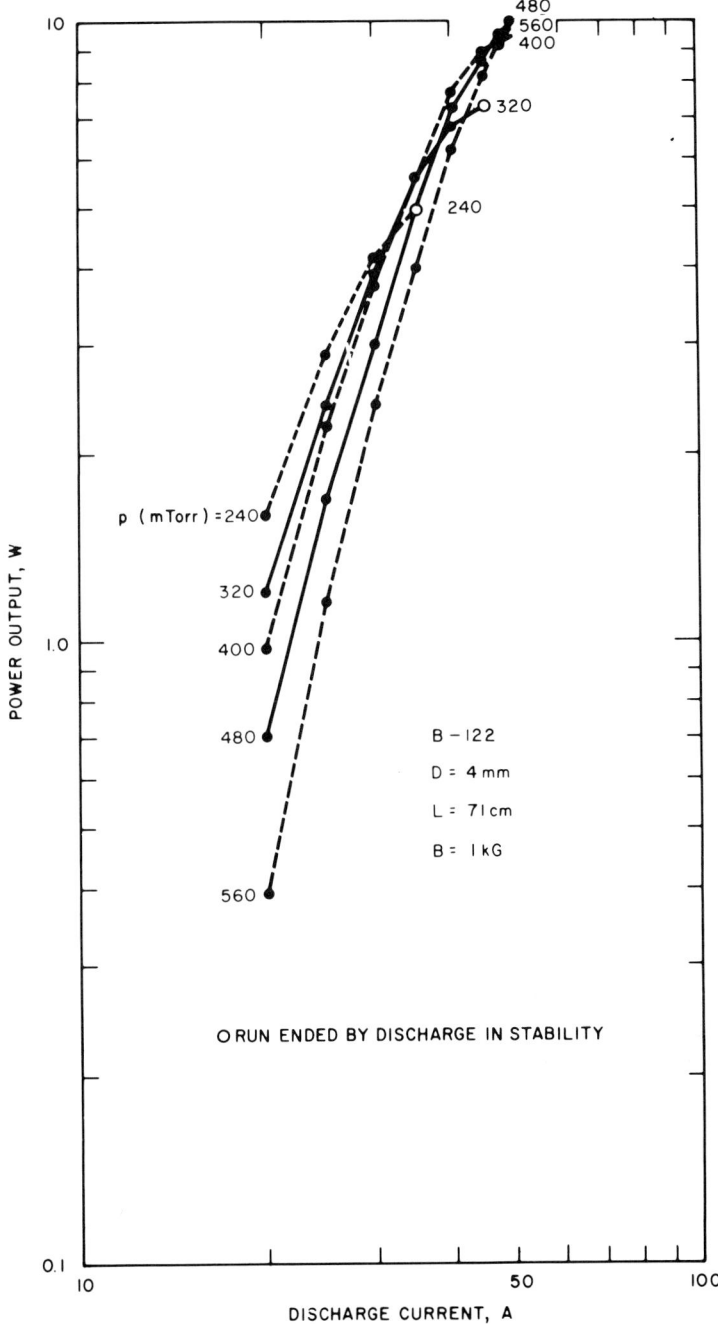

Fig. 13. Variation of the output power with discharge current and gas filling pressure for a cw ion laser (from Bridges et al.,[176] by courtesy of the Hughes Research Laboratories).

typical small diameter cw argon ion laser[176] is shown in Fig. 13. The steep increase with current above threshold was recognized to be significant and suggests a multistep excitation process that depends on some power of the electron density, as proposed by Gordon et al.[163] None of the curves shown in Fig. 13 has the exact form $P \sim J^n$; however, the *envelope* of the set of curves obtained by optimizing the gas pressure at each current is very closely $P \sim J^2$, suggesting a two-electron-collision excitation process. The increase in optimum gas pressure with increasing current is consistent with the idea that the gas is being heated by the discharge and that at higher currents the cold gas filling pressure must be increased to maintain the same number density in the active region. Webb has shown this to be the case by X-ray absorption measurements of the actual number of argon atoms in the discharge.[168] As in the case of pulsed lasers, no strict constant pD relation is observed; Herziger and Seelig report $pD = 50$ mtorr-cm over a diameter range of 7–15 mm,[227] but higher optimum pD products are typically observed in smaller-diameter tubes. The optimum fill pressure for a cw laser is generally about an order of magnitude larger than that of a pulsed laser of the same diameter. When corrected for the higher atom temperature of the cw laser discharge, the optimum atom number density is about the same in both cw and pulsed discharges.

A further complication in describing the gas pressure characteristics of noble-gas ion lasers is raised by the phenomenon of gas pumping. Under typical operating conditions, an ion laser discharge will develop a substantial gradient in gas pressure, with higher pressure at the anode. The gradient may become large enough to extinguish the discharge unless an external gas return path is added to equalize the pressure, as suggested by Gordon and Labuda.[162] Even with a gas return path, significant gradients can exist. Figure 14 shows the pressure differential between anode and cathode for a 3-mm-diameter tube as a function of discharge current and cold-filling pressure, as measured by Halsted[132,160]; a gas return path with a vacuum conductance equal to the laser bore vacuum conductance was employed. Note that the pressure differential is as large as the cold-filling pressure under some circumstances. Experimental and theoretical studies of gas pumping in ion lasers have been made by Halsted[132,160] and Chester,[169,170] although the basic phenomenon was studied much earlier by Langmuir[251] and Druyvesteyn.[252] A qualitative explanation of the basic features can be given simply: Within the positive column of the dis-

---

[251] I. Langmuir, *J. Franklin Inst.* **196**, 751 (1923); also see "The Collected Works of Irving Langmuir," Vol. 4. Pergamon, Oxford, 1961.
[252] M. J. Druyvesteyn, *Physica* **2**, 255 (1935).

2.2. IONS EXCITED BY ELECTRON COLLISION

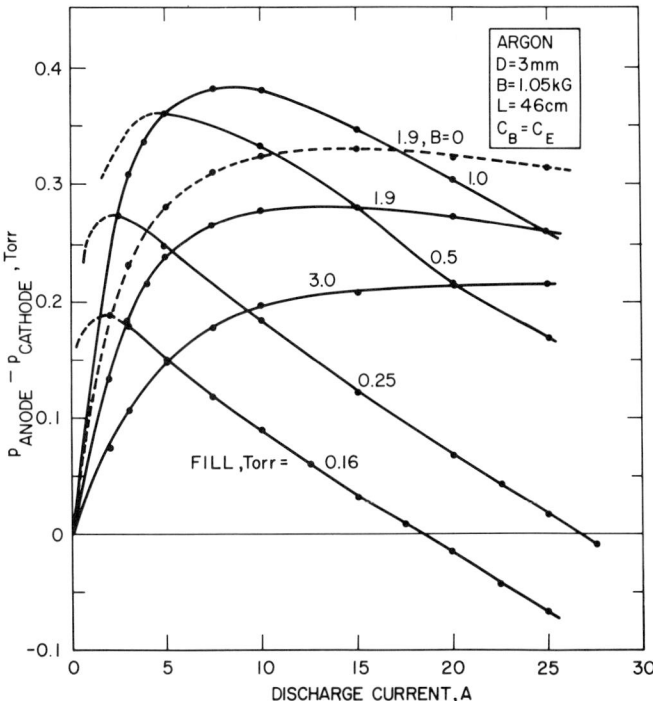

FIG. 14. Variation of the anode-to-cathode pressure differential with discharge current and gas pressure in an argon ion laser (from Bridges and Halsted,[132] by courtesy of the Hughes Research Laboratories).

charge, ions and electrons gain equal and opposite momenta from the longitudinal electric field. However, ions are attracted to the walls by the *radial* electric field (wall sheath) while all but the most energetic electrons are repelled back into the discharge. Thus the ions give up most of their momentum to the walls, while the electrons give theirs up to the gas atoms in the discharge. This creates a net anode-directed flow of neutral gas atoms with a magnitude proportional to the discharge current; this is seen in Fig. 14 as the linearly increasing pressure differential at small currents. At higher currents the cathode-directed drift of the ions themselves makes a significant contribution to the gas flow, since they become atoms upon neutralization; increased ion flow is responsible for the descending portion of the lower-pressure curves in Fig. 14. At some point the ion flow and the electron-collision-driven gas flow can just balance, and the pressure differential goes to zero, as seen for two curves in Fig. 14. Attempts to operate without a gas return path at this particular condition were unsuccessful, however, indicating that it represents an unstable equilibrium.[132] The existence of pressure gradients makes analysis of the plasma properties more difficult, and also complicates any parametric study of the laser, since the operating conditions depend on the nature of the gas return path as well as the other discharge parameters.

A longitudinal dc magnetic field is usually used to increase the output of small-diameter cw ion lasers. The field serves to increase the electron density n_e for a given discharge current density J by decreasing the average longitudinal drift velocity, since

$$n_e = J/v_{\text{drift}}. \tag{2.2.1}$$

The longitudinal electric field is also decreased, as exhibited by a decrease in the overall discharge voltage. Figure 15 shows the increase in spontaneous emission from selected argon upper laser levels and neutral argon levels with increasing magnetic field for a 3-mm-diameter tube.[132] The laser output follows the increasing portion of the spontaneous emission curves but then falls off at fields higher than the value about 1 kG marked B_{opt}, indicating that the lower laser level population is being increased faster than the upper laser level population for still higher magnetic fields. Another contribution to the laser output decrease at higher magnetic fields is the tendency toward circular polarization of the Zeeman-split laser lines and the concomitant loss incurred in the plane-polarizing Brewster's angle windows usually used on discharge tubes, as suggested by Bell[253] and studied experimentally by Labuda et al.[163] However, the laser output is observed to decrease with increasing magnetic field even in

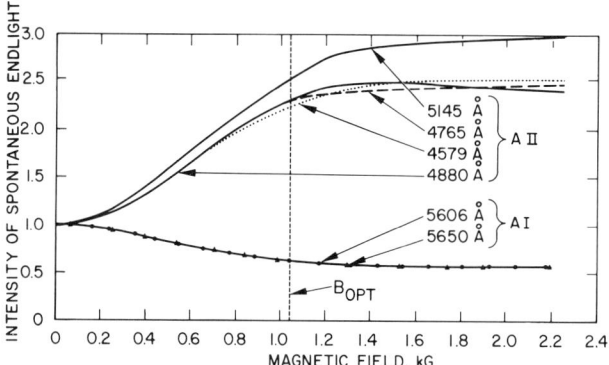

FIG. 15. Variation of the intensity of the spontaneous endlight with axial magnetic field for a number of Ar II ion laser lines and neutral Ar I lines (from Bridges and Halsted,[132] by courtesy of the Hughes Research Laboratories).

[253] W. E. Bell, private communication to E. I. Gordon, as cited in Labuda et al.[163]

tubes without polarizing windows.[163] Larger-diameter ion lasers do not show the same increase in output with magnetic field when operated at higher values of current density. Banse et al.[228] report a factor of 2 increase at values of $JR \leq 50$ A/cm for a 6-mm-diameter tube and a diameter dependence of the optimum field of $B_{\mathrm{opt}}R = 1500$ G-mm. For higher currents, $JR \geq 150$ A/cm, they observed no increase in output with the application of a longitudinal magnetic field of any magnitude. However, small values of magnetic field of the order of tens of gauss are useful in suppressing discharge instabilities, thus allowing the laser to be operated with higher discharge currents, as observed by Lüthi et al.[233]

2.2.2.3. Plasma Properties. Substantial efforts have been made to determine the internal plasma properties of noble gas ion lasers, but the task has been difficult because of the particular operating regime. The high power density of a typical ion laser plasma is the order of 100 W/cm of discharge length, which almost precludes probe measurements. The electron density is typically 10^{13}–10^{14} cm^{-3}, high enough to place the electron plasma frequency in the millimeter or submillimeter wave range, making rf diagnostics difficult. As a consequence, optical linewidth measurements have been used extensively to deduce densities and temperatures of the various plasma species. Theoretical modeling is generally required to make these deductions because of the anisotropic nature of the discharge; linewidths measured parallel and perpendicular to the current flow are not generally the same because of the radial drift velocities resulting from the strong radial electric fields, as described by Kagen and Perel'[254,255] for example. Such modeling is, itself, complicated by the peculiar operating regime, since some of the mean free paths for the various important collision processes are larger than a typical discharge diameter, some smaller and some comparable. Herziger and Seelig[219] have made a detailed treatment of the ion laser plasma and excitation processes based on the Tonks–Langmuir[256] formulation. Lin and Chen[234,237] have attempted to formulate a more general theory and obtain solutions for argon.[239]

Electrons, ions, and neutral atoms in an ion laser discharge have energy distributions that are roughly Maxwellian, but characterized by different temperatures with $T_e > T_i > T_a$. Because of the strong radial electric fields, these are further characterized as *parallel* or *perpendicular* (to the direction of current flow) according to the direction of observation in the optical linewidth measurements.[254,255] Probe and microwave measure-

---

[254] Yu. M. Kagen and V. I. Perel', *Opt. Spectrosc. (USSR)* **2**, 298 (1957).
[255] Yu. M. Kagen and V. I. Perel', *Opt. Spectrosc. (USSR)* **4**, 3 (1958).
[256] L. Tonks and I. Langmuir, *Phys. Rev.* **34**, 876 (1929).

ments average this anisotropy. The results of the measurement of these temperatures and their dependence on the discharge parameters are summarized below. Measurements of the particle densities are also summarized.

2.2.2.3.1. NEUTRAL-ATOM TEMPERATURE. Based on earlier measurements, Chester[170] has suggested an empirical formula for gas temperature in smaller-diameter ion lasers as a function of current density and tube diameter,

$$T_a = 300(1 + 0.02JD^{1/2}) \text{ K}, \qquad (2.2.2)$$

where J is in A/cm² and D is in mm. Kitaeva et al.[187] and Sze and Bennett[214] have measured temperature as a function of J, D, and p in small-diameter tubes. The measurements of Kitaeva et al. generally follow Eq. (2.2.2), with T_a = 1000–1500 K for a 2.8-mm-diameter tube at 100 A/cm² to 3500–4000 K for a 7-mm-diameter tube at 200 A/cm². Kitaeva et al. also determine the atom temperature from line profiles measured perpendicular to the discharge axis ("perpendicular temperature"); these are generally higher than the longitudinal temperature, especially at lower pressures and higher currents (0.2 torr, 200 A/cm²); as the pressure increases and the current decreases (0.6 torr, 100 A/cm²) the two temperatures approach each other. Herziger et al.[227] find a somewhat different dependence on J and D and a generally higher temperature in larger diameter tubes. They find T_a = 9000 K for JD = 200 A/cm and 18,000 K for JD = 300 A/cm.

2.2.2.3.2. ION TEMPERATURES. The ion temperatures measured by Kitaeva et al.[187] and Herziger et al.[227] exhibit variations with J, D, and p similar to their respective measurements of atom temperature, but with somewhat higher values. Kitaeva et al. measure T_i = 1500 K for a 2.8-mm-diameter tube at 100 A/cm² to 6000 K for a 7-mm tube at 200 A/cm². The transverse temperatures are also comparably higher. Herziger et al. found T_i = 12,000 K at JD = 200 A/cm and 25,000 K at 300 A/cm.

2.2.2.3.3. ELECTRON TEMPERATURE. Labuda et al.[164] and Kitaeva et al.[188] have made probe measurements of electron temperature at lower current densities (< 100 A/cm²). Labuda et al. found values of about 3 eV at 0.5 torr pressure, *decreasing* with increasing current density; Kitaeva et al. also found values of about 3 eV but *increasing* with increasing J. Both agree that T_e increases with decreasing pressure. Optical measurements have been made by Kitaeva et al.[177,178] and Sze and Bennett[214] at J > 100 A/cm². Kitaeva et al. again find increasing T_e with increasing J and decreasing p, with T_e = 3 eV at 150 A/cm² and 0.6 torr to 6.5 eV at 250 A/cm² and 0.2 torr; Sze and Bennett found T_e initially increasing with

J, then saturating at an approximately constant value of 3 eV for 150–300 A/cm².

2.2.2.3.4. NEUTRAL ATOM DENSITY. Because of the high neutral-atom temperature, the optimum cold gas-filling pressure must be corrected for the operating conditions as described in the previous section on cw laser characteristics (cf. Fig. 13). The actual number density in the discharge region is reduced by the high gas temperature according to

$$n_a = n_0(300/T_a), \qquad (2.2.3)$$

where n_0 is the number density at room temperature. Webb's X-ray measurements confirm that this relation is approximately true, at least for higher pressures.[168] Kitaeva et al.[188] indicate that Eq. (2.2.3) should be corrected for the electron pressure, which also drives gas atoms out of the active region. Thus

$$n_a = n_0(300/T_a) - n_e(T_e/T_a) \qquad (2.2.4)$$

should be a better approximation. Gas pumping effects produce pressure gradients, which further complicate the determination of n_a, as discussed by Chester[169,170] and Halsted.[132,160]

2.2.2.3.5. ELECTRON DENSITY. Measurements of electron density have been made by Labuda et al.[164] and Kitaeva et al.[178,188] by measuring the Stark broadening of hydrogen impurity lines and also by probe measurements at lower densities. Pleasance and George[257] have used microwave measurements and Sze and Bennett[214] have used Stark broadening of Ar I lines. All results show n_e approximately proportional to current density and gas pressure, but with different constants of proportionality: 2.5 (Kitaeva et al.,[178] $D = 1.6$ mm), 3.3 (Labuda et al.,[164] $D = 2$ mm), 5 (Pleasance and George,[257] $D = 2$ mm), 7.5 (Kitaeva et al.,[188] $D = 10$ mm), and 50 (Sze and Bennett,[214] $D = 2$ mm) 10^{12}cm^{-3}/(torr-A/cm²). Why the values obtained by Sze and Bennett are so high compared with the other four measurements is not known.

2.2.2.3.6 ION DENSITY. The Debye length in a typical argon ion laser discharge is less than 0.01 mm, so that $n_i = n_e$ except very near the discharge walls.

2.2.2.3.7. ELECTRIC FIELD. The longitudinal electric field in the positive column and its variation with p, D, B, and J have been measured by Labuda et al.,[164,170] Bridges and Halsted,[132] Kitaeva et al.,[188] and Herziger and Seelig.[219,220] Labuda et al. found the simple relation $E \cdot D = 12.3$ V-mm/cm, where E is in V/cm and D is in mm for lasers of different diameters operated at optimum discharge conditions. Bridges and

---

[257] L. D. Pleasance and E. V. George, *Appl. Phys. Lett.* **18**, 557 (1971).

Halsted found values of $E \cdot D$ of 11 to 14 V-mm/cm for $1.8 < D < 8$ mm at optimum p and $B = 0$, with values 20–30% lower at $B = 1000$ G. Herziger and Seelig derive a more complicated expression exhibiting the dependence of E on discharge current as well. For small current densities and $B = 0$, their expression reduces to $E \cdot D = 10.6$ V-mm/cm.

2.2.2.4. Excitation Mechanism Models. Various proposals have been made for the mechanisms that excite and de-excite upper and lower laser levels. Four of these proposals are illustrated in Fig. 16. Labuda et al.[163] proposed the process shown in Fig. 16a, involving two successive electron collisions to ionize and then excite the 4p levels from the ion ground state. This model was later modified to include excitation via the ionic metastables as shown in Fig. 16c.[164] Bennett et al.[113] proposed a single electron collision step in which the 4p levels were excited directly from

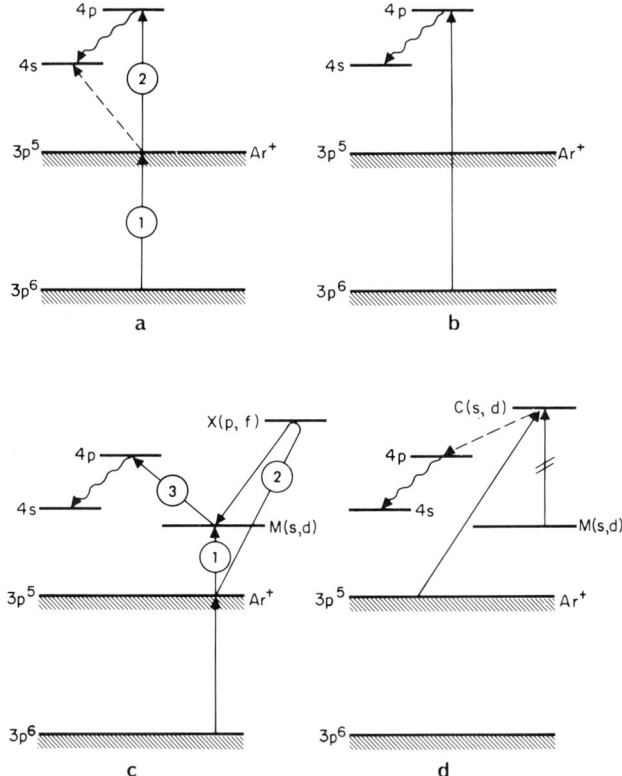

FIG. 16. Energy-level diagrams for the blue and green transitions in Ar II, showing various excitation models (from Bridges and Halsted,[132] by courtesy of the Hughes Research Laboratories).

the neutral ground state as shown in Fig. 16b. By measuring spontaneous emission intensities, Rudko and Tang[172] were able to show that a significant portion of the 4p level population was due to radiative cascade from higher-lying levels as shown in Fig. 16d. Each of these proposals was introduced to explain some of the experimentally observed features of ion laser behavior. As it turns out, all four processes contribute to the operation in varying degrees.

The two-step model of Fig. 16a was introduced to explain the observations that the laser output power varied approximately with the square of the discharge current at currents significantly above threshold, as seen in Fig. 13, for example. The ionization step can actually proceed through many different routes (e.g., via a neutral metastable level) with the same net effect since charge neutrality of the plasma requires that the ion density n_i equals the electron density n_e no matter how the ions are created. (It is tacitly assumed that the majority of ions are in the ion ground state, an assumption that breaks down at high currents.) The second step then gives $N(4p) \sim n_e n_i \sim n_e^2 \sim J^2$, where J is the current density. The 4s lower laser levels are assumed in this model (and the other three models as well) to be depopulated rapidly by radiation to the ion ground state via the strong vuv transitions at 72.3 and 74 nm. The lifetimes of the 4s ^2P levels were calculated by Statz et al.[129,258] and Rubin[183] to be less than 0.4 nsec, or 20 times shorter than the lifetimes of the 4p states. Subsequent measurements have largely confirmed these values,[209,212,259–261] although values of 0.9 nsec[262] and 0.19 nsec[263] have also been measured. In any case, the lifetimes of the 4s states are at least an order of magnitude less than the 4p states.

One objection to the model of Fig. 16a is that on the basis of selection rules for optical transitions, the 4s levels should have much higher electron-collision-excitation cross sections from the ion ground state than the 4p states; in fact, the transition $3p^5$ $^2P_{3/2} \rightarrow 3p^4(^3P)$ 4p is forbidden—both levels have the same parity. However, Zapesochnyi et al.[264] have measured the electron excitation cross section for four of the

---

[258] G. F. Koster, H. Statz, F. A. Horrigan, and C. L. Tang, *J. Appl. Phys.* **39**, 4045 (1968).

[259] F. A. Korolev, V. V. Lebedeva, A. I. Odintsov, and V. M. Salimov, *Radio Eng. Electron. Phys. (Engl. Transl.)* **14**, 1318 (1969).

[260] F. A. Korolev, V. V. Lebedeva, A. E. Novik, and A. I. Odintsov, *Opt. Spectrosc. (Engl. Transl.)* **33**, 435 (1972).

[261] C. E. Webb and R. C. Miller, *IEEE J. Quantum Electron.* **qe-4**, 357 (1968).

[262] A. E. Livingston, D. J. G. Irwin, and E. H. Pinnington, *J. Opt. Soc. Am.* **62**, 1303 (1972).

[263] B. Van der Sijde, J. W. H. Dielis and W. P. M. Graef, *J. Quant. Spectrosc. & Radiat. Transfer* **16**, 1011 (1976).

[264] I. P. Zapesochnyi, A. I. Imre, A. I. Daschenko, V. S. Vukstich, F. F. Danch, and V. A. Kel'man, *Sov. Phys.—JETP (Engl. Transl.)* **36**, 1056 (1973).

4p levels using crossed electron and ion beams and find values of about 1, 3, 4, and 12×10^{-18} cm^2 for 4p $^2P^o_{3/2}$, $^2P^o_{1/2}$, $^2D^o_{3/2}$, and $^2D^o_{5/2}$ levels, respectively. The measured cross sections exhibit a peak near threshold (~ 20 eV) and a second peak at 27–29 eV. A sharp peak near threshold is expected for forbidden transitions (e.g., see Fig. 21 in Section 2.3); the origin of the second peak is not known, although it could indicate cascade contributions from higher-lying levels as illustrated in Fig. 16d. It is interesting that the largest cross section of the four is to the $^2D^o_{5/2}$ level, the upper level of the 0.4880-μm transition laser. Zapesochnyi et al.[264] have made similar measurements on levels in Kr II.

The excitation pathway via the ionic metastables shown in Fig. 16c circumvents the difficulty with the "violation" of selection rules, since the 4p levels are excited by allowed transitions from the 3d and 4s metastables. These metastables can be populated directly from the ground state by electron collision (although forbidden, there is at least a parity change) or by cascade from higher-lying states; their populations can build up to the large values measured by Labuda et al.[164] and Ross.[265] Since metastable levels are destroyed primarily by electron collision at high values of n_e their populations will be tightly coupled to the ground-state ion population n_i. Thus the variation of the ionic metastable population with discharge current will be $N(M) \sim n_i \sim n_e \sim J$, not J^2. A second electron collision (step 3 in Fig. 16c) will yield $N(4p) \sim J^2$, indistinguishable from the process illustrated in Fig. 16a.

The magnitude of the cascade contribution to the 4p population was first determined by Rudko and Tang[131,172,173] by summing the measured spontaneous emission rates of all lines radiating *into* a given upper laser level and comparing this sum with the sum of the rates for all lines *leaving* the same level. They determined that approximately 50% of the 4p $^2D^o_{5/2}$ level population was due to cascade, primarily from the 4d $^2F_{7/2}$ level, in a 1-mm-diameter nonlasing discharge tube operated at 2–4 A.[172] Later measurements[131] in a 2-mm-diameter tube at 5 A give percentages for all the laser lines, e.g., 27.4% for 4p $^2D^o_{5/2}$, 18.5% for 4p $^4D^o_{5/2}$, and 4% for 4p $^2S^o_{1/2}$, and showed an increasing percentage with increasing current for the 4p $^2D^o_{5/2}$ level. Similar measurements by Bridges and Halsted[132] gave values of 23% for the 4p $^2D^o_{5/2}$ level, 22% for the 4p $^4D^o_{5/2}$ level, and less than 5% for the 4p $^2S^o_{1/2}$ level in a 3-mm-diameter discharge. Their measurements indicated that the cascade percentage was approximately independent of current (10–20 A), pressure (300–500 mtorr), and magnetic field (0–1500 G). All the spontaneous emission intensities, both in and out of each level varied as J^2, so the cascade process is inseparable from those in Fig. 16a,c on the basis of current dependence.

[265] J. N. Ross, *J. Phys. D* **6**, 1426 (1974).

The single-step excitation model was proposed by Bennett et al.[113] to explain the observation that the 0.4765-μm laser line was the strongest or the only line that would oscillate in discharges with very short current pulses (< 100 nsec) and high electron energies (i.e., discharges with a high ratio of electric field to gas pressure, E/p). Under the "sudden perturbation" picture of excitation cross section,[113] the most probable final configurations are the $3p^4(^3P)4p\ ^2P$ levels, which, even though they are p states, have a parity opposite to the neutral ground state because an electron has been removed in the ionization. Bennett et al.[210] measured the cross sections for this process for the seven upper laser levels shown in Fig. 9 over the range of 35 eV (threshold) to 110 eV. They obtained peak values of 25, 16, 15, 15, 12, 7, and 5 $\times\ 10^{-19}$ cm^2 for the levels $^2P^o_{3/2}$, $^2D^o_{3/2}$, $^2P^o_{1/2}$, $^2D^o_{5/2}$, $^4D^o_{5/2}$, $^2S^o_{1/2}$, and $^4D^o_{3/2}$, respectively. Hammer and Wen[189] also measured the excitation cross sections for this process to the doublet levels and obtained the same ordering of the cross sections, but with absolute values lower by a factor of two or three: 11, 5.5, 5.3, 5, and 1.5 $\times\ 10^{-19}$ cm^2 for the levels $^2P^o_{3/2}$, $^2D^o_{3/2}$, $^2P^o_{1/2}$, $^2D^o_{5/2}$, $^2S^o_{1/2}$. Similar results were obtained by Latimer and St. John,[266] and Felt'san and Povch.[267] The cross sections measured by Hammer and Wen peaked at electron energies of 100–120 eV. Calculations by Vainshtein and Vinogradov,[268] and Koozekanani[269,270] place the theoretical values in this same range, less than 10^{-18}cm^2, with the theoretical cross sections for the 4p $^2D^o_{5/2}$ level and the quartet levels essentially zero. Excitation from the ion ground state as shown in Fig. 16a has an order of magnitude larger cross section at lower electron energies (20 eV compared to 100 eV); however, the rate from the ion ground state is reduced by the fractional ionization n_i/n_a compared to the process in Fig. 16b. From the observation of which laser lines dominate the output spectrum, it seems clear that the direct process of Fig. 16b dominates for short-pulse (< 100 nsec) discharges with high E/p and that some combination of the processes in Fig. 16a,c,d dominates for pulses of a microsecond out to continuous operation.

2.2.2.5. Radiation Trapping. The lower laser level populations have been ignored thus far because the lower-level radiative lifetimes are 20 times or more shorter than the upper-level lifetimes. However, the lifetimes of these 4s ^2P levels, and hence their populations can be increased greatly by radiation trapping,[11] the reabsorption by the ground state ions of the 72.3- and 74-nm radiation that depopulates the 4s ^2P levels. The ef-

---

[266] I. D. Latimer and R. M. St. John, *Phys. Rev. A* **1**, 1612 (1970).
[267] P. V. Felt'san and M. M. Povch, *Opt. Spectrosc. (Engl. Transl.)* **28**, 119 (1970).
[268] I. A. Vainshtein and A. Vinogradov, *Opt. Spectrosc. (Engl. Transl.)* **23**, 101 (1967).
[269] S. H. Koozekanani, *IEEE J. Quantum Electron.* **qe-2**, 770 (1966).
[270] S. H. Koozekanani, *IEEE J. Quantum Electron.* **qe-3**, 206 (1967).

fects of this trapping process can be seen in the temporal and spatial properties of the output in pulsed ion lasers; at higher currents and pressures a "dead" or negative gain interval develops near the beginning of the output pulse,[116,166] and the laser may also oscillate in annular or "ring" modes, indicating the gain near the walls is larger than on the discharge axis.[248] Continuous lasers also exhibit ring modes at higher pressures and magnetic fields as shown by Gorog and Spong.[190] Figure 12 shows the typical development of an absorbing interval in the output of a 0.4880-μm pulsed ion laser. The absorbing interval widens with increases in current, pressure, or magnetic field, eventually "swallowing up" one or both pulses; this occurs first for radiation along the tube axis, then progresses radially outward toward the tube walls, giving rise to oscillation in annular cavity modes.

The explanation for these phenomena in terms of radiation trapping on vuv lines from the lower laser levels to the ion ground state was first given by E. I. Gordon in June 1964,[271] and is summarized briefly below, following the treatment in Bridges and Halsted[132] and Klein.[272] (See the Thesis of M. B. Klein[273] for a more complete development.) If both upper and lower laser levels are populated primarily by two-electron processes, then the small-signal gain coefficient may be written

$$g_0 = K_1 J^2 [1 - (K_2/\gamma)], \qquad (2.2.5)$$

where γ is Holstein's trapping factor[11] by which the effective A coefficient is reduced, and K_2 is independent of p, J, and B. With some approximation

$$\gamma \sim \Delta\nu_D / N_0 R \qquad (2.2.6)$$

where $\Delta\nu_D$ is the Doppler width of the vuv line, N_0 the ion ground state population, and R the discharge radius. If we further assume that most of the ions are in the ground state (or at least $N_0 \sim n_i$) then

$$N_0 \sim n_i \sim n_a n_e \sim pJR \qquad (2.2.7)$$

and

$$g_0 = K_1 J^2 [1 - (K_3 pJR^2/\Delta\nu_D)]. \qquad (2.2.8)$$

The shape of the pulsed laser waveform in Fig. 12 can now be understood from this equation. The gain rises as J^2 until the second term in parenthe-

---

[271] E. I. Gordon, private communication (1964).
[272] M. B. Klein, *Appl. Phys. Lett.* **17**, 29 (1970).
[273] M. B. Klein, Ph.D. Dissertation, University of California, Berkeley (1969) (unpublished).

ses becomes significant. The gain drops to zero when the current is high enough to make this term unity and then becomes negative for higher currents. Increasing the gas pressure (actually, the neutral number density) makes this occur at lower currents. However, the ion temperature increases during the pulse due to an increase in electron temperature and gas temperature according to the theory of Kagen and Perel'[254,255] and shown experimentally by Klein,[272,273] so that the Doppler linewidth $\Delta\nu_D$ increases to make the second term less than unity again, and the net gain becomes positive. Of course, it is always possible to increase p sufficiently so that no increase in $\Delta\nu_D$ will suffice.

Typically, cw ion lasers operate with an order of magnitude higher gas pressure than pulsed ion lasers, say 200 vs. 20 mtorr. The high average power of the discharge heats the gas and reduces the atom number density to values similar to those in pulsed lasers both by heating according to Eq. (2.2.3) or (2.2.4) and by gas pumping processes. Radaelli and Sona[274] observed the discharge starting transient under pressure and current conditions appropriate to cw ion laser operation. The spontaneous emission on the laser lines began promptly with the current, but the oscillation was delayed several milliseconds. The delay time increased with increasing fill pressure and increasing discharge current, again consistent with the radiation-trapping model result given by Eq. (2.2.8) but at higher pressures. The pressure and magnetic field observations by Gorog and Spong[190] are also consistent with the radiation-trapping model given here. A longitudinal magnetic field of the magnitude ordinarily employed to increase the output power of cw ion lasers, 500–1000 G, serves to increase the electron density n_e and hence the ion density n_i for a given current by reducing the electron drift velocity as in Eq. (2.2.1) and is thus equivalent to an increase in current as far as radiation trapping is concerned.

Despite the relatively good agreement between experimental observations and the simple theory given here, it is clear that the actual situation is more involved. Some of the approximations made in explaining the behavior of Ar II lines break down when the excitation rate to higher ionization states becomes significant; the singly ionized ground state is depleted by excitation to higher states. Levinson et al.[275] have proposed that the saturation in output power of pulsed ion lasers with increasing current is caused by such depletion. Ring-mode formation has also been observed in higher ionization state lasers Ar III and Xe IV by Cheo and Cooper.[248] It may be possible to extend the radiation-trapping model to these higher ionization states even after it has become invalid for the

---

[274] G. Redaelli and A. Sona, *Alta Freq.* **36**, 150 (1967).
[275] G. R. Levinson, V. F. Papulovskii, and V. P. Tychinskii, *Radio Eng. Electron. Phys.* (*Engl. Transl.*) **13**, 578 (1968).

lower-lying levels. Skurnick and Schacter[276] have proposed that electron-collision destruction of the upper laser level is the saturation mechanism at higher values of J, p, and B. Cottrell[277] and Davis and King[278] have proposed that the behavior of pulsed lasers can be explained by a relative increase in lower-level population at higher currents if the upper levels are excited in a single step and the lower levels excited by two steps. Boscher et al.[224] and Lüthi and Seelig[229] conclude that radiation trapping is the power-limiting mechanism in large-diameter cw ion lasers operated at high current densities, including the cw, doubly ionized lines.

2.2.2.6. Ion Laser Excitation Techniques. Many different methods of creating an intensely ionized plasma have been investigated in an attempt to improve the output, efficiency, and operating life of ion lasers. Simple hot- or cold-cathode dc discharges in dielectric-wall tubes were the first employed and are still widely used in commercial lasers. Segmented metal walls and metal disks have also been used to confine the plasma with little difference in output characteristics. The performances obtained with these techniques are discussed in the next section. Other excitation techniques that have successfully produced ion laser oscillation are described in this section.

Radio-frequency excitation has been used to produce ion laser plasmas. The high-power, low-impedance nature of the ion laser plasma generally requires a different coupling method than the capacitive or E-field electrodes commonly employed with low-power, high-impedance discharges such as the helium–neon laser. *Inductive* or B-*field* coupling was first successfully demonstrated by Bell,[204] who made the laser optical path one of the long legs in a rectangular ring of plasma. The ring formed a lossy single turn secondary of a transformer with a ferrite core and a multiturn primary driven by millisecond-duration pulses of 1 MHz rf. Continuous generation was later demonstrated with 40-MHz excitation and no core.[206] Zarowin and Williams[279] produced quasi-cw operation (90% duty cycle pulses) with a similar ring laser excited by square waves at 2.5 kHz applied through a tape-wound permendur transformer core. Goldsborough and Hodges[280] demonstrated a different inductively coupled excitation arrangement by placing the laser discharge tube on the axis of an rf-

---

[276] E. Skurnick and H. Schacter, *J. Appl. Phys.* **43**, 3393 (1972).
[277] T. H. E. Cottrell, *IEEE J. Quantum Electron.* **qe-4**, 435 (1968).
[278] C. C. Davis and T. A. King, *Phys. Lett. A* **36**, 169 (1971).
[279] C. B. Zarowin and C. K. Williams, *Appl. Phys. Lett.* **11**, 47 (1967).
[280] J. P. Goldsborough and E. B. Hodges, International Electron Devices Meeting, p. 88. IEEE, New York, 1967.

2.2. IONS EXCITED BY ELECTRON COLLISION

excited solenoid. Ling et al.[281] discuss the design considerations for this arrangement and report 1 W output from a tube 1 cm in diameter and 20 cm long. Goldsborough[207] has also obtained ion laser oscillation with cyclotron resonance excitation of the plasma; a periodic longitudinal rf electric field at 2.45 GHz was applied by placing the discharge tube in the fringing field of an interdigital microwave slow-wave structure. Maximum excitation occurred at the predicted cyclotron resonance with an applied transverse magnetic field of 1.3 kG. Katsurai and Sekiguchi[282] have also made a study of this mode of excitation. Paik and Creedon[283] have produced ion laser oscillation with high-power pulsed microwaves at 1.26 GHz and no magnetic field by placing a short laser discharge across a waveguide in the high-field region near a reflective short. Powers of 100 kW were required to reach threshold for the 0.4880-μm line because of the inefficient coupling to the plasma. Kato and Shimizu[284] have obtained much lower thresholds at 9 GHz by placing the discharge along the axis of a long cylindrical waveguide cavity. Oscillation threshold was 7.5 kW, and 5 W output was obtained at 18 kW in a 5-mm-diameter by 41-cm-long discharge.

A salient advantage of all of these rf excitation methods is that chemically reactive vapors can be used in the discharge tube without unfavorable reactions with any metallic electrodes, etc. Bell[204] operated his pulsed ring with sulfur vapor and Goldsborough et al.[206] produced continuous oscillation on lines of Cl II and Br II. Akirtava et al.[285] have studied the O II and O III ultraviolet lines with a pulsed rf-excited ring discharge similar to Bell's.

Very high longitudinal currents in a plasma produce electromagnetic forces that can compress the plasma radially, producing still higher current densities. The electron temperatures and densities developed in these *self-compressed* or *Z-pinched* discharges can be very high. Extensive studies of such plasmas have been made for controlled thermonuclear fusion, and they have also been used successfully for ion lasers. Kulagin et al.[286,287] were the first to produce oscillation on the Ar II

---

[281] H. Ling, J. Colombo, and C. L. Fisher, *Rev. Sci. Instrum.* **41**, 1436 (1970).

[282] M. Katsurai and T. Sekiguchi, *Electron. Commun. Jpn.* **54-B**, 61 (1971).

[283] S. F. Paik and J. E. Creedon, *Proc. IEEE* **56**, 2086 (1968).

[284] I. Kato and T. Shimizu, *Electron. Commun. Jpn.* **55-C**, 108 (1972).

[285] O. S. Akirtava, A. M. Bogus, V. L. Dzhikiya, and Yu. M. Oleinik, *Sov. J. Quantum Electron.* (*Engl. Transl.*) **3**, 519 (1974).

[286] S. G. Kulagin, V. M. Likhachev, E. V. Markuzon, M. S. Rabinovich, and V. M. Sutovskii, *JETP Lett.* (*Engl. Transl.*) **3**, 6 (1968).

[287] S. G. Kulagin, V. M. Likhachev, M. S. Rabinovich, and V. M. Sutovskii, *J. Appl. Spectrosc.* (*Engl. Transl.*) **5**, 398 (1966).

0.4765-μm line in a 25-mm-diameter tube 100 cm long with a 15-kA self-compressing pulse. They reported an output of 20–25 kW with a self-terminating duration of 200 nsec. Only the 0.4765-μm line was observed to oscillate, consistent with the *sudden perturbation* excitation model proposed by Bennett *et al.*[113] and discussed in the previous section. Vasil'eva *et al.*[288] later obtained oscillation also on the Ar III 0.3511- and 0.3638-μm lines, and measured electron densities of 10^{15} cm^{-3} and electron temperatures of 14–20 eV. Illingworth[289–291] has also studied the Z-pinch argon ion laser, obtaining oscillation on the Ar II 0.4765-μm and Ar III 0.3511-μm lines; he has also measured electron densities of 10^{15} cm^{-3} at the peak of the compression. Papayoanou *et al.*[292–294] have used the Z-pinch discharge to produce high power oscillation on many of the visible and uv lines in Xe III and Xe IV with 1.5–4-kA discharge currents. Outputs of 100–500 W/line were typical, with pulse durations of 200–350 nsec. Time-resolved electron densities as high as 10^{16} cm^{-3} at the peak compression were measured interferometrically.[294] Hashino *et al.*[149,295] used peak currents of 30 kA in a 100-cm-long tube to produce oscillation on many lines in the spectra Ne I–III, Ar II–IV, N III–IV, and O II–V. Output pulse widths were also typically 200 nsec for all lines. Katzenstein and Lovberg[296] have given a theoretical discussion of some of the features observed in the Z-pinch experiments and describe the applicable scaling laws and impedance matching considerations.

Beam-generated plasmas have also been used to obtain ion laser oscillation. Tkach *et al.*[297,298] produced peak powers of 100 W for 30 μsec on 8 of the 11 lines listed in Table V by injecting a 3–30-kA beam of 10–45-kV electrons into a tube containing argon at about 1 mtorr pressure. Electron temperatures of 90 eV and ion temperatures of about 1 eV were measured, substantially higher than in other argon ion laser excitation

[288] A. N. Vasil'eva, V. M. Likhachev, and V. M. Sutovskii, *Sov. Phys.—Tech. Phys.* (*Engl. Transl.*) **14**, 246 (1969).
[289] R. Illingworth, *J. Phys. D* **3**, 924 (1970).
[290] R. Illingworth, *J. Phys. D* **5**, 686 (1972).
[291] R. Illingworth, *J. Phys. D* **8**, 1956 (1975).
[292] A. Papayoanou and I. M. Gunmeiner, *Appl. Phys. Lett.* **16**, 5 (1970).
[293] A. Papayoanou and I. M. Gunmeiner, *J. Appl. Phys.* **42**, 1914 (1971).
[294] A. Papayoanou, R. G. Buser, and I. M. Gunmeiner, *IEEE J. Quantum Electron.* **qe-9**, 580 (1973).
[295] Y. Hashino, Y. Katsuyama, and K. Fukuda, *Jpn. J. Appl. Phys.* **11**, 907 (1972).
[296] J. Katzenstein and R. H. Lovberg, *Appl. Phys. Lett.* **26**, 113 (1975).
[297] Yu. V. Tkach, Ya. B. Fainberg, L. I. Bolotin, Ya. Ya. Bessarab, N. P. Gadetskii, Yu. N. Chernen'kii, and A. K. Berezin, *JETP Lett.* (*Engl. Transl.*) **6**, 371 (1967).
[298] Yu. V. Tkach, Ya. B. Fainberg, L. I. Bolotin, Ya. Ya. Bessarab, N. P. Gadetskii, I. I. Magda, A. V. Bogdanovich, and Yu. N. Chernen'kii, *Ukr. Fiz. Zh.* (*Russ. Ed.*) **14**, 1470 (1969).

methods. The optimum longitudinal magnetic field in these experiments was about 1.5 kG.

2.2.3. Ion Laser Performance

Noble-gas ion lasers have been in commercial production for the past 13 years. The output power of available argon units now exceeds 20 W summed over the blue-green lines. Table V lists the wavelength distribution of the output for typical commercial argon ion lasers. These lasers use small diameter (less than 5 mm) discharge tubes and magnetic fields of the order of 2 kG. For higher power generation, the trend has been toward larger-diameter discharges (greater than 10 mm) operating at very high currents without magnetic fields. Table VI gives the multiline performance for blue-green Ar II lasers of this type. (The first entry is typical of the earlier small bore lasers and is included in the table for comparison.) While the output power has increased 50 times, the efficiency has increased only a little over twice; the 500-W laser demonstrated by Alferov et al.[247] requires almost a quarter of a megawatt cw input. All of the later higher-power argon ion lasers utilize discharge tubes with water-cooled, segmented-metal walls. Figure 17 has been adapted and updated from Bridges et al.[160] and provides a comparison between the performance of the earlier, small-bore lasers (dotted or shaded curves) and the later, large-bore lasers (solid curves and circles).

Much less information has been published on the Kr II and Xe II visible laser performance compared to Ar II, primarily because these gases produce substantially less power overall and are more susceptible to discharge instability. Table VII lists the outputs for two relatively recent commercial krypton ion lasers, including both Kr II and Kr III lines. No xenon lasers are commercially available at the present time. Boersch et al.[221] have reported the relative output power of different Kr II lines in a larger diameter tube but give no absolute numbers. Boscher et al.[224] report saturated power less than 0.5 W for the various Xe II visible lines.

Table VIII lists the data on ultraviolet ion lasers published within the last decade. Increases in power per unit length have been especially dramatic during the last two years with the work of Lüthi et al.[231-233] and Tio et al.[240] As in the singly ionized laser, larger-diameter discharges without magnetic fields produce better results. However, one other key factor peculiar to the ultraviolet performance has been the development of lower-loss uv multilayer dielectric coatings and techniques for reducing the optical losses in Brewster's angle windows. As is evident by comparing Tables VII and VIII, the Ar III and Kr III uv lasers are generally the same discharge tube structures as the ones used for the higher power Ar II lasers. This is true of the commercial uv ion lasers also.

TABLE VI. Performance of High Power cw Blue–Green Argon Ion Lasers

Ref.	Date	Operating parameters							Performance			Bore type
		D (mm)	L (cm)	I (A)	V (V)	B (G)	P (torr)	JR (A/cm)	P (W)	P/L (W/m)	η (%)	
a	1967	4	70	50	219	900	0.5	80	10	14.3	0.09	fused silica
b	1970	6.35	254	90	600	600	0.38	90	93	35.4	0.16	graphite disks
c	1969	12	200	(320)	(280)	0	(0.4)	170	120	(60)	0.13	fused silica
d	1970	12	85	340	—	0	1.1	180	90	105	—	Al_2O_3 on Al segments
e	1972	11	55	320	—	0	1	(185)	45	(82)	} 0.16	Al_2O_3 on Al segments
e	1972	16	135	420	255	0	0.7	(168)	175	(130)		
f	1972	12	85	(377)	—	0	—	200	(47)	40	—	W segments
g	1973	12	85	300	(150)	0	—	(159)	25	(29)	} 0.055	W segments
g	1973	12	185	360	(430)	0	—	(191)	125	(68)	0.08	
h	1973	16	150	390	(<640)	0	0.4	(155)	500	(333)	0.2	Al_2O_3 on Al segments

[a] W. B. Bridges, P. O. Clark, and A. S. Halsted, "High Power Gas Laser Research," Tech. Rep. AFAL-TR-66-369. Hughes Res. Lab., Malibu, California, 1967 (unpublished). (Available from National Technical Information Service, Accession Number AD807363).

[b] J. R. Fendley, Jr. and J. J. O'Grady, "Development, Construction, and Demonstration of a 100 W cw Argon-Ion Laser," Tech. Rep. ECOM-0246-F. RCA, Lancaster, Pennsylvania, 1970 (unpublished). (Available from National Technical Information Service.)

[c] G. Herziger and W. Seelig, Z. Phys. 219, 5 (1969).

[d] H. Boersch, J. Boscher, D. Hoder, and G. Schäfer, Phys. Lett. A 31, 188 (1970).

[e] V. I. Donin, Sov. Phys.—JETP (Engl. Transl.) 35, 858 (1972).

[f] C. P. Wang and S.-C. Lin, J. Appl. Phys. 43, 5068 (1972).

[g] C. P. Wang and S.-C. Lin, J. Appl. Phys. 44, 4681 (1973).

[h] G. N. Alferov, V. I. Donin, and B. Ya. Yurshin, JETP Lett. (Engl. Transl.) 18, 369 (1973).

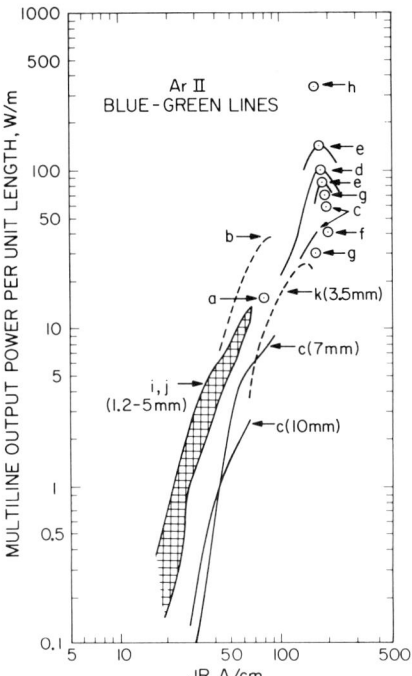

FIG. 17. Summary of the multiline output powers per unit length as a function of the product JR observed by various investigators. The solid curves and points are larger-bore, zero-magnetic-field lasers given in Table VI with the corresponding reference. The dotted and shaded curves are smaller-bore tubes with magnetic files of 0.5–1.5 kG, from i,[163] j,[132] and k.[201]

Pulsed noble gas ion lasers were studied in research and development laboratories worldwide and were also available commercially for a few years after their discovery in 1964, but they have now all but disappeared from the scene. The exception to this is the highly ionized xenon laser. Peak powers are typically observed on these xenon lines that are one to two orders of magnitude greater than argon operated under the same discharge conditions. The energy levels from which these lines originate are not known, but they are thought to originate in the Xe IV spectrum.[139–144] Table IX lists these lines and the output powers observed by Simmons and Witte,[299] Hoffmann and Toschek,[300] Papayoanou et al.,[294] Marling,[125] and Schearer.[301] A study of the coherence properties has been made by

[299] W. W. Simmons and R. S. Witte, *IEEE J. Quantum Electron.* **qe-6**, 466 (1970).
[300] V. Hoffmann and P. Toschek, *IEEE J. Quantum Electron.* **qe-6**, 757 (1970).
[301] L. D. Schearer, *IEEE J. Quantum Electron.* **qe-11**, 935 (1975).

TABLE VII. Performance of Commercial Krypton Ion Lasers

Krypton		Output power (W)		
Wavelength (μm)	Ion	CR3000K[a]	SP171-01[b]	SP921-01[b,c]
0.3374	III	⎫	⎫	⎫
0.3507	III	⎬ 2.0	⎬ 1.1	⎬ 2.5
0.3564	III	⎭	⎭	⎭
0.4067	III	0.9	— ⎫	— ⎫
0.4131	III	1.5	0.5 ⎬ 1.3	2.4 ⎬ 3.0
0.4154	III	0.1	— ⎪	— ⎪
0.4226	III	—	— ⎭	— ⎭
0.4680	II	0.5	0.3	0.8
0.4762	II	0.4	0.4	1.0
0.4825	II	0.4	0.4	1.0
0.5208	II	0.7	0.7	2.0
0.5309	II	1.5	1.4	3.0
0.5682	II	1.1	1.0	2.5
0.6471	II	3.5	3.5	7.0
0.6764	II	0.9	0.9	2.5
0.7525	II	1.2	1.1	3.0
0.7931	II	⎫ 0.3	—	—
0.7993	II	⎭	0.25	1.2
Line current @460 V 3ϕ (A)		70	60	60 × 2

[a] Date source: Anonymous brochure, "Super Graphite Ion Lasers and Accessories." Coherent Radiation, Palo Alto, California, dated 12/76.

[b] Data source: Anonymous brochure, "Spectra-Physics High Power Ion Lasers." Spectra-Physics, Mountain View, California, dated 5/77.

[c] This laser is essentially two SP171-01 discharge tubes in a common optical cavity.

Moskalenko et al.[302] and of the output power for different excitation conditions and pulse lengths by Jarrett and Barker,[303] Davis and King,[141] Hoffmann and Toschek,[143] and Razmadze et al.[304] Harper and Gundersen[305] have demonstrated the record peak output power of 80 kW with 0.25 W average at a 36-pps repetition rate. The Soviet commercial laser LGI-37 operates at 2 kW peak and 0.4 W average power.[306] The parame-

[302] V. F. Moskalenko, E. P. Ostapchenko, S. V. Pechurina, V. A. Stepanov, and Yu. M. Tsukanov, Opt. Spectrosc. (Engl. Transl.) **30**, 201 (1971).

[303] S. M. Jarrett and G. C. Barker, IEEE J. Quantum Electron. qe-5, 166 (1969).

[304] N. A. Razmadze, L. L. Gol'dinov, and Z. D. Chkuaseli, J. Appl. Spectrosc. (Engl. Transl.) **22**, 621 (1975).

[305] C. D. Harper and M. Gundersen, Rev. Sci. Instrum. **45**, 400 (1974).

[306] Anonymous, "ПГИ.37 High Power Pulsed Optical Quantum Oscillator," Specification sheet ca. 1970.

ters of many of these lasers are listed in Table X. The blue and green lines have also been operated continuously but at only ~0.5 W output power.[144]

A recent development that seems to hold considerable promise for the future is the demonstration of quasi-cw, high-power oscillation on visible lines of Hg III by Burkhard et al.[307] Using a 10-mm diameter by 120-cm-long fused-silica tube at currents up to 300 A they obtained outputs of 40 W on the 0.6501-μm line, 25 W on the 0.4797-μm line, and 5 W on the 0.5210-μm line. The Hg III 0.4797-μm line was first observed in pulsed operation by Gerritsen and Goedertier[118] and the Hg III 0.6501-μm line by Lüthi et al.[308] The silica discharge tube here was essentially the same apparatus used in earlier argon experiments[220] and was limited to pulsed operation by thermal dissipation; however, discharge current pulses of 25 μsec to 4 msec produced laser output pulses of 10 μsec to 1 msec, indicating that cw operation should be possible. Mercury vapor pressures of 1–20 mtorr were employed, and buffer gases helium and argon were investigated over the range 1–100 mtorr. Best operation was obtained with pure mercury vapor at 2.5 mtorr. Operation on ultraviolet lines in the 0.31- to 0.385-μm range was predicted.

2.3. Neutral Atom Lasers Excited by Collisions with Atoms

Only a relatively few lasers are known in which oscillation in the neutral atom is excited by collisions with other neutral atoms. However, the most widely used helium–neon laser is among these. Substantial efforts have been made to understand the excitation mechanisms quantitatively and to determine the optimum operating parameters for this laser. The results of these studies are reviewed in this section. A few other atomic lasers are known in which neutral atom collisions are, or are thought to be, the principal excitation means; these are discussed briefly at the end of the section.

2.3.1. Helium–Neon Lasers

Laser oscillation in a mixture of helium and neon with excitation transfer from helium metastable to neon upper laser levels was predicted theoretically by Javan[6] in 1959. In 1961, Javan et al.[309] were successful in

---

[307] P. Burkhard, H. R. Lüthi, and W. Seelig, *Opt. Commun.* **18**, 485 (1976).
[308] H. R. Lüthi, W. Seelig, and A. Stadler, *IEEE J. Quantum Electron.* **qe-12**, 317 (1976).
[309] A. Javan, W. R. Bennett, Jr., and D. R. Herriott, *Phys. Rev. Lett.* **6**, 106 (1961).

TABLE VIII. Performance of cw Ultraviolet Noble-Gas Ion Lasers

Ref.	Date	Ion $\lambda(\mu m)$	D (mm)	L (cm)	I (A)	V (V)	B (G)	P (torr)	JR (A/cm)	P (W)	P/L (W/m)	η (%)	Bore type, comment
a	1968	Ne II 0.3324 (0.75) 0.3378 (0.25)	1.7	34	18	510	1100	0.7	67	0.13	0.38	0.0014	graphite segments
		Ar III 0.3511 (0.5) 0.3638 (0.5)			20	315	1200	0.33	75	0.1	0.35	0.0016	
		Kr III 0.3507 (0.7) 0.3564 (0.3)			20	270	1200	0.35	75	0.14	0.41	0.0025	
		Xe III 0.3781 (0.8) 0.3454 (0.2)			18	240	1100	0.25	67	0.075	0.22	0.0017	
b	1968	Ar III 0.3511 0.3638	12	200	280	—	0	0.3	150	1.5	0.75	—	fused silica
		Kr III 0.3507								1.3	0.65	—	
c	1968	Ar III 0.3511 (0.2) 0.3638 (0.8)	4	90	135	—	840	0.85	214	1.7	1.9	—	W disks
		Kr III 0.3507			120	—			185	0.44	0.5	—	
d	1969	Ar III 0.3511 0.3638	2.3	46	64	390	1500	0.8	177	2.3	5	0.01	W disks
		Kr III 0.3507 0.3564			57	—	1500	—	158	0.38	0.83	—	
e	1970	Ar III 0.3511 0.3638	12	85	(475)	—	0	—	250	3	3.5	—	Al_2O_3 on Al segments

ref	year	Species/Wavelength											Notes
		Ne II 0.3324			400	—	0	0.8–1	212	0.15	0.08	—	
		Ar III 0.3511 (0.45) / 0.3638 (0.55)			485	—	0	0.7–1.4	180	16	8.6	—	} W segments
f	1976	Kr III 0.3507 (0.8) / 0.3564 (0.2)	12	185	430	—	0	0.6–1.2	228	7	3.8	—	
		Xe III 0.3781 (0.9) / 0.3746 (0.1)			340	—	0	0.8–1.1	257	1.8	1	—	
g	1977	Ar III 0.3511 / 0.3638	12	170	490	480	25	1.2	260	40	24	0.017	Al$_2$O$_3$ on Al segments
		Ar III 0.3511 / 0.3638								61	36	0.03	
		Ar III 0.33 triplet								17	10		
h	1977	Ar III 0.30 triplet	12	170	480	—	~30	1.3	255	4	2.3		} Al$_2$O$_3$ on Al segments
		Ar III 0.275								0.4	0.23		
		Kr III 0.3507 / 0.3564								19	11.2		
		Kr III 0.32 triplet								4.5	2.7		

[a] J. R. Fendley, *IEEE J. Quantum Electron.* **qe-4**, 627 (1968)

[b] K. Banse, G. Herziger, G. Schäfer, and W. Seelig, *Phys. Lett. A* **27**, 682 (1968).

[c] I. D. Latimer, *Appl. Phys. Lett.* **13**, 333 (1968).

[d] W. B. Bridges and G. N. Mercer, "Ultraviolet Ion Laser," Rep. ECOM-0229-F. Hughes Res. Lab., Malibu, California, 1969. (Unpublished) (Available from National Technical Information Service, Accession Number AD861927.)

[e] H. Boersch, J. Boscher, D. Hoder, and G. Schäfer, *Phys. Lett. A* **31**, 188 (1970).

[f] T. K. Tio, H. H. Luo, and S.-C. Lin, *Appl. Phys. Lett.* **29**, 795 (1976).

[g] H. R. Lüthi, W. Seelig, J. Steinger, and W. Lobsiger, *IEEE J. Quantum Electron.* **qe-13**, 404 (1977).

[h] H. R. Lüthi, W. Seelig, and J. Steinger, *Appl. Phys. Lett.* **31**, 670 (1977).

[i] Numbers in parentheses indicate fraction of the output at the wavelength listed.

TABLE IX. Comparison of Power (W) Observed on Xe IV Lines

λ (μm)	Date: 1970[a]	1973[b]	1973[c]	1975[d]	1975[e]
0.2232	—	—	—	—	10–50
0.2315	—	—	—	—	1400
0.3079	—	—	—	—	32
0.3247	—	—	weak	—	—
0.3331	—	—	100	—	330
0.3645	—	—	400	—	3600
0.4306	—	—	240	50	1000
0.4954	155	—	525	250	
0.5008	150	—	180	—	
0.5159	85	—	weak	250	6500 Total
0.5260	160	110	350	680	
0.5353	180	60	570	800	
0.5395	170	120	560	480	
0.5956	—	100	120	110	

[a] W. W. Simmons and R. S. Witte, *IEEE J. Quantum Electron.* **qe-6**, 466 (1970).

[b] V. Hoffmann and P. Toschek, *IEEE J. Quantum Electron.* **qe-6**, 757 (1970).

[c] A. Papayoanou, R. G. Buser, and I. M. Gunmeiner, *IEEE J. Quantum. Electron.* **qe-9**, 580 (1973).

[d] L. D. Schearer, *IEEE J. Quantum Electron.* **qe-11**, 935 (1975).

[e] J. B. Marling, *IEEE J. Quantum Electron.* **qe-11**, 822 (1975).

TABLE X. Performance of Blue–Green Pulsed Xenon Ion Lasers

Ref.	Year	Discharge parameters					Output		
		D (mm)	L (cm)	I (kA)	Rep. rate (Hz)	Press (mtorr)	P_{peak} (kW)	τ (μsec)	P_{avg} (W)
a	1970	2.3	152	0.1–0.6	—	8–28	0.9	0.5–5	—
b	1970	5	175	1.8	10	—	0.4	0.6–1.5	—
c	1970	4	<150	1–1.2	100–200	—	2	0.3	(0.4)
d	1973	7–12	90	1.5–4	—	16–20	2	0.2	—
e	1974	17	300	2	10–36	6	82	0.3	0.1–0.25
f	1975	7	150	10	600	11	6.5	0.2–0.3	—
g	1975	4	120	—	30	—	4	0.35	0.02

[a] W. W. Simmons and R. S. Witte, *IEEE J. Quantum Electron.* **qe-6**, 466 (1970).

[b] V. Hoffmann and P. Toschek, *IEEE J. Quantum Electron.* **qe-6**, 757 (1970).

[c] Anonymous, "ПГИ.37 High Power Pulsed Optical Quantum Oscillator," Specification sheet ca. 1970.

[d] A. Papayoanou, R. G. Buser, and I. M. Gunmeiner, *IEEE J. Quantum Electron.* **qe-9**, 580 (1973).

[e] C. D. Harper and M. Gundersen, *Rev. Sci. Instrum.* **45**, 400 (1974).

[f] J. B. Marling, *IEEE J. Quantum Electron.* **qe-11**, 822 (1975).

[g] L. D. Schearer, *IEEE J. Quantum Electron.* **qe-11**, 935 (1975).

obtaining oscillation in a helium–neon discharge at five wavelengths around 1.1 μm. This was the first observation of laser oscillation in any gas discharge. Many additional lines in neutral neon were later observed by others; some of these lines required or were greatly enhanced by helium collisions. White and Rigden[310] were the first to demonstrate the now-ubiquitous 0.6328-μm red laser in 1962. Other visible He–Ne laser transitions were obtained by Bloom[311] and White and Rigden.[312] Operation at 3.39 μm was first reported in 1963 by Bloom et al.[313] High-power pulsed operation on the 1.1-μm transitions was first obtained by Boot and Clunie.[314]

Since 1961, hundreds of papers have appeared dealing with the helium–neon laser system. In this section the spectroscopy, excitation mechanisms, and performance are reviewed briefly. For a more detailed treatment see the book by Willett[2] or the long article by Arrathoon.[315]

2.3.1.1. Spectroscopy of the Helium–Neon Laser. Partial energy level diagrams of neutral helium and neon are shown in Fig. 18, with the strongest laser transitions indicated explicitly. The older Paschen notation is used for the neon levels for simplicity, although the intermediate-coupling model and notation introduced by Racah[65] is generally preferred. The correspondence between these two notations is given in Table XI for some levels. The 3s and 2s neon levels are populated by collision with the 2s $^1S_0$ and 2s $^3S_1$ helium metastable levels. Several of the possible transitions in each of the 3s → 3p, 3s → 2p, and 2s → 2p multiplets oscillate in addition to those shown in the figure. Table XI lists all of the 3s → 2p and 2s → 2p transitions observed to date. All of the $3s_2$ → 2p transitions allowed by the ΔJ selection rules oscillate, but none from the other 3s upper levels oscillate. Of the 2s → 2p transitions, all but one of the allowed transitions oscillate, although five of these have been obtained observed only in pulsed Ne–H_2 hollow-cathode discharges by Chebotayev and Vasilenko,[316] and some of the cw lines were obtained only in a 10-m-long He–Ne discharge by Zitter.[317] Many of these transitions share levels, so that competition for population will prevent oscillation on the weaker lines unless an intracavity wavelength-selective ele-

---

[310] A. D. White and J. D. Rigden, *Proc. IEEE* **50**, 1697 (1962).
[311] A. L. Bloom, *Appl. Phys. Lett.* **2**, 101 (1963).
[312] A. D. White and J. D. Rigden, *Appl. Phys. Lett.* **2**, 211 (1963).
[313] A. L. Bloom, W. E. Bell, and R. C. Rempel, *Appl. Opt.* **2**, 317 (1963).
[314] H. A. H. Boot and D. M. Clunie, *Nature* (*London*) **197**, 173 (1963).
[315] R. Arrathoon, *Lasers* **4**, 119 (1976).
[316] V. P. Chebotayev and L. S. Vasilenko, *Sov. Phys.—JETP* (*Engl. Transl.*) **21**, 515 (1965).
[317] R. N. Zitter, *J. Appl. Phys.* **35**, 3070 (1964).

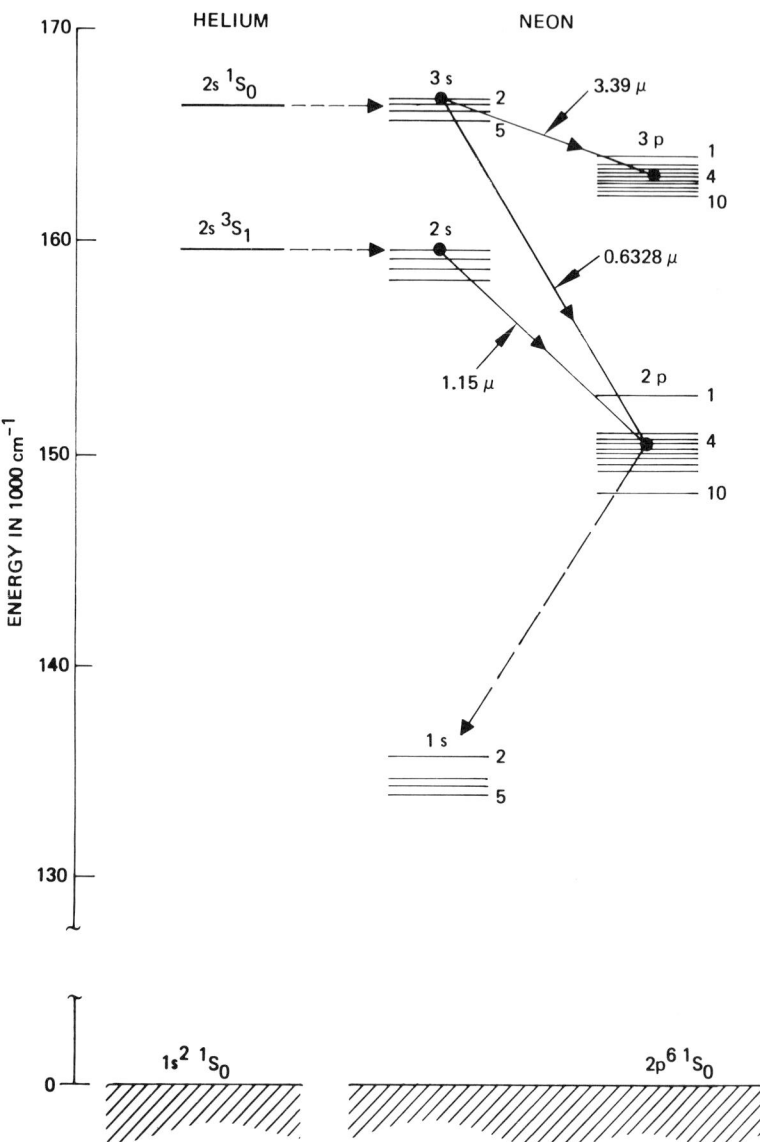

FIG. 18. Partial energy level diagrams of helium and neon showing the three strongest neon laser transitions. Paschen notation has been used for simplicity to designate the neon excited states.

TABLE XI. Helium–Neon Laser Transitions in the 3s → 2p and 2s → 2p Multiplets[a]

Lower level	Upper level 3s$_2$ 5s'[1/2]$_1^o$	2s$_2$ 4s'[1/2]$_1^o$	2s$_3$ 4s'[1/2]$_0^o$	2s$_4$ 4s[3/2]$_1^o$	2s$_5$ 4s[3/2]$_2^o$
2p$_1$ 3p'[1/2]$_0$	0.7305	1.5230	×	1.7162[b]	×
2p$_2$ 3p'[1/2]$_1$	0.6401	1.1766	1.1984	(1.2887)	1.3219[c]
2p$_3$ 3p[1/2]$_0$	0.6352	1.1602	×	1.2689	×
2p$_4$ 3p'[3/2]$_2$	0.6328	1.1523	×	1.2594[c]	1.2912
2p$_5$ 3p'[3/2]$_1$	0.6294	1.1409	1.1614	1.2459[b]	1.2769[c]
2p$_6$ 3p[3/2]$_2$	0.6118	1.0844	×	1.1788[b]	1.2066
2p$_7$ 3p[3/2]$_1$	0.6046	1.0621	1.0798	1.1525	1.1790[d]
2p$_8$ 3p[5/2]$_2$	0.5939	1.0295[b]	×	1.1143	1.1391
2p$_9$ 3p[5/2]$_3$	×	×	×	×	1.1178
2p$_{10}$ 3p[1/2]$_1$	0.5433	0.8865[b]	0.8988[b]	0.9486[c]	0.9665[c]

[a] ×, Forbidden by ΔJ selection rules; ( ), no oscillation observed.
[b] Oscillation observed only in 10-m-long He–N$_2$ discharge by R. N. Zitter, *J. Appl. Phys.* **35**, 3070 (1964).
[c] Pulsed oscillation only observed in a Ne–H$_2$ hollow cathode discharge by V. P. Chebotayev and L. S. Vasilenko, *Sov. Phys.—JETP (Engl. Transl.)* **21**, 515 (1965).
[d] Oscillation observed only in a Ne–H$_2$ hollow cathode discharge by V. P. Chebotayev, *Radio Eng. Electron. Phys. (Engl. Transl.)* **10**, 316 (1965).

ment such as a prism is used. This is particularly true of the 3s$_2$ → 2p transitions; without any intracavity prism, only the 0.6328-μm line will oscillate.[311,312] (An exception to this is the 0.6401-μm line. Anomalous dispersion from a nearby absorption line creates a lens effect at 0.6401 μm not felt by the 0.6328-μm line; both lines will oscillate if the optical cavity is arranged to be marginally stable at 0.6328 μm[317a,317b].)

In addition to the lines listed in Table XI, there are 13 laser lines in the 2.42–4.22-μm range belonging to the 3s → 3p multiplet. The gain on the 3s$_2$ → 3p$_4$ transition is exceptionally high, so that an intracavity prism and

[317a] A. L. Bloom and D. L. Hardwick, *Phys. Lett.* **20**, 373 (1966).
[317b] L. A. Schlie and J. T. Verdeyn, *IEEE J. Quantum. Electron.* **qe-5**, 21 (1969).

other means of spoiling the high gain at 3.39 μm is often employed to prevent quenching of the visible transitions.[311-313]

2.3.1.2. Excitation Processes in the Helium–Neon Laser.
As indicated in Fig. 18, the neon upper laser levels are populated by near-resonant collisions of the second kind with helium atoms. Excitation is transferred selectively from the helium $2s\,^3S_1$ and $2s\,^1S_0$ metastable atoms primarily to the neon $3s_2$ and $2s_2$ states by the processes

$$He(2^3S_1) + Ne(^1S_0) \longrightarrow He(^1S_0) + Ne(2s_2), \qquad (2.3.1)$$

$$He(2^1S_0) + Ne(^1S_0) \longrightarrow He(^1S_0) + Ne(3s_2). \qquad (2.3.2)$$

The reactions given in Eqs. (2.3.1) and (2.3.2) also proceed in the reverse direction as well, but the high ratio of helium to neon in the typical laser discharge and the rapid destruction of the neon excited states via the laser output assure that the rate in the direction indicated is much greater than the reverse rate.

Both $2s_2$ and $3s_2$ states have spin 0 (singlets) in the LS coupling model. Thus the Wigner spin rule, which requires the total spin of the system to be conserved in a collision, is violated for reaction (2.3.1) and obeyed for reaction (2.3.2). However, the cross section for resonant excitation transfer also depends strongly on the difference (defect) in energy between the interacting levels. The details of this dependence are complicated, but a variation of the form $\exp(-\Delta E/kT)$ is expected, where ΔE is the magnitude of the energy defect and kT the kinetic energy of the colliding atoms. Figure 19 shows the relative positions of the levels in more detail. For reaction (2.3.1) the $2s_2$ level lies just below the $He(2^3S_1)$ level and is the closest member of the multiplet. This energetically favored position evidently outweighs the spin conservation violation, and the $2s_2$ is preferentially populated. Existence of laser oscillation from the other three 2s levels, for which the spin rule is obeyed, suggests that the spin rule does have an influence on the transfer cross sections. The branching of the excitation among the 2s states is not known quantitatively. The situation is quite different for reaction (2.3.2). The neon $3s_2$ level lies above the helium metastable and is not the closest level to it. Yet the $3s_2$ level is observed to be preferentially populated, even to the point that no laser transitions are observed from the other 3s levels. Evidently the spin conservation outweighs the unfavorable energetic position of the $3s_2$ in this case.

Helium metastables are created by collision with the discharge electrons by several paths: (a) directly, (b) by radiative cascade from higher-lying levels created by electron collision, (c) by ionization, recombination, and then radiative cascade to the metastable levels. Process (a) is dipole-forbidden (which is why the helium 2s levels are metastable),

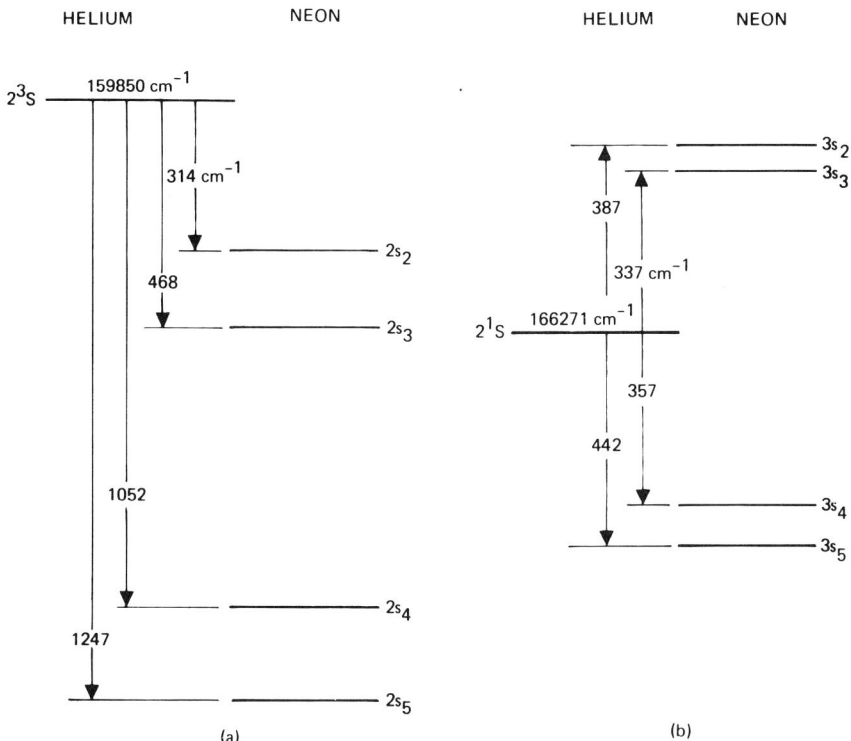

FIG. 19. Energy level relationship between the helium metastables and the 2s, 3s neon levels.

and the cross sections for direct electron excitation from the ground state are small and highly peaked just above their thresholds around 20 eV. Nevertheless, there is a relatively large contribution from this source since the discharge electron energy distribution falls off rapidly above 20 eV. Wada and Heil[318] have measured electron distributions in helium–neon discharges that fall off substantially faster than Maxwellian as shown in Fig. 20. Using their measured electron energy distribution for $pD = 3.0$ torr-mm and published excitation cross sections as shown in Fig. 21, they calculate the following rates:

$$\text{He}(1s^2\ ^1S_0) + \textit{fast}\ e \longrightarrow \text{He}(2s\ ^1S_0) + \textit{slow}\ e = 1.2, \quad (2.3.3)$$

$$\text{He}(1s^2\ ^1S_0) + \textit{fast}\ e \longrightarrow \text{He}(2p\ ^1P_1^{\text{o}}) + \textit{slow}\ e = 1.5, \quad (2.3.4)$$

$$\text{He}(1s^2\ ^1S_0) + \textit{fast}\ e \longrightarrow \text{He}^+ + 2\ \textit{slow}\ e = 4.6, \quad (2.3.5)$$

[318] J. Y. Wada and H. Heil, *IEEE J. Quantum Electron.* **qe-1,** 327 (1965).

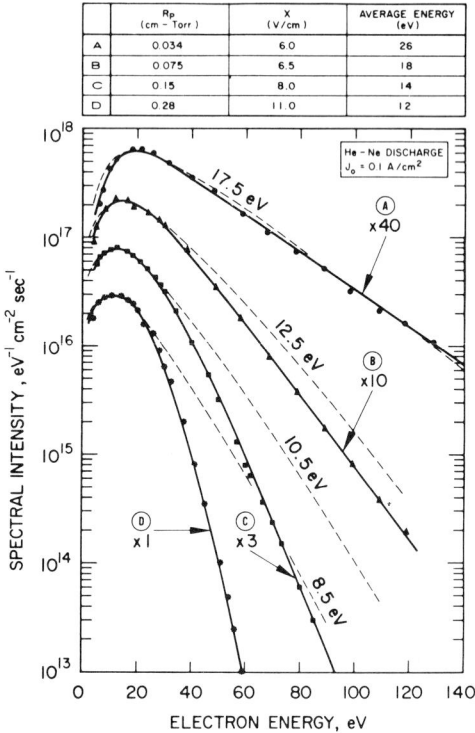

FIG. 20. Electron energy distributions determined by Wada and Heil[318] for different valuse of the tube radius pressure product R_P. The dashed curves are Maxwellian distributions for the temperatures indicated in kT/e; X is the longitudinal elective field in the discharge column (by courtesy of the IEEE).

where the rates are given as excitations per second per unit ground state atom density. Relation (2.3.3) gives the contribution to the excitation via the direct path. Relation (2.3.4) gives the excitation rate to one of the important higher-lying levels. The reverse process to relation (2.3.4), vuv radiation to the ground states is allowed, but the radiation is highly trapped,[11] so that the only effective radiative path from the np levels is to the metastable levels. The contribution from the higher n^1P states via path (b) alone is probably negligible since their electron excitation cross sections are substantially lower, but may be significant via path (c).

Although the relative contributions from these three paths are not known quantitatively, all three should be directly proportional to electron density and thus discharge current. White and Gordon[319] have measured the He($2s\ ^1S_0$) density by absorption and find a linear increase with dis-

[319] A. D. White and E. I. Gordon, *Appl. Phys. Lett.* **3**, 197 (1963).

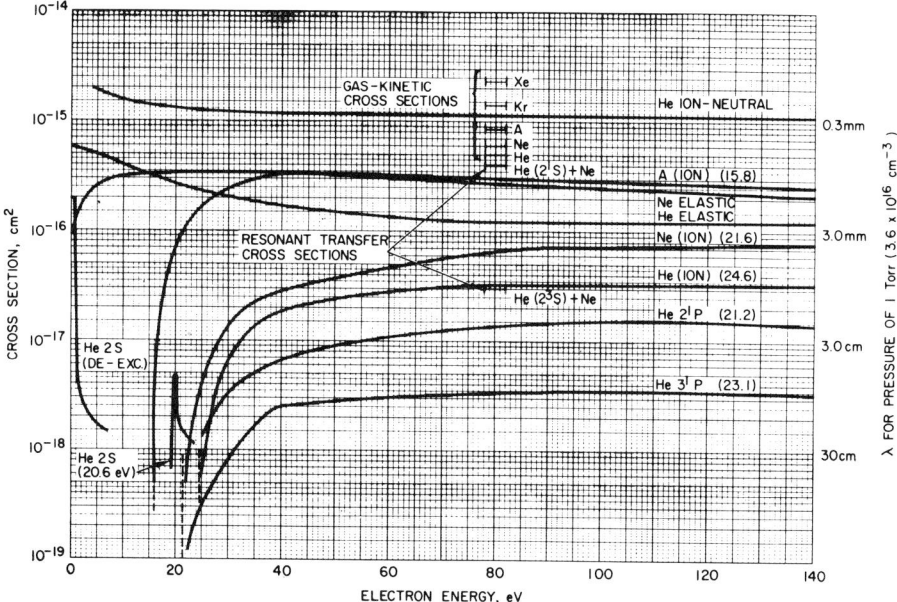

FIG. 21. Various collision cross sections compiled by Wada and Heil[318] from references given in their paper. Note the sharply peaked nature of the He 2s cross section (by courtesy of the IEEE).

charge current at low currents, followed by a saturation to a constant value of 2.5×10^{11} cm^{-3}, as shown by the open circles in Fig. 22. Such a variation is consistent with a creation rate $K_1 I$ and a destruction rate $(K_2 + K_3 I)$, where I is the discharge current density. The constant K_2 represents metastable destruction by diffusion to the walls and $K_3 I$ the destruction by electron collision. The form of the measured metastable density curve in Fig. 22 gives an excellent fit to $K_1 I/(K_2 + K_3 I)$. White and Gordon assume that the current-dependent destruction mechanism $K_3 I$ is the result of superelastic collisions with slow electrons. There are two such processes for the 2s $^1S_0$ metastable and one for the 2s $^3S_1$:

$$\text{He}(2s\ ^1S_0) + slow\ e \longrightarrow \text{He}(1s^2\ ^1S_0) + fast\ e, \qquad (2.3.6)$$

$$\text{He}(2s\ ^1S_0) + slow\ e \longrightarrow \text{He}(2s\ ^3S_1) + fast\ e, \qquad (2.3.7)$$

$$\text{He}(2s\ ^3S_1) + slow\ e \longrightarrow \text{He}(1s^2\ ^1S_0) + fast\ e. \qquad (2.3.8)$$

The cross sections for processes (2.3.6) and (2.3.7) can be derived from Schulz' measurements[320]; the cross-section curve for process (2.3.8) is

[320] G. J. Schulz, *Phys. Rev.* **116**, 1141 (1959).

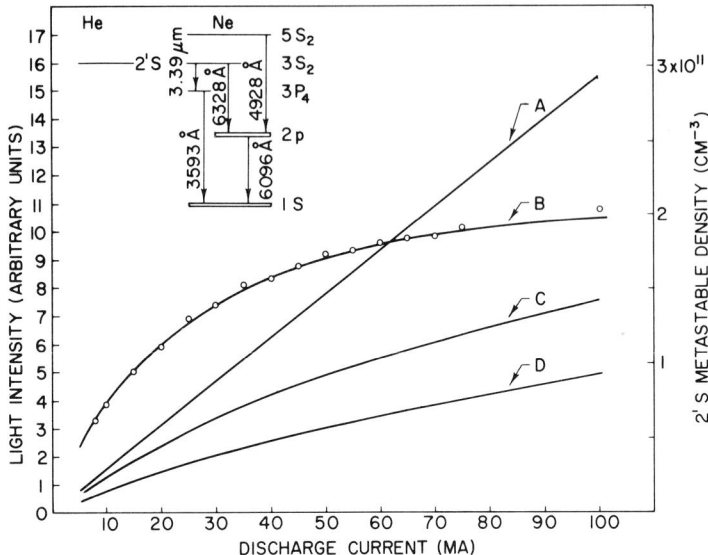

FIG. 22. Light intensity and 2s $^1S_0$ helium metastable density (in arbitrary units) in an He–Ne discharge plotted as a function of discharge current. (A) 0.4928-μm Ne($5s_2 \rightarrow 2p_2$); (B) 0.6328-μm Ne ($3s_2 \rightarrow 2p_4$); circled points, He 2s $^1S_0$ metastable density; (C) 0.6096-μm Ne ($2p_4 \rightarrow 1s_4$); (D) 0.3593-μm Ne ($3p_4 \rightarrow 1s_2$) (from White and Gordon[319] by courtesy of American Institute of Physics).

shown on Fig. 21. It is sharply peaked at zero electron energy, so that the rate for this process is sensitive only to the number of very slow electrons, not the high-energy portion of the curve. Process (2.3.7) has a cross section even larger than (2.3.6) or (2.3.8) according to Phelps.[321] However, Wada and Heil[318] calculate that the rates due to these processes are insufficient by *orders of magnitude* to explain the observed saturation at a metastable density of 2.5×10^{11} cm^{-3} in Fig. 22. Arrathoon[315] has considered the possibility that helium metastables are destroyed by excitation to higher-lying states, specifically He (2p ^{3}P) and He(2p ^{1}P) and calculated a plausible rate out of the metastable levels. However, no mechanisms are known that would deplete the P levels at a sufficient rate to prevent the radiative cascade back into the metastable levels. Browne and Dunn in Ref. (440) have made measurements of the triplet and singlet metastable densities in a pure helium discharge, obtaining a saturation value of 2×10^{12} cm^{-3} for the singlet metastables. They obtain good agreement between their experimental results and a model that includes both process (2.3.6) and destruction by electron collisional ionization:

$$\text{He}(2s\ ^1S_0) + e \longrightarrow \text{He}^+ + 2e \qquad (2.3.9)$$

[321] A. V. Phelps, *Phys. Rev.* **99**, 1307 (1955).

Since ions can recombine only at the walls, where they also return to the ground state, there is no reverse process to (2.3.9). Their model indicates that process (2.3.9) is about an order of magnitude more likely than (2.3.6).

The exact coincidence in shape between the measured current variation of the neon $3s_2$ spontaneous emission (curve B) and helium $2s\ ^1S_0$ metastable population (open circles) shown in Fig. 22 provides convincing evidence that collisional transfer is the dominant population mechanism. Additional measurements by Young et al.[322] show that electron collisions contribute only 1–2% to the $3s_2$ population and 10–20% to the other 3s levels. The measurement corresponding to Fig. 22 for transfer from the helium $2s\ ^3S_1$ to the neon $2s_2$, for which the spin rule is violated, have not been made. Some contribution to the neon 2s levels by electron collision is expected, since many of the 2s → 2p transitions will oscillate in a pure neon discharge. Bennett suggests[323a] that the 2s levels are about equally populated by helium and electron collision. Measurements by Young et al.[322] give values to 3–20%, depending on the particular neon 2s level and the discharge current.

The chain of processes responsible for populating and depopulating the lower laser levels is also not completely determined. Depopulation occurs radiatively via the strong transitions responsible for the usual bright reddish-orange color of the typical neon discharge and the pulsed laser transitions discussed in Section 2.1.1.2. Transitions from the $1s_3$ and $1s_5$ levels to the ground state are dipole-forbidden by the selection rules on the total angular momentum J; they are metastable and are destroyed only by collisions or by diffusion to the walls. The $1s_2$ and $1s_4$ levels are radiatively coupled to the neon ground state, but they become effectively metastable because their radiation to the ground state is highly trapped. Additional processes in the gas act to counter this depopulation route for the 2p and 3p levels. Figure 22 shows that the populations of the neon $2p_4$ and $3p_4$ levels vary approximately linearly with discharge current and hence with electron density. This variation suggests that the 2p and 3p levels are populated by a single-electron collision process, for example, direct excitation from the ground state or radiative cascade from higher-lying levels that are excited from the ground state. The first process is dipole-forbidden by parity, and should have an electron excitation cross section similar to that for the helium $2s^3S_1$ metastable, sharply peaked just above 18.5 and 20.5 eV. While such cross sections are small, they occur at the peak of the measured electron energy distribution as

[322] R. T. Young, Jr., C. S. Willett, and R. T. Maupin, *J. Appl. Phys.* **41**, 2936 (1970).
[323] E. F. Labuda, Ph.D. Dissertation, Polytechnic Institute of Brooklyn, New York (1967)
[323a] W. R. Bennett, Jr., *Appl. Opt.*, Suppl., 24 (1962).

shown in Fig. 20. A two-electron process that should have higher cross section was proposed by Javan[6]:

$$Ne(2p^6\ ^1S_0) + fast\ e \longrightarrow Ne(1s) + slow\ e, \quad (2.3.10)$$

$$Ne(1s) + fast\ e \longrightarrow Ne(2p) + slow\ e, \quad (2.3.11)$$

and similarly for the 3p levels. Such processes should result in a quadratic variation with discharge current for the 2p and 3p level populations, which is observed in pure neon discharges but not in the He–Ne discharges.[319] Radiation trapping on process (2.3.10) could produce saturation with current in the 1s population similar to that for the helium metastables shown in Fig. 22. Then the electron collision of process (2.3.11) would yield an overall linear dependence on discharge current. No direct measurement of the 1s population variation with current has been published as it has for the helium metastable. Arrathoon[315] states that his unpublished measurements show no saturation in the 1s levels with current over the range of typical current values. On the basis of the similar current dependencies of the different $2p \to 1s$ transitions, White and Gordon[319] also conclude that trapping of the radiation to the ground state from the 1s levels is insignificant. They conclude instead that the almost linear variation with current of the 2p and 3p levels results from population by radiative cascade from higher-lying states such as the strongly populated 2s and 3s levels. Whatever the cause, the linear current dependence is well verified.

2.3.1.3. Performance of the Helium–Neon Laser. The observed current dependencies of upper and lower laser level spontaneous emission may be combined to give the variation of small-signal gain. White and Gordon[319] derive the gain variation expected for a 3.39-μm $3s_2 \to 3p_4$ laser as

$$\text{gain}\ (\text{dB}) = \frac{2.14I}{1 + 0.04I} - 0.194I, \quad (2.3.12)$$

for a particular tube diameter, length, and gas fill. The current I is in mA. Figure 23 shows Eq. (2.3.12) compared to experimentally measured points (solid circles). The higher gain available from the lighter isotope ^{3}He is shown by the open circles, a consequence of the $(4/3)^{1/2}$ higher collision frequency for this atom. Willett[9] has combined the measurements of White and Gordon[319] with those of their co-worker Labuda[323] to deduce the corresponding relation for the 0.6328-μm $3s_2 \to 2p_4$ laser:

$$\text{gain}\ (\text{nepers/m}) = \left(\frac{0.594I}{1 + 0.04I} - 0.0572I\right) \times 10^{-2}. \quad (2.3.13)$$

This expression is compared with Labuda's measured values for a 6-

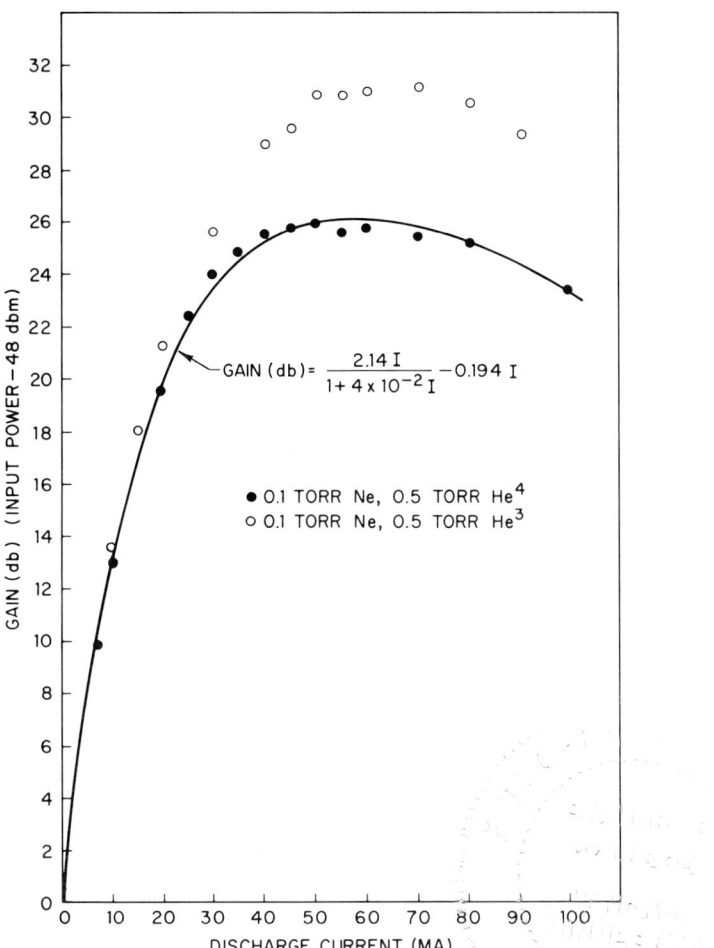

FIG. 23. Small-signal gain at 3.39-μm Ne ($3s_2 \to 3p_4$) as a function of discharge current. The solid circles are measured values; the solid curve is obtained by fitting the data presented in Fig. 22. The gain with ^{3}He substituted for ^{4}He is given by the circles (from White and Gordon[319] by courtesy of the American Institute of Physics).

mm-diameter tube in Fig. 24. The corresponding comparison for the 1.15-μm $2s_2 \to 2p_4$ laser has not been made.

Several workers have measured the optimum gas filling conditions for helium–neon lasers. For the 0.6328-μm laser, the optimum values are a He:Ne ratio of 5:1 or 6:1 and pD = 3.6 to 4.0 torr-mm. The values for the 3.39-μm laser are expected to be similar, while the 1.15-μm laser[324] prefers a He:Ne ratio of 8:1 or 10:1 and pD = 14 to 17 torr-mm.

[324] C. K. N. Patel, *J. Appl. Phys.* **33**, 3194 (1962).

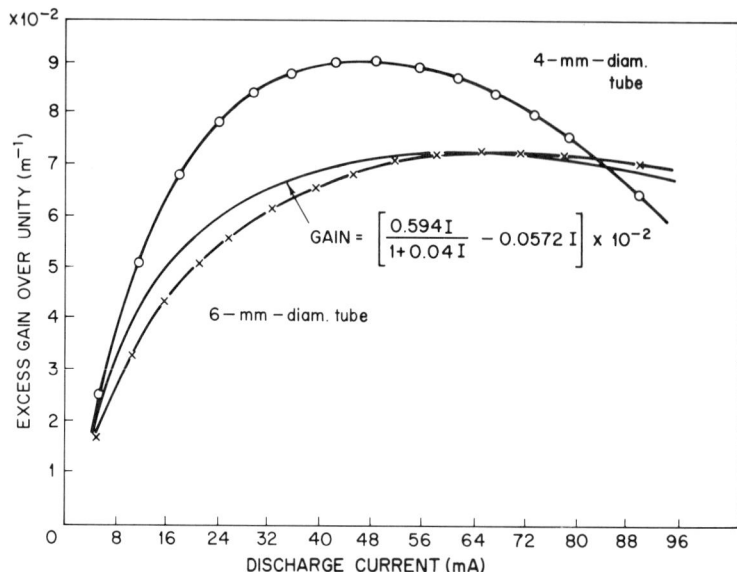

FIG. 24. Small-signal gain at 0.6328-μm Ne($3s_2 \to 2p_4$) as a function of discharge current for 7:1 He:Ne mixture, at $pD = 3.6$ torr-mm. The curve without data points was obtained by fitting the 0.6328- and 0.3593-μm curves of Fig. 22 (from data by Labuda[323] as cited by Willett,[2] by courtesy of Pergamon Press).

The fact that the optimum gas pressure can be specified by a single value of pD over a wide range of discharge diameters indicates that it is the *electron temperature* that is being optimized. The elementary theory of the positive column[325] with charge neutrality, ambipolar diffusion, and a mean-free path less than the discharge diameter yields a Maxwellian electron energy distribution whose temperature depends only on the product pD. Even more realistic theories[68] that allow non-Maxwellian electron energy distributions have pD as the parameter controlling the width of the distribution. By measuring the microwave noise radiation from discharges, Labuda and Gordon[326] have confirmed the dependence of the electron temperature on the product pD for 5:1 helium–neon mixtures in tubes 2–6 mm in diameter. At the optimum value of $pD = 3.6$–4 torr-mm, they find radiation temperatures of 80–87,000 K, corresponding to an average electron energy of 10–11 eV if a Maxwellian distribution is assumed (the average electron energy is 1.5 kT/e for a Maxwellian). Wada and Heil's measurements (Fig. 20), although indicating a non-

[325] Cf. A. von Engel, "Ionized Gases," 2nd ed. Oxford Univ. Press (Clarendon), London and New York, 1955.
[326] E. F. Labuda and E. I. Gordon, *J. Appl. Phys.* **35**, 1647 (1964).

Maxwellian distribution, give a value of about 13 eV for the average electron energy at $pD = 3.6-4$ torr-mm, in good agreement with Labuda and Gordon.[326] Labuda and Gordon's measurements also indicate an exactly linear relation between electron density and discharge current.

The small-signal gain of helium–neon lasers is observed experimentally to vary as D^{-1}, provided pD is kept constant. A simple explanation follows directly from the result just cited: at constant pD the electron temperature is constant, and thus the integral involving the excitation cross section and the electron energy distribution function is also a constant. Hence, all electron-collision excitation processes scale simply as the number of atoms available to excite. Since both upper and lower laser levels are ultimately populated by electron-collision processes, their populations and hence the laser gain is directly proportional to pressure, or to D^{-1} at constant pD. In fact, this relationship is probably better thought of as the *pressure* dependence of the gain rather than the *diameter* dependence. Note that this relation breaks down if other processes that are not electron-collision dominated come into play, for example, destruction of metastables by diffusion to the walls. Gordon and White[327] have formulated the gain expression in more general terms and show that at pressures much below 1 torr (large-diameter tubes) the gain should vary as p^2 or D^{-2}. At the other extreme, if the pressure is sufficiently high that the gain linewidth becomes pressure broadened rather than Doppler broadened, for example, in a very small bore waveguide laser,[328] then the gain should become independent of p or D. The relation $g \sim p$ or $g \sim D^{-1}$ has been experimentally verified over the range 1–10 mm in diameter. The numerical value for the small-signal gain coefficient found by Smith[329] for the 0.6328-μm line at optimum pressure, mixture, and current is

$$g = 0.3D^{-1} \quad \text{nepers/m},$$

or

$$= 1.3D^{-1} \quad \text{dB/m}, \quad (2.3.14)$$

when D is in millimeters.

The output power of a laser depends not only on the small-signal gain, but also the saturation parameter, the optical losses, the mode filling factor, and the output coupling. Smith[330,331] has measured the saturation

---

[327] E. I. Gordon and A. D. White, *Appl. Phys. Lett.* **3**, 199 (1963).
[328] R. L. Abrams and W. B. Bridges, *IEEE J. Quantum Electron.* **qe-9**, 940 (1973).
[329] P. W. Smith, *IEEE J. Quantum Electron.* **qe-2**, 77 (1966).
[330] P. W. Smith, *IEEE J. Quantum Electron.* **qe-2**, 62 (1966).
[331] P. W. Smith, *J. Appl. Phys.* **37**, 2089 (1966).

TABLE XII. Comparison of He–Ne 0.6328 μm Measured and Predicted Output Power

Diameter D (mm)	Length l (m)	Field observed[a] P/Dl (mW/mm-m)	Smith predicted[b] P/Dl (mW/mm-m)	Optimum transmission[b] t_{opt} (%)	Transmission used[a] t (%)
1.5	0.125	1.05	0.76	1.2	1.1
3	0.55	1.55	1.86	2.7	1.0
5	0.65	0.87	1.38	1.9	1.0
8	2.00	0.92	2.28	3.5	1.1

[a] Values from R. L. Field, Jr., *Rev. Sci. Instrum.* **38**, 1720 (1967).
[b] Predicted from Eqs. 2.3.16 and 2.3.17, derived from P. W. Smith, *IEEE J. Quantum Electron.* **qe-2**, 62 (1966).

parameter for the 0.6328-μm laser in modest-size discharge tubes. He finds an expression for the single-ended output power of

$$P \text{ mW} = \frac{\pi D^2}{5} S_0 G[1 - (a/G)^{1/2}]^2, \quad (2.3.15)$$

where D is in mm, S_0 is the empirically determined saturation flux density = 300 mW/(mm)^2, a is the optical loss per pass, and $G = gl$ is the optical gain per pass. Inherent in Eq. (2.3.15) are the assumptions that the optical mode occupies 1/5 the physical tube cross-sectional area (appropriate to TEM$_{00}$ mode operation), and that the single-ended output coupling t is at its optimum value

$$t_{opt} = 2G[(a/G)^{1/2} - (a/G)]. \quad (2.3.16)$$

Equations (2.3.14) and (2.3.15) can be combined in the form

$$P/D = 56.6[1 - (a/G)^{1/2}]^2 \text{ mW/mm-m}. \quad (2.3.17)$$

Field[332] has measured the output power of helium–neon lasers with a range of diameters and lengths as a function of gas pressure and mixture. Table XII compares Field's results with the predictions of Smith's equation. The comparison is reasonably good considering that Field did not optimize the optical coupling for each tube and operated the lasers with high-order transverse modes rather than TEM$_{00}$. Field does not give a value for the optical loss per pass a; a value of 1% has been assumed in the calculations of Table XII.

2.3.1.4. Pulsed Operation of Helium–Neon Lasers. Much higher output powers have been obtained on a pulsed basis from some of the

[332] R. L. Field, Jr., *Rev. Sci. Instrum.* **38**, 1720 (1967).

2s → 2p neon transitions as first demonstrated by Boot et al.[314,333] This mode of operation is enhanced by much higher helium pressures (~200 torr) and much higher discharge currents (10–100 A) than those appropriate to cw He–Ne operation. At such high currents both upper and lower neon laser levels are excited by electron collision during the current pulse and no strong laser oscillation is observed after the initial transient characteristic of cyclic laser operation (see Section 2.1). However, after the current pulse decays, the plasma electrons recombine rapidly, but the helium 2s $^3S_1$ metastable atoms remain for times on the order of 100 μsec and are very effective in exciting the 2s levels in neon. Since there are no longer electrons around to excite the 2p lower levels, very high inversion densities are obtained at high helium pressures.

Boot and Clunie[314] originally reported 1 W peak output power on the $2s_2 \to 2p_4$ line at 1.1523 μm, occurring 30–80 μsec after the decay of the 10-μsec exciting pulse in a mixture of 70 mtorr neon and 1 torr helium. Both dc and 3-GHz microwave pulses produced similar results. Further optimization[333] increased the output to 84 W peak in a tube 20 mm in diameter and 150 cm long excited with 1-μsec, 35-A current pulses. The optimum pressures were 240 torr helium and 5 torr neon. Four transitions were observed:

$2s_5 \longrightarrow 2p_9$ 1.1178 μm
$2s_2 \longrightarrow 2p_4$ 1.1523 μm
$2s_3 \longrightarrow 2p_5$ 1.1614 μm
$2s_5 \longrightarrow 2p_6$ 1.2066 μm

An average power of 224 mW was obtained at a 2.5-kHz repetition rate, limited by the power supply. Shtyrkov and Subbes[334] have observed the $2s_2 \to 2p_7$ 1.0621-μm line in addition to the four reported by Boot et al.[333] in a discharge 11 mm in diameter and 80 cm long. Studies of the excitation of the various 2s → 2p laser lines in helium–neon mixtures have been performed by Javan et al.,[309] Petrash and Knyazev,[335] and particularly at the higher pressures appropriate to pulsed laser operation, by Shtyrkov and Larionov[336] and Perchanok et al.[337] Clunie and Rock[338] have measured the gain saturation of the 1.1178- and 1.1523-μm transitions in a

---

[333] H. A. H. Boot, D. M. Clunie, and R. S. A. Thorn, *Nature* (London) **198**, 773 (1963).
[334] Ye. I. Shtyrkov and E. V. Subbes, *Opt. Spectrosc.* (Engl. Transl.) **21**, 143 (1966).
[335] G. G. Petrash and I. N. Knyazev, *Sov. Phys.—JETP* (Engl. Transl.) **18**, 571 (1964).
[336] Ye. I. Shtyrkov and N. P. Larionov, *Sov. Phys.—Tech. Phys.* (Engl. Transl.) **13**, 248 (1968).
[337] T. M. Perchanok, V. M. Russov, and S. A. Fridrikhov, *Sov. Phys.—Tech. Phys.* (Engl. Transl.) **11**, 1633 (1967).
[338] D. M. Clunie and N. H. Rock, *Phys. Lett.* **13**, 213 (1964).

multipass laser amplifier 45 mm in diameter and 150 cm long. A small-signal gain of 19 in a 12-m folded path was observed at 1.1523 μm, and saturated output powers of 38 and 30 W were obtained at 1.1178 and 1.1523 μm, respectively, with 4 W input power. The pressures used in these measurements were 5 torr neon and 155 torr helium.

2.3.2. Other Neutral Atom Lasers Excited by Atomic Collisions

2.3.2.1. Argon–Chlorine Laser. The only resonant-transfer laser analogous to the He–Ne laser presently known is the Ar–Cl laser investigated by Dauger and Stafsud.[339] Argon 1s metastables (Paschen notation) excite the chlorine upper laser levels via the reaction

$$Ar^*(1s_2) + Cl(^2P_{3/2}) \longrightarrow Ar(^1S_0) + Cl^*(5p\ ^2D^o_{5/2}) \quad (2.3.18)$$

for chlorine atoms in the $^2P_{3/2}$ ground state or

$$Ar^*(1s_3) + Cl(^2P_{1/2}) \longrightarrow Ar(^1S_0) + Cl^*(5p\ ^2D^o_{5/2}) \quad (2.3.19)$$

for chlorine atoms in the $^2P_{1/2}$ level just 881 cm^{-1} above the ground state. The energy defect is 4 cm^{-1} for reaction (2.3.18) and 39 cm^{-1} for (2.3.19). The chlorine atoms then oscillate on the Cl I 5p $^2D^o_{5/2}$ → 5s $^2P_{3/2}$ transitions at 3.067 μm. Dauger and Stafsud[339] report a cw output power of 8 mW from a 2.5-cm-diameter tube 187 cm long at a discharge current of 125 mA with a gas mixture of 0.09 torr Cl$_2$, 2.1 torr Ar, and 17 torr He. No further development of this laser has been reported.

2.3.2.2. Neon–Oxygen Laser. Oscillation at 0.8446 μm in atomic oxygen was observed by Bennett et al.[340] in a neon–oxygen discharge and studied further by Kolpakova and Redko.[341] The excitation is thought to proceed via the reactions

$$Ne^*(1s) + O_2 \longrightarrow Ne(^1S_0) + O_2^*, \quad (2.3.20)$$

$$O_2^* \longrightarrow O + O^*(3p\ ^3P_{0,1,2}), \quad (2.3.21)$$

in which the neon metastables excite ground-state molecules to an as yet unidentified repulsive-potential excited state, which decays to a ground-state atom and an atom in one of the 3p ^3P states. The oxygen then oscillates on the 3p $^3P_2$ → 3s $^3S_1$ transition at 0.8446 μm. Resonance radiation depopulating the 3s $^3S_1$ state is highly trapped if a large number of ground-state oxygen atoms are produced, as implied by reaction (2.3.21); this spoils the inversion on the laser line and allows oscillation only in one wing of the laser transition.[340]

---

[339] A. B. Dauger and O. M. Stafsud, *IEEE J. Quantum Electron.* **qe-6,** 572 (1970).
[340] W. R. Bennett, Jr., W. L. Faust, R. A. McFarlane, and C. K. N. Patel, *Phys. Rev. Lett.* **8,** 470 (1962).
[341] I. V. Kolpakova and T. P. Redko, *Opt. Spectrosc. (Engl. Transl.)* **23,** 351 (1971).

2.3.2.3. Argon–Oxygen Laser. Discharges in argon–oxygen mixtures exhibit oscillation on the same transition as neon–oxygen, but the population processes are different.[340,342] The oxygen upper levels are thought to result from the three steps

$$Ar^*(1s) + O_2 \longrightarrow O_2^* + Ar \qquad (2.3.22)$$

$$O_2^* \longrightarrow O + O^*(^1S_0), \qquad (2.3.23)$$

$$Ar^*(1s) + O \longrightarrow Ar + O^*(3p\ ^3P_{0,1,2}). \qquad (2.3.24)$$

The oscillation line structure of the resulting inversion is somewhat different from that of Ne–O_2 because the amount of kinetic energy left over in reaction (2.3.24) is larger than that in reactions (2.3.20) and (2.3.21); as a consequence the 3p $^3P_{0,1,2}$ states are Doppler broadened more in Ar–O_2 than in Ne–O_2.[343]

2.3.2.4. High-Pressure Noble-Gas Mixtures. Chapovsky et al.[66] have reported oscillation at several wavelengths in neutral Ar, Kr, and Xe using approximately 1% of the lasing gas in a buffer of 0.4–7 atmospheres of He, Ne, or Ar. High-voltage transverse excitation was employed in a laser cell 6 × 15 mm in cross section and 60 mm long. Peak output powers as high as 10 kW were observed on some of the $ns \to np$ lines listed in Table III and greater than 100 kW was seen on the Ar I 1.79144-μm line ($3d[1/2]_1^0 \to 4p[3/2]_2$ or $3d[1/2]_0^0 \to 4p[3/2]_1$). The particular lines that oscillate depend on the choice of buffer gas and are generally not the same lines that are strong in the pure noble gases (Table IV). It is possible that the excitation proceeds via the formation of excimers, e.g., (HeAr)*, which then dissociate with one partner in a specific excited state. Further study is obviously required to trace out the excitation pathways in these high-powered lasers.

High pulsed output powers at high pressures have been observed previously in He–Xe mixtures, but on the same lines that are generally observed in a low-power cw operation (Table IV). Schwarz et al.[344] used a transverse discharge with a helical arrangement of pin electrodes to obtain 1 kW peak output on the Xe 2.027-, 3.508-, and 3.652-μm (Xe I $7p[1/2]_1 \to 7s[3/2]_2^0$) lines from a tube 25 mm in diameter and 200 cm long. The optimum gas pressure was 250 torr of 1:20 xenon:helium mixture. Targ and Sasnett[345,346] obtained 20 kW peak power and greater than 5 W average power by using a transverse discharge with transverse gas flow at 30 m/sec to allow high repetition rates (1.4 kHz). A gas mixture

---

[342] L. N. Tunitskii and E. M. Cherkasov, *Opt. Spectrosc. (Engl. Transl.)* **26**, 344 (1969).
[343] L. N. Tunitskii and E. M. Cherkasov, *J. Opt. Soc. Am.* **56**, 1783 (1966).
[344] S. E. Schwarz, T. A. DeTemple, and R. Targ, *Appl. Phys. Lett.* **17**, 305 (1970).
[345] R. Targ and M. W. Sasnett, *Appl. Phys. Lett.* **19**, 537 (1971).
[346] R. Targ and M. W. Sasnett, *IEEE J. Quantum Electron.* **qe-8**, 166 (1972).

TABLE XIII. Helium–Fluorine Laser Transitions

λ (μm)	Transition	Observers[a]
0.6349	$3p\ ^4S^o_{3/2} \to 3s\ ^4P_{3/2}$	b
0.6966	$3p\ ^2P^o_{1/2} \to 3s\ ^2P_{3/2}$	c,d
0.7037	$3p\ ^2P^o_{3/2} \to 3s\ ^2P_{3/2}$	c–g
0.7128	$3p\ ^2P^o_{1/2} \to 3s\ ^2P_{1/2}$	b–g
0.7202	$3p\ ^2P^o_{3/2} \to 3s\ ^2P_{3/2}$	c,d,f,g
0.7311	$3p\ ^2S^o_{1/2} \to 3s\ ^2P_{3/2}$	b–e
0.7399	$3p\ ^4P^o_{5/2} \to 3s\ ^4P_{5/2}$	b
0.7489	$3p\ ^2S^o_{1/2} \to 3s\ ^2P_{1/2}$	c
0.7552	$3p\ ^4P^o_{5/2} \to 3s\ ^4P_{3/2}$	b
0.7755	$3p\ ^2D^o_{5/2} \to 3s\ ^2P_{3/2}$	c
0.7800	$3p\ ^2D^o_{3/2} \to 3s\ ^2P_{1/2}$	c,f

[a] Source of fluorine atoms: a, (SF_6, CF_4, C_2F_6); b, (HF); c, e, f, (F_2); d, (NF_3).

[b] T. R. Loree and R. C. Sze, *Opt. Commun.* **21**, 255 (1977).

[c] L. O. Hocker and T. B. Phi, *Appl. Phys. Lett.* **29**, 493 (1976).

[d] I. J. Bigio and R. F. Begley, *Appl. Phys. Lett.* **28**, 263 (1976).

[e] A. E. Florin and R. J. Jensen, *IEEE J. Quantum Electron.* **qe-7**, 472 (1971).

[f] W. Q. Jeffers and C. E. Wiswall, *Appl. Phys. Lett.* **17**, 444 (1970).

[g] M. A. Kovacs and C. J. Ultee, *Appl. Phys. Lett.* **17**, 39 (1970).

of 1–2 torr xenon and 300 torr helium was excited by 8000-A pulses at 20 kV to obtain an efficiency of 0.12% in a discharge 240 cm long and 50 × 50 mm in cross section. Targ and Sasnett attribute the high output power to the higher plasma electron energies at the high E/p (13 V/cm-torr) resulting from the high helium pressure. Whether the high power outputs result from higher electron energies or from (HeXe)* excimer formation, helium was essential to the laser operation.

2.3.2.5. Helium–Fluorine. Atomic fluorine exhibits oscillation on 11 red and near-infrared lines of the 3p → 3s multiplet in mixtures of helium and F_2, HF, or other fluorine-containing vapors[347–352] as listed in Table XIII. The stronger lines typically have sufficiently high gain to be superradiant. The exact route by which the transfer of excitation from helium to fluorine occurs is still subject to some question, but helium is definitely required. The first observations were made by Kovacs and Ultee,[347] who

[347] M. A. Kovacs and C. J. Ultee, *Appl. Phys. Lett.* **17**, 39 (1970).

obtained the 0.7037-, 0.7128-, and 0.7202-μm lines from mixtures of helium and SF_6, CF_4, C_2F_6 at relatively low pressures (2–10 torr He, 0.03–1 torr fluorine source gas). They produced 150-W peak output pulses 1–2 μsec in duration in tubes 25 mm in diameter by 100–300 cm long. At higher pressures the output occurred in annular or "ring" modes, and they conclude that radiation trapping significantly reduces the rate at which the lower laser levels are depopulated. Jeffers and Wiswall [348] observed the same three lines and a fourth at 0.7800 μm in a flowing He–HF mixture also at relatively low pressures (0.3 torr He, 0.05 torr HF). They obtained 15-W pulses 20 μsec long from a tube 100 mm in diameter and 300 cm long. No oscillation could be obtained by substituting Ar, N_2, O_2, H_2, or CO_2 for helium. Jeffers and Wiswall conclude that the upper levels of atomic fluorine are excited by dissociative collisions between the helium singlet metastables and HF ground state molecules:

$$He^*(2s\ ^1S) + HF \longrightarrow He + H + F^*(3p) + K.E. \qquad (2.3.25)$$

The energy defect is less than 1200 cm^{-1} for this process, including the dissociation energy of HF. Florin and Jensen[349] obtained an additional line at 0.7311 μm by using a flowing mixture of He and F_2 at 5 torr in a smaller tube, 12 mm in diameter by 40 cm long.

The highest outputs reported to date were obtained by Bigio and Begley,[350] 70 kW peak output on five lines using a 1:100 mixture of NF_3:He at 100–150 torr total pressure. Two different transverse discharge configurations were successfully employed: a pin laser with 350-nsec current pulses gave 100-nsec output pulses and a faster, Blumlein-type laser with 10-nsec current pulses produced output pulses 30 nsec in duration. The Blumlein laser electrodes were 50 cm long and spaced 2.5 cm, giving an estimated specific output power of 500 W/cm^3. The longer pulses obtained from the pin laser at approximately the same specific output lead Bigio and Begley to conclude that radiation trapping is not a significant limitation in lower-level depopulation, in contrast to the conclusions of Kovacs and Ultee.[347] Replacing helium by N_2, Ne, or Xe failed to produce oscillation on the fluorine lines. Hocker and Phi[351] have observed two additional lines in a 1:100 mixture of F_2:He in a 6-mm-diameter by 60-cm longitudinal discharge, and they have studied the variation of the output power of six lines with pressure from 1 to 50 torr. They conclude that the excitation process is more involved than simple dissociative collisions that leave the fluorine atom in an excited state and

[348] W. Q. Jeffers and C. E. Wiswall, *Appl. Phys. Lett.* **17**, 444 (1970).
[349] A. E. Florin and R. J. Jensen, *IEEE J. Quantum Electron.* **qe-7**, 472 (1971).
[350] I. J. Bigio and R. F. Begley, *Appl. Phys. Lett.* **28**, 263 (1976).
[351] L. O. Hocker and T. B. Phi, *Appl. Phys. Lett.* **29**, 493 (1976).

that the HeF excimer is probably involved as an intermediary. Loree and Sze[352] have obtained a still-different set of oscillating lines, as indicated in Table XIII. They used a commercial TEA laser with a mixture of $0.8\% F_2 : 99.2\%$ He at relatively high pressures, 3–30 psia (250–2500 torr). From the particular lines that oscillate under these conditions and their intensity variation with pressure Loree and Sze conclude that the upper-level excitation process is

$$He^*(2s\ ^3S_1) + F_2 \longrightarrow He + F + F^*(3p\ doublets) + 2.5 eV, \qquad (2.3.26)$$

for the doublet levels and

$$F^*(3p\ doublets) + F \longrightarrow F^*(3p\ quartets) + F, \qquad (2.3.27)$$

for the quartets as the pressure is increased. They have obtained an output of 0.6 mJ/pulse (75 kW in 8 nsec) at 30 psia from a TEA laser 60 cm long with 15 mm electrode spacing.

2.4. Ion Lasers Excited by Collisions with Atoms or Ions

Figure 1 shows that there are a number of laser transitions produced in ions by exciting collisions with other atoms or ions. The Penning ionization reaction[353]

$$A^* + B \longrightarrow A + B^{+*} + e \qquad (2.4.1)$$

and the asymmetric thermal charge-exchange reaction[354–354b]

$$A^+ + B \longrightarrow A + B^{+*} \qquad (2.4.2)$$

are the two most prominent processes. Charge transfer is somewhat less resonant than the neutral–neutral processes exemplified by the He–Ne laser; energy defects of 0.1 to as much as 2 eV are typical. The Penning reaction, on the other hand, seems to be quite nonresonant, the only requirement being that the energy defect is positive.[354c]

Many lasers have been discovered utilizing processes (2.4.1) or (2.4.2)

---

[352] T. R. Loree and R. C. Sze, *Opt. Commun.* **21**, 255 (1977).
[353] F. M. Penning, *Physica* **1**, 1028 (1934); A. A. Kruithof, and F. M. Penning, *ibid.* **4**, 430 (1937).
[354] O. S. Duffendack and K. Thomson, *Phys. Rev.* **43**, 106 (1933).
[354a] J. H. Manley and O. S. Duffendack, *Phys. Rev.* **47**, 56 (1935).
[354b] O. S. Duffendack and W. H. Gran, *Phys. Rev.* **51**, 804 (1937).
[354c] C. E. Webb, *in* "Proceedings of the Summer School on the Physics and Technology of High Power Gas Lasers"(E.R. Pike,ed.) Vol. 29, pp.1–28 Institute of Physics Conf. Series, 1976.

2.4. IONS EXCITED BY ATOMIC COLLISION

in which reactant A is a noble gas, usually helium or neon. Often different lasing lines can be excited in the same ion B by changing the noble gas; for example, singly ionized copper exhibits different sets of lasing lines when excited by helium and neon. In other cases different laser lines are produced in the same ion by the two processes (38) and (39), for example in the He–Cd$^+$ system. The lasing ion may originate as a permanent gas or as a vaporized solid. Brief descriptions are given in this section for those systems known to date; pairs of noble gases: He–Kr$^+$, Ne–Xe$^+$; atomic vapors obtained by simple self-heating discharges: Hg$^+$, Cd$^+$, Zn$^+$, I$^+$, Se$^+$, Te$^+$, As$^+$, Mg$^+$, Sn$^+$, Pb$^+$, Tl$^+$, and Be$^+$; and atoms obtainable by high-temperature evaporation or physical sputtering: Cu$^+$, Ag$^+$, Au$^+$, Al$^+$. Predictions of additional laser lines in these and other ions excited by collisions with noble gases have been made by Green and Webb[355,356] and Collins.[357]

2.4.1. Noble-Gas Ion Lasers Excited by Atomic Collisions

Two examples of atom-excited noble-gas ion lasers are known: the He–Kr$^+$ system and the Ne–Xe$^+$ system. Both systems oscillate on $(n + 2)$s $\rightarrow (n + 1)$p transitions as listed in Table XIV, and together they make up the eight of the nine transitions indicated on Fig. 7. The He–Kr$^+$ system was first investigated by Laurès *et al.*[358,359] Their studies conclude that excitation of the upper laser levels occur via the two steps

$$\text{He}^*(2s\ ^3S_1) + \text{Kr}(^1S_0) \longrightarrow \text{He}(^1S_0) + \text{Kr}^+(^2P_3) + e, \quad (2.4.3)$$

$$\text{He}^*(2s\ ^3S_1) + \text{Kr}^+(^2P_{3/2}) \longrightarrow \text{He}(^1S_0) + \text{Kr}^+(6s\ ^4P^o_{3/2,5/2}). \quad (2.4.4)$$

Process (2.4.3) is a Penning reaction creating ground state krypton ions, while (2.4.4) is a resonant excitation transfer more like that in the He–Ne laser. The Wigner spin rule is satisfied by reaction (2.4.4), with the total spin of 3/2 being conserved. Laurès *et al.*[358] obtained oscillation in a 4-mm-diameter discharge tube 1 m in length with 1–3 mtorr of krypton in 10 torr of helium and 1-μsec current pulses of 40–150 A. Peak output powers of the order of 5 W in pulses of a few to 15 μsec duration were obtained at a 50-Hz repetition rate. These pulses occurred in the discharge afterglow, 3–15 μsec after the discharge current had decreased to zero. Breton[360] has made further measurements of the operating parameters and

---

[355] J. M. Green and C. E. Webb, *J. Phys. B* **7**, 1698 (1974).
[356] J. M. Green and C. E. Webb, *J. Phys. B* **8**, 1484 (1975).
[357] G. J. Collins, *J. Appl. Phys.* **44**, 4633 (1963).
[358] P. Laurès, L. Dana, and C. Frapard, *C. R. Hebd. Seances Acad. Sci.* **258**, 6363 (1964).
[359] J. Dana and P. Laurès, *Proc. IEEE* **53**, 78 (1965).
[360] L. Breton, *IEEE J. Quantum Electron.* **qe-9**, 854 (1973).

TABLE XIV. Noble-Gas Ion Lasers Excited by
Collisions with Noble-Gas Atoms

λ (μm)	Transition	References
	Krypton II	
0.4319	6s $^4P_{5/2}$ → 5p $^4P^o_{5/2}$	a,c,d,k
0.4388	6s $^4P_{5/2}$ → 5p $^4P^o_{3/2}$	a,c
0.4584	6s $^4P_{3/2}$ → 5p $^4D^o_{5/2}$	a,c,h
0.4696	6s $^4P_{5/2}$ → 5p $^4D^o_{7/2}$	a,c,d,e,f,g,h,i,j,k
0.5127	6s $^4P_{3/2}$ → 5P $^4D^o_{3/2}$	a,c
	Xenon II	
0.4864	7s $^4P_{5/2}$ → 6p $^4P^o_{5/2}$	b,c,j
0.5315	7s $^4P_{5/2}$ → 6p $^4D^o_{7/2}$	b,c,h,j
0.5729	5d′ $^2S_{1/2}$ → 6p $^4P^o_{3/2}$	c
0.6095	7s $^4P_{3/2}$ → 6p $^4D^o_{3/2}$	c

[a] P. Laurès, L. Dana, and C. Frapard, *C. R. Hebd. Seances Acad. Sci.* **258,** 6363 (1964).
[b] P. Laurès, L. Dana, and C. Frapard, *C. R. Hebd. Seances Acad. Sci.* **259,** 745 (1964).
[c] L. Dana and P. Laurès, *Proc. IEEE* **53,** 78 (1965).
[d] J. Breton, *IEEE J. Quantum Electron.* **qe-9,** 854 (1973).
[e] I. Kato, N. Seki, and T. Shimizu, *Jpn. J. Appl. Phys.* **11,** 1236 (1972).
[f] I. Kato, T. Satake, M. Nakaya, and T. Shimizu, *J. Appl. Phys.* **46,** 5051 (1975).
[g] I. Kato, M. Nakaya, T. Satake, and T. Shimizu, *Jpn. J. Appl. Phys.* **14,** 2001 (1975).
[h] W. B. Bridges unpublished (1975). Pulsed laser oscillation in a hollow-cathode discharge.
[i] M. Jánossy, L. Csillag, K. Rózsa, and T. Salamon, *Phys. Lett A* **46,** 379 (1974).
[j] K. Rózsa, M. Jánossy, J. Bergou, and L. Csillag, *Opt. Commun.* **23,** 15 (1977).
[k] Y. Pacheva, M. Stefanova, and P. Pramatarov, *Opt. Commun.* **27,** 121 (1978).

delay in the He–Kr+ laser in a 10-mm-diameter tube 1 m long, obtaining 20-W peak output on the Kr II 0.4694-μm line. The optimum discharge current was 500 A at optimum pressures of 30 mtorr krypton and 10 torr helium. The delay of the laser pulse with respect to the current pulse reached a minimum value of about 2 μsec at the optimum current. Kato et al.[361–368] have studied the properties of the 0.4696-μm laser in a pulsed

---

[361] I. Kato, N. Seki, and T. Shimizu, *Jpn. J. Appl. Phys.* **11,** 1236 (1972).
[362] I. Kato, T. Satake, M. Nakaya, and T. Shimizu, *J. Appl. Phys.* **46,** 5051 (1975).

microwave discharge. A peak output power of 3.5 W was obtained from a 4.5-mm-diameter tube 90 cm long in a microwave cavity excited by an 18-kW pulse at 9 GHz. The optimum pressures were 50 mtorr krypton and 10 torr helium. Jánossy et al.[364] have observed the 0.4696-μm line in continuous oscillation in a hollow cathode discharge 4 mm in diameter by 50 cm long. A cw output power of 15 mW was obtained at 6 A discharge current, with the output still increasing steeply at the maximum current. The optimum gas pressures in the hollow cathode arrangement were 50 mtorr krypton and 40 torr helium. Pacheva et al.[368a] have also obtained cw oscillation on the 0.4694-μm line in a hollow cathode discharge 3 mm in diameter and 60 cm in length, with optimum operating parameters similar to Jánossy et al.[364] They also observed cw oscillation on an additional line at 0.4319 μm. Pulsed operation of both 0.4696- and 0.4584-μm lines was obtained by Bridges[365] in a 3-mm by 50-cm hollow cathode discharge at similar gas pressures, but at currents of 20–200 A. Rózsa et al in Ref. (481) have obtained cw operation of the 0.4694 μm line in an unusual type of hollow cathode discharge which they term a hollow-anode-cathode (HAC) discharge. They employ a hollow cathode 10 mm in diameter with a coaxial wire grid anode 8 mm in diameter running the entire 40 cm length of the cathode. The maximum laser output from this discharge was about twice that of a conventional HCD of similar dimensions, and the dc discharge impedance was several times higher.

The Ne–Xe$^+$ laser was also first demonstrated by Laurès et al.[366] The excitation mechanism is presumed to be the same as the He–Kr$^+$ laser for three of the four xenon lines listed in Table XIV:

$$\text{Ne*}(1s_5) + \text{Xe}(^1S_0) \longrightarrow \text{Ne}(^1S_0) + \text{Xe}^+(^2P_{3/2}) + e, \qquad (2.4.5)$$

$$\text{Ne*}(1s_5) + \text{Xe}^+(^2P_{3/2}) \longrightarrow \text{Ne}(^1S_0) + \text{Xe}^+(7s\ ^4P^o_{3/2,5/2}). \qquad (2.4.6)$$

The Paschen notation $1s_5$ for the neon metastable level gives no indication of the spin, but Shortley (as cited in Moore[135]) has assigned the LS designation 3s $^3P_2^o$ to this level, so that again the total spin of 3/2 is conserved for reaction (2.4.6). The fourth xenon laser line observed by Dana and Laurès[359] originates from the 5d' $^2S^o_{1/2}$, a level lying 0.67 eV below the neon metastable. The exact excitation route for this relatively weak line is not known. The optimum neon pressure for the Ne–Xe$^+$ system was observed to be 4 torr; otherwise the performance parameters were the same as those of the He–Kr$^+$ laser given above.[358] Pulsed operation at

---

[363] I. Kato, M. Nakaya, T. Satake, and T. Shimizu, Jpn. J. Appl. Phys. **14**, 2001 (1975).
[364] M. Jánossy, L. Csillag, K. Rózsa, and T. Salamon, Phys. Lett. A **46**, 379 (1974).
[365] W. B. Bridges, unpublished (1975).
[366] P. Laurès, L. Dana, and C. Frapard, C. R. Hebd. Seances Acad. Sci. 259, 745 (1964).

0.5315 μm was obtained in a hollow-cathode discharge by Bridges,[365] with a xenon pressure of about 100 mtorr and neon pressures of 10–20 torr. Oscillation occurred in the decaying portion of the 40-μsec current pulse and in the immediate afterglow. Rózsa, et al. in Ref. (481) have obtained cw oscillation on the 0.5314- and 0.4863-μm lines in their hollow-anode-cathode discharge, with 100 μW output on the 0.5314-μm line. A He:Ne:Xe mixture of 7:4:0.046 torr was found to be optimum, although the role played by the helium is not clear; oscillation was much weaker without helium.

2.4.2. Mercury Ion Laser

Mercury was the first gaseous ion to exhibit laser oscillation. Bell[109] obtained pulsed oscillation on the green 0.5677- and red 0.6150-μm lines and also on the 0.7346- and 1.0583-μm infrared lines of singly ionized mercury in a helium–mercury discharge late in 1963. Additional infrared wavelengths were reported by Bloom et al.[24] (with the classifications of some of these lines later corrected by Bockasten et al.[25]) and also by Byer et al.[367] and Goldsborough and Bloom[368]. A single ultraviolet line at 0.3984 μm was reported in pulsed oscillation in a neon–mercury discharge by Isaev and Petrash.[369] Figure 25 shows all of the Hg II lines presently known to oscillate. Most of these lines were originally observed in the positive columns of pulsed discharges, but a few have also been made to oscillate in pulsed hollow-cathode discharges.[367,368,370] The 0.6150-, 0.7945-, and 1.5554-μm lines originating from the 7p $^2P^o$ levels require helium, but lines originating from other levels have been obtained with other buffer gases; for example, the 0.5677-μm lines from the 5f $^2F^o_{7/2}$ level will oscillate with neon[24,371] or argon.[372,373] The lines requiring helium have also been operated continuously in positive-column discharges[374,375] and in hollow-cathode discharges.[376,377]

---

[367] R. L. Byer, W. E. Bell, E. Hodges, and A. L. Bloom, *J. Opt. Soc. Am.* **55**, 1598 (1965).
[368] J. P. Goldsborough and A. L. Bloom, *IEEE J. Quantum Electron.* **qe-5**, 459 (1969).
[368a] Y. Pacheva, M. Stefanova, and P. Pramatarov, *Opt. Commun.*, **27**, 121 (1978).
[369] A. A. Isaev and G. G. Petrash, *J. Appl. Spectrosc. (Engl. Transl.)* **12**, 835 (1970).
[370] H. Wieder, R. A. Myers, C. L. Fisher, C. G. Powell, and J. Colombo, *Rev. Sci. Instrum.* **38**, 1538 (1967).
[371] W. B. Bridges, unpublished (1964).
[372] W. E. Bell, unpublished (1964).
[373] H. G. Heard and J. Peterson, *Proc. IEEE* **52**, 1049 (1964).
[374] A. Ferrario, *Opt. Commun.* **7**, 376 (1973).
[375] H. Kano, T. Goto, and S. Hattori, *J. Phys. Soc. Jpn.* **38**, 596 (1975).
[376] W. K. Schuebel, *IEEE J. Quantum Electron.* **qe-7**, 39 (1971).
[377] J. A. Piper and C. E. Webb, *Opt. Commun.* **13**, 122 (1975).

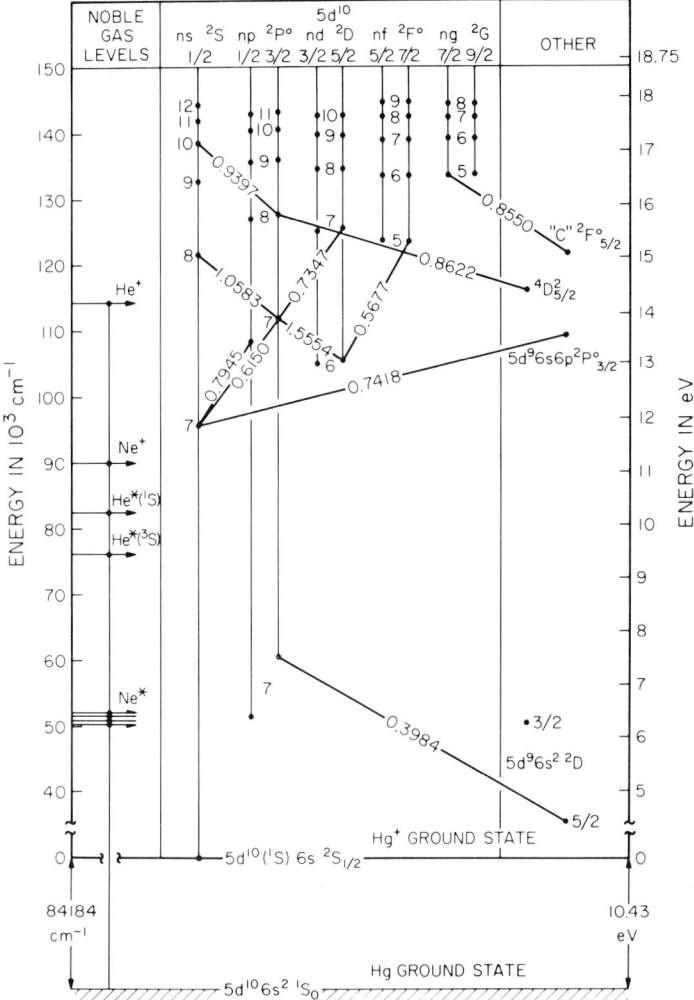

FIG. 25. Energy level diagram for singly ionized mercury, showing all of the Hg II laser lines observed to date. The energies of the helium and neon metastables and ground-state ions are also shown with respect to the ground state of the Hg neutral atom.

Early studies of excitation mechanisms in the He–Hg$^+$ laser were made by Dyson[378] and Suzuki.[379] On the basis of his measured variations in spontaneous emission intensities and decay rates with mercury vapor pressure, Dyson concluded that the thermal charge exchange reaction

$$\text{He}^+(^2S_{1/2}) + \text{Hg}(^1S_0) \longrightarrow \text{He}(^1S_0) + \text{Hg}^+(7p\ ^2P^o_{3/2}) + \text{KE} \qquad (2.4.7)$$

[378] D. J. Dyson, *Nature (London)* **207**, 361 (1965).
[379] N. Suzuki, *Jpn. J. Appl. Phys.* **4**, 452 (1965).

was responsible for the excitation of the 0.6150-μm line. He derived a cross section of 1.3×10^{-14} cm^2 for process (2.4.7). Suzuki, from similar measurements in a smaller-diameter He–Hg discharge, proposed that a Penning-like process between the metastables of both mercury and helium was responsible:

$$\text{He}^*(2s\ ^3S_1) + \text{Hg}^*(6p\ ^3P_0) \longrightarrow \text{He}(^1S_0) + \text{Hg}^+(7p\ ^2P_{3/2}) + \text{KE} \qquad (2.4.8)$$

Later studies by Aleinikov,[380] Willett,[381] Ferrario,[374] Piper and Webb,[377] Littlewood et al.[382] and Kano et al.[382a,b] establish quite conclusively that the thermal charge exchange reaction (2.4.7) is responsible for the excitation of the 7p $^2P^0_{3/2,1/2}$ levels. Note that the Wigner spin rule is obeyed for reaction (2.4.7) but not for reaction (2.4.8). Using measurement techniques similar to Dyson's, Aleinkov and Ushakov[383] derive a cross section of 1.4×10^{-14} cm^2 for reaction (2.4.7), in excellent agreement with Dyson's value. Similarly, Kano et al.[382a] obtained a value of 1.3×10^{-14} cm^2, also in excellent agreement with Dyson. On the other hand, Bogdanova et al.[384] determine a much lower value of 1.6×10^{-16} cm^2 from measurements in an electron-beam-excited discharge. The reason for the large difference in the two values is not known.

In the first publication on the He–Hg$^+$ laser, Bell[109] reported peak output powers of 40 W for the 0.6150-μm line in a discharge 3.5 mm in diameter and 300 cm long. Output pulses were about 5 μsec in duration and exhibited a double-peaked form similar to those later attributed to radiation trapping in argon ion lasers (see Section 2.2.2.5). The average output power was observed to increase linearly with repetition rate, reaching 4 mW at 120 pps. Pressures of about 1 mtorr Hg and 0.5 torr He were used in these first pulsed positive-column discharges. Bell did not give a value for the output power of the green line but other experiments show comparable powers on the two lines.[371,372] Bell observed a complex-shaped laser output pulse coincident with the discharge current pulse. In contrast, Bridges observed simple pulses of a few microseconds in duration but delayed from the end of the current pulse (also a few microseconds in duration) by 3–8 μsec. The red and green lines exhibited different delays (red longer) so that the two wavelengths did not oscillate

---

[380] V. S. Aleinkov, *Opt. Spectrosc. (Engl. Transl.)* **28**, 15 (1970).

[381] C. S. Willett, *IEEE J. Quantum Electron.* **qe-6**, 469 (1970).

[382] I. M. Littlewood, J. A. Piper, and C. E. Webb, *Opt. Commun.* **16**, 45 (1976).

[382a] H. Kano, T. Shay, and G. J. Collins, *Appl. Phys. Lett.* **27**, 610 (1975).

[382b] T. Shay, H. Kano, S. Hattori, and G. J. Collins, *J. Appl. Phys.* **48**, 4449 (1977).

[383] V. S. Aleinkov and V. V. Ushakov, *Opt. Spectrosc. (Engl. Transl.)* **33**, 116 (1972).

[384] I. P. Bogdanova, V. D. Marusin, and V. E. Yakhontova, *Opt. Spectrosc. (Engl. Transl.)* **37**, 365 (1974).

2.4. IONS EXCITED BY ATOMIC COLLISION

simultaneously.[371] The tube used for these measurements was 5 mm in diameter and 130 cm in length with 0.5 torr He and 1–10 mTorr Hg. Square current pulses 2–8 μsec up to 30 A were employed. Heard and Peterson also report observations of oscillation in the discharge afterglow.[373] Another feature of the output observed by Bloom et al.[24] and Bridges[371] was the tendency for oscillation to occur at 0.6150 μm in a narrow annulus near the tube walls and at 0.5677 μm in the central region, with no spatial overlap. A color photograph of this striking effect taken by Bloom is reproduced as the frontispiece in Willett's book.[2]

Oscillation on the 0.3984-μm line was observed in a positive-column discharge 1.3 mm in diameter and 20 cm long by Isaev and Petrash.[369] The superradiant output occurred only for about 40 nsec on the leading edge of a high-current pulse. Optimum pressures were 25–100 mTorr Hg (60–80°C reservoir temperature) and 0.06–2 torr Ne. Repetition rates of 5–8 pps were employed, with the output decreasing as the repetition rate was increased. Oscillation was observed on three of the isotope hyperfine components, with the strongest corresponding to the ^{202}Hg component. The mechanisms responsible for inversion on this transition are not known. Isaev and Petrash speculate that electron collisions with the Hg$^+$ ground state are responsible because of the similarity of the energy level structure to the cyclic Cu I and Au I lasers. Population directly from the neutral ground state by "sudden perturbation"[113] would favor the 7s $^2S_{1/2}$ state. The fact that only neon will produce this line is also curious, although no reasonable coincidences for Penning or charge-exchange processes are evident from Fig. 25. The depopulation of the $6s^2$ level undoubtedly occurs by diffusion to the walls, since the transitions to the ion ground state are optically forbidden just as in the Cu I and Au I lasers. This is consistent with the low repetition rate limitation observed by Isaev and Petrash.

Continuous oscillation on the 0.6150-μm line in a positive-column discharge was first obtained by Ferrario[374] in a tube 3 mm in diameter and 100 cm long. He obtained 10 mW output at 370 mA discharge current and an optimum pressure of 30 mTorr Hg (63°C reservoir temperature) and 4.5 Torr He. Ferrario observed a linear increase in spontaneous emission at 0.6150 μm with discharge current up to about 300 mA, then saturation to a constant value beyond about 400 mA. Laser oscillation occurred over the range 200–650 mA, with a clearly defined maximum at around 400 mA. The spontaneous emission at 0.5677 μm, by contrast, increased quadratically up to 300 mA, then subquadratically at higher currents, giving a clear demonstration that different excitation mechanisms are responsible for these two lines. Continuous oscillation on the 0.7945-μm line in a positive column was obtained by Kano et al.[375] in a tube 3.5 mm in diameter by 123 cm long, with 6 mW output at 400 mA and pressures of 6 mtorr Hg (40°C reservoir) and 5 torr He.

Operation of the He–Hg$^+$ laser in a hollow-cathode discharge was first obtained by Byer et al.[367] on the 0.6150- and 0.7945-μm lines. Their operating parameters are summarized in Table XV along with hollow-cathode results obtained subsequently by others. Only those lines originating from the 7p $^2P^o_{3/2,1/2}$ levels and the 5d^{9}6s6p $^2P^o_{1/2}$ level that lies between the two 7p levels have oscillated in HCDs, again indicating a different mechanism for these lines from all the others. Byer et al.[367] obtained simultaneous oscillation on each of the four even isotope components 198–204 in natural mercury, spanning a range of about 3000 MHz. The four components were clearly resolved since their measured Doppler linewidths are about 500 MHz for the 0.6150-μm line in a HCD. As Byer et al. suggest, higher gains can be obtained if monoisotopic mercury is used. Some results obtained with ^{202}Hg are given in Table XV. One interesting feature of pulsed HCD He–Hg$^+$ lasers is the high gains obtainable in relatively large diameter lasers. This feature was exploited by Myers et al.[385] in the development of a laser oscillator whose transverse mode structure could be controlled to produce images. Continuous oscillation in HCDs was first demonstrated by Schuebel,[376] even before continuous operation was obtained in positive-column discharges. Piper and Webb[377] have continued the optimization of the cw HCD to produce the largest average output power observed to date in the He–Hg$^+$ laser, 80 mW at 0.6150 μm, as listed in Table XV.

2.4.3. Cadmium Ion Laser

The He–Cd$^+$ laser provides a good example of a single atom in which both Penning and charge-exchange collisions are operative, each producing its own set of laser transitions. Figure 26 shows a simplified Grotrian diagram of Cd$^+$ indicating all of the Cd II laser transitions observed to date. The corresponding positions of relevant noble gas levels are also shown with energies measured from the neutral Cd ground state for the Penning and charge-exchange processes:

$$\text{He}^*(2s\ ^3S_1) + \text{Cd} \longrightarrow \text{He} + \text{Cd}^+(5s^2\ ^2D) + e + \text{K.E.}, \tag{2.4.9}$$

$$\text{He}^+(^2S_{1/2}) + \text{Cd} \longrightarrow \text{He} + \text{Cd}(9s, 6f, 6g) + \text{K.E.} \tag{2.4.10}$$

2.4.3.1. Spectroscopy of the Cadmium Ion Laser.
The two green laser lines at 0.5337 and 0.5378 μm were the first ion laser transitions to be observed in cadmium, obtained by Fowles and Silfvast[386] in a 5-mm-

---

[385] R. A. Myers, H. Wieder, and R. V. Pole, *IEEE J. Quantum Electron.* **qe-2**, 270 (1966).
[386] G. R. Fowles and W. T. Silfvast, *IEEE J. Quantum Electron.* **qe-1**, 131 (1965).

TABLE XV. Parameters of Hollow-Cathode He–Hg$^+$ Ion Lasers

Ref.	Date	Dimensions		Output		Excitation				Gas			Comment
		D (mm)	L (cm)	λ (μm)	P (W)	I (A)	V (kV)	τ (μsec)	rate (kHz)	Hg (°C)	He (torr)		
a	1965	25	30	0.6150	2–3	—	5	2	1	30	1–2	$g = 2.2$ dB/m
				0.7945	—	—	—	2	2	30	1–2	—
b	1966	25	20	0.6150	—	30	5	—	1	—	—	^{202}Hg; $g = 7$ dB
c	1967	25	15	0.6150	1.8	10	2	1–3	30	—	—	^{202}Hg; $g = 0.5$ dB/cm
				1.5554	—	—	—	—	—	—	—	—
d	1969	24	53	0.6150	—	60	—	2	—	80–100	10	^{202}Hg
				0.7418	—	60	—	—	—	—	—	^{202}Hg
				0.7945	7	60	—	—	—	—	30	^{202}Hg; $g = 0.6\%$ cm^{-1}
e	1971	5.6	50	0.6150	—	0.7^g	—	cw	—	60–80	4–9	
				0.7945	—	0.24^g	—	cw	—	60–80	4–9	
f	1975	3	50	0.6150	0.08	10	—	1200	—	115	16	
				0.6150	0.02	3.5	—	cw	—	115	16	
				0.7945	0.0005	4	—	cw	—	115	10	

^{a} R. L. Byer, W. E. Bell, E. Hodges, and A. L. Bloom, *J. Opt. Soc. Am.* **55**, 1598 (1965).
^{b} R. A. Myers, H. Wieder, and R. V. Pole, *IEEE J. Quantum Electron.* **qe-2**, 270 (1966).
^{c} H. Wieder, R. A. Myers, C. L. Fisher, C. G. Powell, and J. Colombo, *Rev. Sci. Instrum.* **38**, 1538 (1967).
^{d} J. P. Goldsborough and A. L. Bloom, *IEEE J. Quantum Electron.* **qe-5**, 459 (1969).
^{e} W. K. Schuebel, *IEEE J. Quantum Electron.* **qe-7**, 39 (1971).
^{f} J. A. Piper and C. E. Webb, *Opt. Commun.* **13**, 122 (1975).
^{g} Threshold current; power not measured.

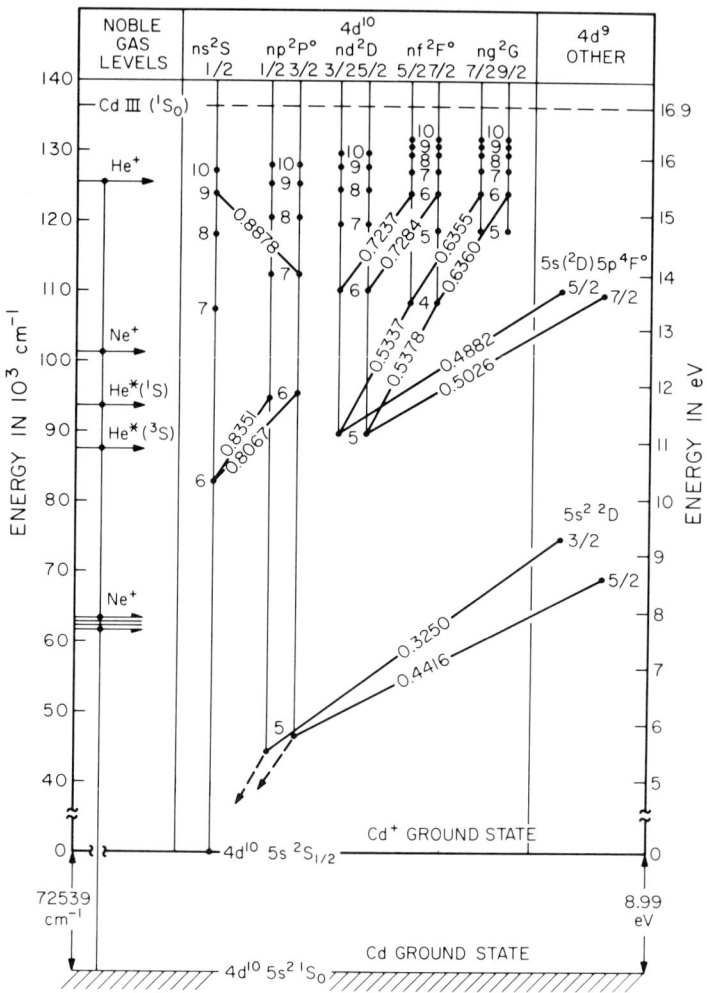

FIG. 26. Energy level diagram for singly ionized cadmium, showing all the Cd II laser lines observed to date. The energies of the helium and neon metastables and ground-state ions are also shown with respect to the ground state of the Cd neutral atom.

diameter by 80-cm-long pulsed positive-column discharge. Helium or neon at a few torr could be used as a buffer gas. The blue 0.4416-μm line was obtained about a year later by Silfvast et al.[387] in a somewhat larger tube (6 mm by 100 cm). The vapor pressure of cadmium employed in these experiments was in the range of 1–100 mtorr, corresponding to a

[387] W. T. Silfvast, G. R. Fowles, and B. D. Hopkins, *Appl. Phys. Lett.* **8**, 318 (1966).

tube wall temperature of 200–320°C. Little attention was given to the helium–cadmium system until Fowles and Hopkins[388] reported quasi-cw oscillation on the 0.4416-μm line in a He–Cd discharge. Their tube was excited by a neon-sign transformer, with oscillation persisting for 5 msec out of each 8-msec half-cycle. Enriched cadmium isotopes were used to narrow the line width and increase the gain. Continuous oscillation in a dc-excited discharge tube was reported by Silfvast,[389] who obtained 50 mW at 0.4416 μm from a tube 5 mm in diameter by 100 cm long at 120 mA discharge current. Monoisotopic ^{114}Cd was used to increase the gain coefficient by approximately four times over natural Cd. The optimum vapor pressure of Cd was estimated by Silfvast to be 10 mtorr and the optimum helium pressure about 2 torr. In his paper, Silfvast proposed the Penning process (2.4.9) to be responsible for the excitation, and ruled out electron excitation because of the high optimum helium pressures observed. He also speculated that the Cd II 0.3250-μm line (among others) might oscillate, and subsequently demonstrated it,[390] obtaining 6 mW from a 4-mm by 100-cm tube. Operation on this ultraviolet line was also obtained independently by Goldsborough.[391] Additional weak cw lines at 0.4882 and 0.5026 μm were reported by Bloom and Goldsborough.[392]

Hollow-cathode discharges were used by other groups to obtain the additional transitions shown in Fig. 26. Schuebel[393,394] and Sugawara *et al.*[395,396] independently obtained cw oscillation on the original green lines at 0.5337 and 0.5378 μm and on new lines at 0.6355, 0.6360, 0.7237, and 0.7284 μm. Karabut *et al.*[397] first reported operation at 0.8067, and later Sugawara *et al.*[396] obtain this line and 0.8531 and 0.8879 μm, all in cw HCDs. These transitions all result from the thermal charge-exchange process (2.4.10) or cascade from levels populated by (2.4.10).

2.4.3.2. Excitation Mechanisms in the He–Cd$^+$ Laser. The first discussion of population mechanisms was that given by Silfvast,[389] who, following a suggestion by Webb, proposed the Penning process (2.4.9) for the excitation of the $5s^2$ ^2D upper laser levels, and fast radiative decay via the

---

[388] G. R. Fowles and B. D. Hopkins, *IEEE J. Quantum Electron.* **qe-3**, 419 (1967).
[389] W. T. Silfvast, *Appl. Phys. Lett.* **13**, 169 (1968).
[390] W. T. Silfvast, *Appl. Phys. Lett.* **15**, 23 (1969).
[391] J. P. Goldsborough, *IEEE J. Quantum Electron.* **qe-5**, 133 (1969).
[392] A. L. Bloom and J. P. Goldsborough, *IEEE J. Quantum Electron.* **qe-6**, 164 (1970).
[393] W. K. Schuebel, *Appl. Phys. Lett.* **16**, 470 (1970).
[394] W. K. Schuebel, *IEEE J. Quantum Electron.* **qe-6**, 574 (1970).
[395] Y. Sugawara, Y. Tokiwa, and T. Iijima, *Jpn. J. Appl. Phys.* **9**, 588 (1970).
[396] Y. Sugawara, Y. Tokiwa, and T. Iijima, *Jpn. J. Appl. Phys.* **9**, 1537 (1970).
[397] E. K. Karabut, V. S. Mikhalevskii, V. F. Papakin, and M. F. Sém, *Sov. Phys.—Tech. Phys. (Engl. Transl.)* **14**, 1447 (1970).

highly allowed uv transitions to the ion ground state as the depopulating process for the 5p ^{2}P lower laser levels. This speculation was subsequently verified by several workers. Schearer and Padovani[398] measured the decay rate of the helium 2s $^3S_1$ metastable population as a function of cadmium pressure in a pulsed He–Cd discharge; they showed that the decay rate varied linearly with cadmium pressure, and further that the spontaneous emission at 0.4416 and 0.3250 μm exhibited exactly the same variation. Collins et al.[399] obtained the same result in a similar measurement of decay rate; they also measured the decay of the green, red, and some of the near-infrared laser lines and found that these rates also varied linearly with cadmium pressure, but with a smaller constant of proportionality. This suggested population by collisions with a different long-lived species, and Collins et al.[399,399a] argued that the species must be ground-state helium ions. A dramatic and direct demonstration of these different excitation routes was provided by Webb et al.,[400] who employed a flowing helium system in which the populations of He(2s $^3S_1$) metastables and He$^+(^2S_{1/2})$ ions in the flow could be modulated independently. By observing the modulation of the spontaneous radiation from the various levels shown in Fig. 26 when cadmium vapor was injected into this helium flow, the dominant process responsible for the population of a particular level could be determined. As a result of these measurements it was clear that the 0.4416- and 0.3250-μm lines were excited by process (2.4.9) and the other laser lines by process (2.4.10) or by cascade from levels excited by (2.4.10). These results were given in more detail in a later paper.[401]

Silfvast[402] measured the density of both He(2s $^3S_1$) and He(2s $^1S_0$) metastables as a function of cadmium pressure in a cw He–Cd discharge. Using these data he calculated the variation of the 0.4416-μm spontaneous emission with cadmium pressure assuming only Penning processes were present; his calculated values followed his measured variation of spontaneous emission very closely. Silfvast also measured the cadmium ion ground-state density and the discharge parameter E/p as functions of the cadmium vapor pressure and argued that their variations were consistent with Penning ionization as the dominant mechanism. Unfortunately, Silfvast's measurements were made with other discharge parameters (10 mm diameter, 10 mA current) that were not typical of He–Cd$^+$ lasers.

[398] L. D. Schearer and F. A. Padovani, J. Chem. Phys. **52**, 1618 (1970).
[399] G. J. Collins, R. C. Jensen, and W. R. Bennett, Jr., Appl. Phys. Lett. **19**, 125 (1971).
[399a] G. J. Collins, "Cw Oscillation and Charge-Exchange Excitation in the Zinc-Ion Laser," Thesis, Yale University, 1970.
[400] C. E. Webb, A. R. Turner-Smith, and J. M. Green, J. Phys. B **3**, L135 (1970).
[401] A. R. Turner-Smith, J. M. Green, and C. E. Webb, J. Phys. B **6**, 114 (1973).
[402] W. T. Silfvast, Phys. Rev. Lett. **27**, 1489 (1971).

2.4. IONS EXCITED BY ATOMIC COLLISION

Measurements of metastable density in He–Cd discharges have also been made by Dyatlov et al.,[403] Browne and Dunn,[404] and Miyazaki et al.[405], and Goto et al.[405a] The values obtained by Browne and Dunn for discharges typical of He–Cd$^+$ lasers show triplet metastable densities of the order of 5×10^{12} cm^{-3} at the optimum conditions, about half the value for an identical pure helium discharge. The singlet metastable density was about five times smaller under the same conditions. The corresponding cadmium neutral density was about 2×10^{13} cm^{-3}, and the discharge E/p about 6 V/torr-cm. Similar values for the densities were found by Miyazaki et al.[405] and Goto et al.[405a] Other measurements of the discharge parameters including electron density and temperatures have been made by Goto et al.[406] Ivanov and Sém,[407], Dunn,[408] and Mori et al.[408a] Measurements of ion densities and helium excited state populations have also been made by Mori et al.[408a,b]

Early studies by Csillag et al.[409-411] cast some doubt on the simple picture of the Penning process (2.4.9) as the sole excitation mechanism for the 0.4416-μm line. They observed a complex behavior of output power with discharge current[409]; they also observed pulsed oscillation of 0.4416-μm laser in a neon–cadmium discharge. The output power in Ne–Cd was much lower than it was with helium in the same size discharge tube. It is obvious from Fig. 26 that the neon metastables are not energetic enough to populate the 5s^2 ^{2}D levels by a Penning process analogous to (2.4.9) [although the ground-state neon ions could do so by a charge-exchange process analogous to (2.4.10)]. Wang and Siegman[412] subsequently obtained cw oscillation of both 0.4416- and 0.3250-μm laser transitions in a hollow-cathode discharge with Ne–Cd, Ar–Cd, and Xe–Cd mixtures. None of the other cadmium lines could be obtained, however. They observed that the laser output generally *increased* as the noble gas pressure *decreased,* even to pressures as low as a few mtorr. They speculate that excitation of the cadmium lines occurs by direct electron excitation for

[403] M. K. Dyatlov, E. P. Ostapchenko, and V. A. Stepanov, *Opt. Spectrosc.* (*Engl. Transl.*) **29**, 539 (1970).
[404] P. G. Browne and M. H. Dunn, *J. Phys. B* **6**, 1103 (1973).
[405] K. Miyazaki, Y. Ogata, T. Fujimoto, and K. Fukuda, *Jpn. J. Appl. Phys.* **13**, 1866 (1974).
[405a] T. Goto, M. Mori, and S. Hattori, *Appl. Phys. Lett.* **29**, 358 (1976).
[406] T. Goto, A. Kawahara, G. J. Collins, and S. Hattori, *J. Appl. Phys.* **42**, 3816 (1971).
[407] I. G. Ivanov and M. F. Sém, *Sov. Phys.—Tech. Phys.* (*Engl. Transl.*) **17**, 1234 (1973).
[408] M. H. Dunn, *J. Phys. B* **5**, 665 (1972).
[408a] M. Mori, T. Goto, and S. Hattori, *J. Phys. Soc. Japan* **43**, 662 (1977).
[408b] M. Mori, K. Takasu, T. Goto, and S. Hattori, *J. Appl. Phys.* **48**, 2226 (1977).
[409] L. Csillag, M. Jánossy, K. Kántor, K. Rózsa, and T. Salamon, *J. Phys. D* **3**, 64 (1970).
[410] L. Csillag, M. Jánossy, and T. Salamon, *Phys. Lett. A* **31**, 532 (1970).
[411] L. Csillag, V. V. Itagi, M. Jánossy, and K. Rózsa, *Phys. Lett. A* **34**, 110 (1971).
[412] S. C. Wang and A. E. Siegman, *Appl. Phys.* **2**, 143 (1973).

those conditions. In this connection it is interesting that Aleinkov and Ushakov[413] measured an exceptionally high cross section, 1.5×10^{-16} cm², for electron excitation of the $5s^2\ ^2D_{5/2}$ from the neutral Cd ground state by 100 eV electrons. This process may also be the explanation for the Ne–Cd laser obtained by Csillag et al.[409]

Mori et al.[413a] have recently modeled the He–Cd⁺ laser processes using their measured values of the various species in the discharge.[405a,408a,b,413b] They find they must invoke stepwise excitation of both upper and lower laser levels in addition to Penning excitation of the upper laser level to explain their observations of saturation at 0.4416 μm.

Many other studies have been made of particular facets of the He–Cd⁺ laser excitation mechanisms. Hodges[414] measured the spontaneous emission intensity from many ultraviolet transitions including those which Silfvast had suggested might oscillate.[389] Hodges' results indicate that inversions occur only on the 0.4416- and 0.3250-μm lines; no inversion exists on the 0.2573-, 0.2749-, and 0.3536-μm uv lines. Goldsborough had already reported the failure of 0.3536 μm to oscillate in an otherwise high-performance He–Cd⁺ laser.[391] Watanabe et al.[415] have recently modeled the various processes in the He–Cd⁺ laser and speculate that high-gain, high-power pulsed operation should be possible at atmospheric helium pressures.

2.4.3.3. Cataphoresis in He–Cd⁺ Lasers. It was evident from the earliest work with cw He–Cd⁺ lasers that there were difficulties in maintaining a stable discharge and a uniform distribution of cadmium vapor in the presence of cataphoresis,[98] discharge self-heating, and oven temperature gradients. In addition, there was the practical matter of keeping the cadmium vapor from condensing on the optical surfaces of the laser. Answers to these problems had to be found before any serious attempts at parameter optimization could be undertaken.

The key to practical positive-column He–Cd⁺ lasers is to *use* cataphoresis to produce a continuous flow of Cd atoms through the discharge tube rather than to fight the process. This concept was suggested and demonstrated independently by Goldsborough,[416] Fendley et al.,[417] and

[413] V. S. Aleinkov and V. V. Ushakov, *Opt. Spectrosc. (Engl. Transl.)* **29**, 111 (1970).

[413a] M. Mori, M. Murayama, T. Goto, and S. Hattori, *IEEE J. Quantum Electron.*, **qe-14**, 427 (1978).

[413b] M. Mori, T. Goto, and S. Hattori, *J. Phys. Soc. Japan* **44**, 1715 (1978).

[414] D. T. Hodges, *Appl. Phys. Lett.* **17**, 11 (1970).

[415] S. Watanabe, K. Kuroda, and I. Ogura, *J. Appl. Phys.* **47**, 4887 (1976).

[416] J. P. Goldsborough, *Appl. Phys. Lett.* **15**, 159 (1969).

[417] J. R. Fendley, Jr., I. Gorog, K. G. Hernqvist, and C. Sun, *RCA Rev.* **30**, 422 (1969).

Sosnowski.[418] A source of cadmium in a temperature-controlled sidearm or bulge in the discharge positive column is located near the anode, and a cool region in a similar side-arm or bulge is provided near the cathode as shown in Fig. 27a. If the helium discharge heats the tube walls sufficiently the cadmium vapor will be swept toward the cathode by cataphoresis and condense on the walls in the appropriate cool region. The short length of pure helium discharge between the anode and Cd source serves to protect the anode end window from Cd deposition since the direction of the cataphoretic flow in this discharge is opposed to that for diffusion, as suggested by Fendley *et al.*[417] At the cathode end of the tube the cold region must contain the cadmium vapor sufficiently to keep it from reaching the cathode end window. An auxiliary discharge from the cathode to an anode near the window can also be used to keep cadmium vapor from the window, as shown in Fig. 27b. Alternatively, the discharge tube can be constructed with the cathode and cadmium sink in the center of the tube and an anode and cadmium source at each end as shown in Fig. 27c. Using a tube design based on the cataphoretic flow concept, Goldsborough[416] obtained 200 mW output at 0.4416 μm and 20 mW at 0.3250 μm in a tube 2.4 mm in diameter and 143 cm long at 110 mA discharge current and 3.4 torr helium pressure. The cadmium source temperature was 220°C, corresponding to a vapor pressure of about 2 mtorr. Other tube designs based on cataphoretic flow were described by Hernqvist and Pultorak[419] and Silfvast and Szeto.[420]

Simple cataphoretic flow does not result in a completely uniform distribution of cadmium vapor along the bore as shown by Sosnowski.[418] In a second paper Silfvast and Szeto[421] describe an improved tube design in which small washers of cadmium separated by short sections of glass tubing 6–8 cm long are slipped inside a glass vacuum envelope as shown in Fig. 27d. A simple condenser near the cathode end was employed. The periodically spaced cadmium sources provide a more uniform distribution of vapor along the bore. In addition, the slip-fit glass tubing inserts allow the inner wall next to the discharge to operate at a higher temperature than the cadmium washers, which are in contact with the outer glass vacuum envelope. The exterior of the tube is surrounded with thermal insulation, and only the discharge power was used to heat the tube. Silfvast and Szeto obtained 10 mW at 0.4416 μm and 2.5 mW at 0.3250 μm in a tube 2 mm in diameter and 26 cm long at 67 mA and 6 torr helium pressure.

[418] T. P. Sosnowski, *J. Appl. Phys.* **40**, 5138 (1969).
[419] K. G. Hernqvist and D. C. Pultorak, *Rev. Sci. Instrum.* **41**, 696 (1970).
[420] W. T. Silfvast and L. H. Szeto, *Appl. Opt.* **9**, 1484 (1970).
[421] W. T. Silfvast and L. H. Szeto, *Appl. Phys. Lett.* **19**, 445 (1971).

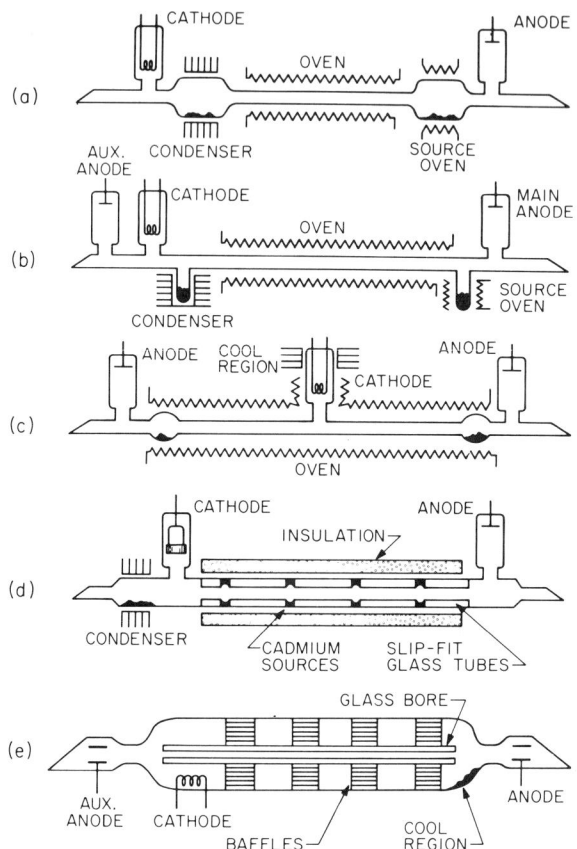

FIG. 27. Discharge tube configurations used in cataphoretic-flow He–Cd⁺ lasers. (a) Simple unidirectional flow; (b) unidirectional flow between side-arms with an auxiliary anode to protect the cathode-end window; (c) symmetrical discharge with condensing region behind the cathode; (d) unidirectional flow with Cd sources distributed along the bore and thermally insulated walls; (e) continuous-circulation laser with metal baffles in the vapor return path.

Helium cleanup is observed in He–Cd⁺ lasers employing cataphoretic flow to a condensing region located near the discharge itself. Sosnowski and Klein[422] demonstrated that this loss rate is two orders of magnitude greater than the rate of helium permeation through the hot glass envelope. They were further able to demonstrate that the loss was not by simple burial of helium atoms under layers of condensing cadmium atoms; rather, the loss is caused by helium ions that are accelerated across the

[422] T. P. Sosnowski, and M. B. Klein, *IEEE J. Quantum Electron.* **qe-7**, 425 (1971).

discharge wall sheath (≈ 20 eV) into the cadmium-covered walls of the condenser region. They suggest a tube design with the condenser region removed from the vicinity of the discharge, such as that shown in Fig. 27c.

Tubes utilizing cataphoretic flow between source and condenser have a finite operating life limited by the amount of cadmium originally in the source. This limit becomes especially important if the relatively expensive monoisotopic ^{114}Cd is used to increase the tube gain. Fendley et al.[417] suggest a symmetrical tube in which the role of source and condenser can be reversed periodically when the cadmium has moved from one end to the other. Hernqvist[423-425] has suggested tube designs in which the cadmium vapor is recirculated by diffusion instead of being condensed at one end. In one version a gas return path is added from cathode to anode in which the cadmium vapor can flow by diffusion.[423,424] To allow proper equilization this return path must have a much larger cross-sectional area than the discharge bore, since flow by diffusion is much slower than the cataphoretic flow in a given diameter tube.[417] Baffles in the return path are required to prevent electrical breakdown. Hernqvist gives the performance of several sizes of tubes with coaxial return paths and honeycomb kovar baffles as shown schematically in Fig. 27e. Outputs at 0.4416 μm ranged from 2 mW in a 1-mm by 10-cm tube to 160 mW in a 2.2-mm by 180-cm tube. All tubes used ^{3}He–^{114}Cd fills. In a later paper, Hernqvist[425] describes a modification of this design in which holes are added periodically to the discharge bore to allow local circulation through the coaxial return path as well as from cathode to anode.

2.4.3.4. Noise in the He–Cd$^+$ Laser. Despite their similarity to He–Ne lasers, cw He–Cd$^+$ lasers typically exhibit orders of magnitude greater noise modulation of the light output than the He–Ne lasers. Fluctuation amplitudes of 15% in the 10–100 kHz range were reported by Fendley et al.[417] in simple cataphoretic flow lasers. The magnitude of the noise can be reduced to the order of 1% by careful adjustment of pressure and wall temperature in the laser, as reported by Silfvast and Szeto[421] for a unidirectional-flow configuration and by Hernqvist[424,425] for the recirculating configurations. However, 1% is still two or more orders of magnitude greater noise modulation than that of typical He–Ne lasers. Additional measurements of the noise spectrum were made by Brown and Ginsburg,[426] which exhibited several peaks at discrete frequencies in the 0–450-kHz range.

[423] K. G. Hernqvist, *Appl. Phys. Lett.* **16**, 464 (1970).
[424] K. G. Hernqvist, *IEEE J. Quantum Electron.* **qe-8**, 740 (1972).
[425] K. G. Hernqvist, *RCA Rev.* **34**, 401 (1973).
[426] D. C. Brown and N. Ginsburg, *Appl. Phys. Lett.* **24**, 287 (1974).

The exact nature of the noise-producing mechanism is not yet understood, but it is clear that the process involves the Penning ionization process and fluctuations in the helium metastable density. Fluctuations in cadmium vapor density appear to be relatively unimportant. Silfvast[402] attributes the noise to instability in the helium metastable density through creation of additional metastables by the Penning electrons liberated during process (2.4.9). Jánossy[427] has observed such an increase in a novel manner; he observed that the 1.15-μm neon laser intensity would increase in a He–Ne–Cd discharge when the simultaneously oscillating 0.4416-μm laser line became noisy.

Johnston and Kolb[428] have explained the noise generation process in terms of the rapid growth of striation waves. They show that the striation-wave gain exhibits a strong increase in the same region of He–Cd$^+$ operating parameters that produces noisy operation. Johnston and Kolb also found that ^{3}He produced 25% more noise than ^{4}He.

2.4.3.5. **Optimum Parameters for the He–Cd$^+$ Laser.** Perhaps the most complete parametric study of the He–Cd$^+$ laser has been made by Johnston and Kolb.[428] Several self-heating tubes were studied, all of the glass-insert design described by Silfvast and Szeto[421] but with only a single cadmium source. The results for 1.0- and 1.6-mm-diameter tubes are shown in Figs. 28 and 29 as contour plots of output power vs. helium pressure and cadmium pellet temperature. Values of rms noise are shown as dotted contours. These plots show that the pressures can be adjusted to reduce the rms noise from 4% to lower values with only modest sacrifices in output power. The similarity between the two plots when expressed in terms of pD suggests that the scaling laws pD = const hold for these lasers. Figure 30 shows the contours of output power and noise as a function of discharge current and helium pressure. Table XVI gives a summary of Johnston and Kolb's results for the optimum tube parameters and performance over the diameter range 1–1.6 mm. Natural cadmium and ^{4}He were used in obtaining these results.

2.4.3.6. **Other Excitation Methods for He–Cd$^+$ Lasers.** Hollow-cathode discharges have been used successfully to yield new cw laser transitions as outlined previously.[393–397] Power levels of a few mW per line for the green, red, and near-infrared lines was reported by Schuebel[393,394] in a slotted HCD with a cathode 5.6 mm in diameter and 45 cm long. Currents of 150–450 mA at 300–400 V were used. Piper and Webb[429] have obtained outputs on 10 Cd lines, including the 0.4416-

---

[427] M. Jánossy, *Phys. Lett. A* **47**, 409 (1974).
[428] T. F. Johnston, Jr. and W. P. Kolb, *IEEE J. Quantum Electron.* **qe-12**, 482 (1976).
[429] J. A. Piper and C. E. Webb, *J. Phys. D* **6**, 400 (1973).

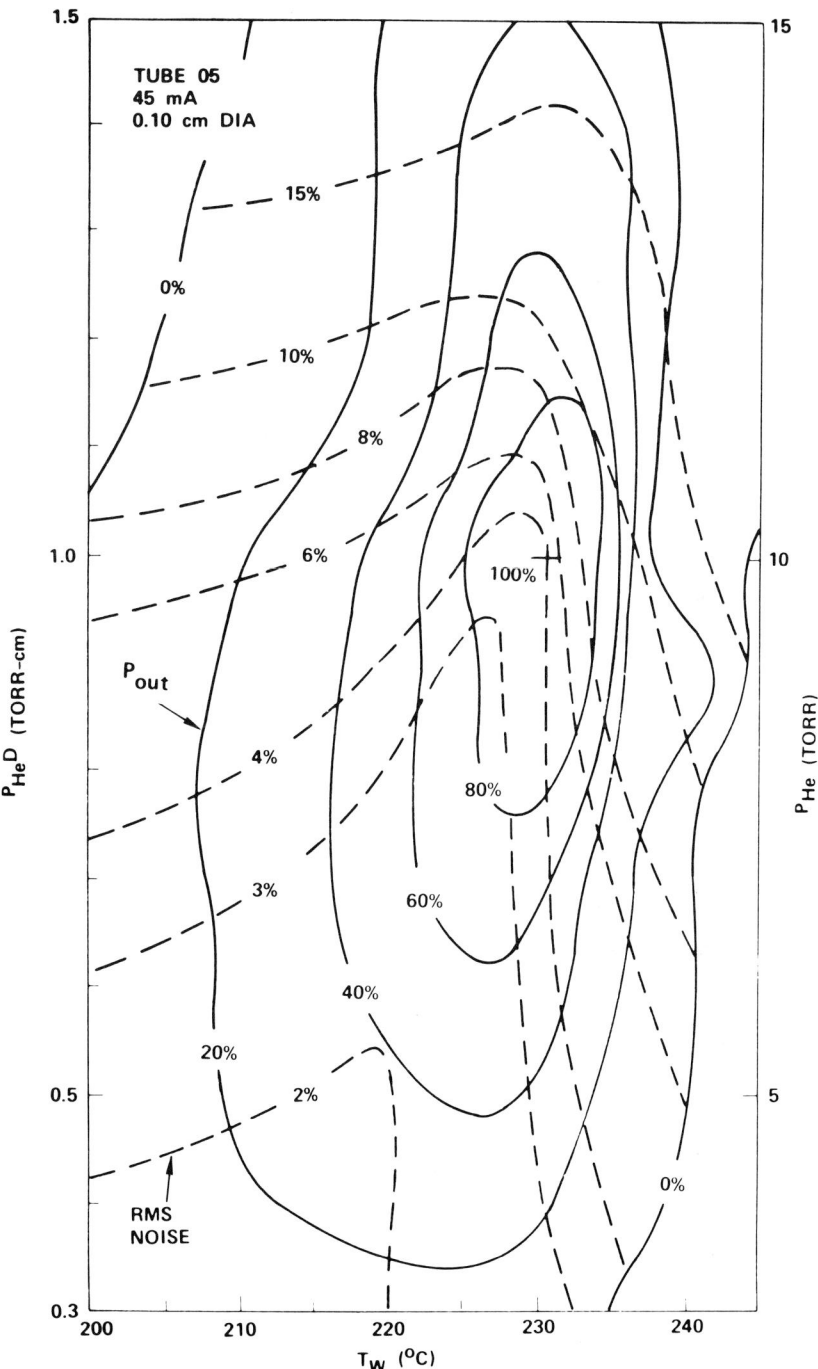

FIG. 28. Power output and beam noise contour plot for a 45-mA current in a 1.0-mm-diameter He–Cd$^+$ laser; the normalization power (100%) is 9.0 mW. (From Johnston and Kolb,[428] by courtesy of the IEEE.)

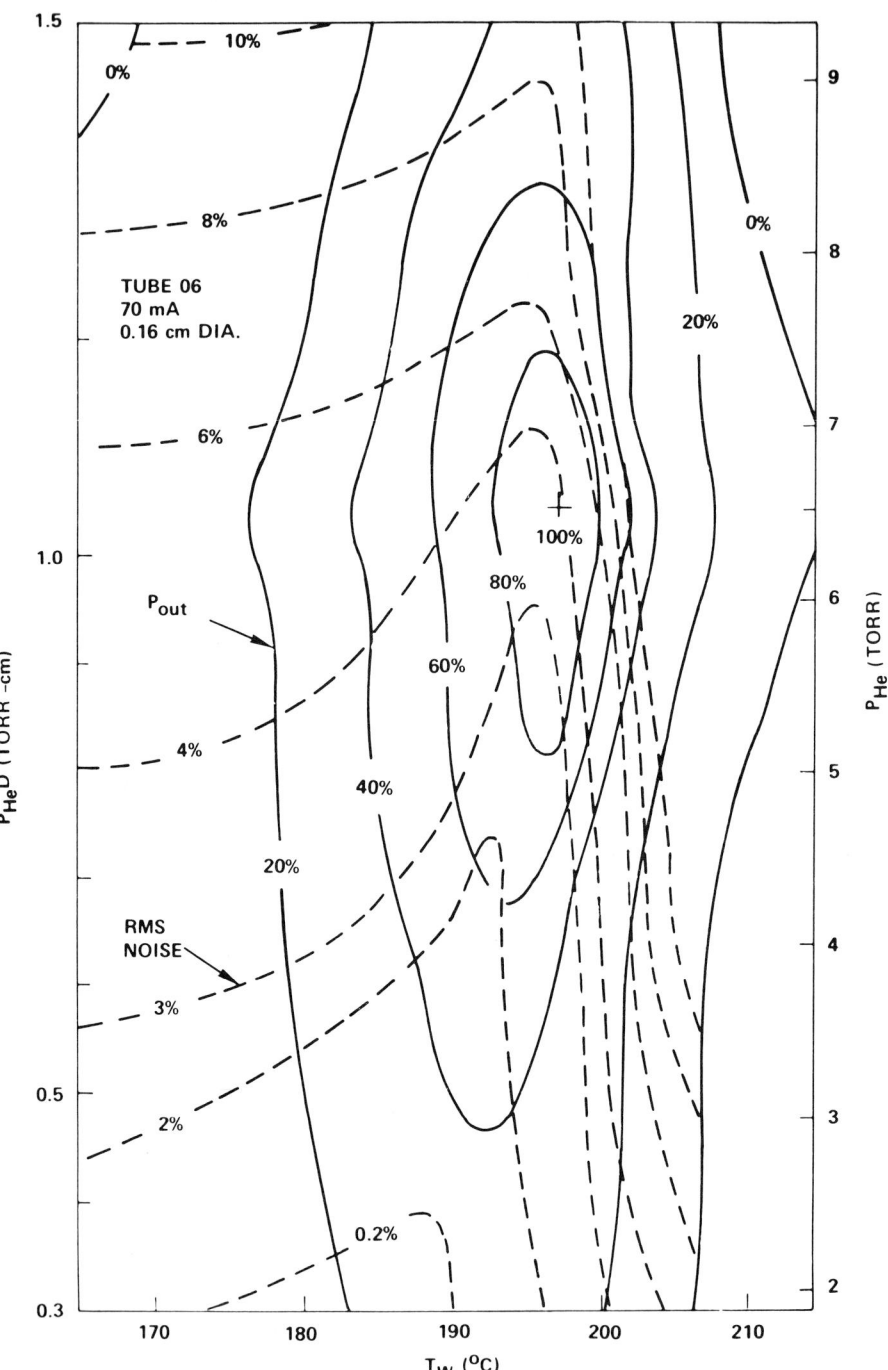

FIG. 29. Power output and beam noise contour plot for a constant 70-mA current in a 1.6-mm-diameter He–Cd$^+$ laser; the normalization power (100%) is 14.5 mW. (From Johnston and Kolb,[428] by courtesy of the IEEE.)

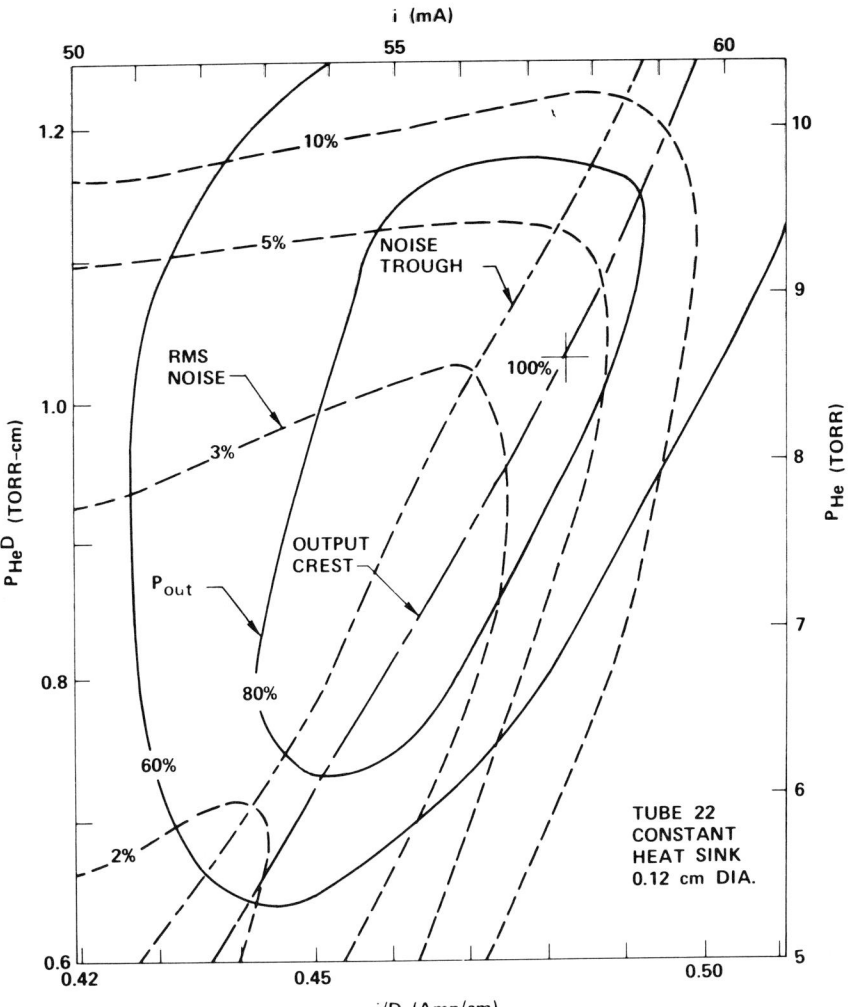

FIG. 30. Power output and beam noise contour plot for a coaxial tube of 1.2 mm diameter with a constant heat sink; the normalization power (100%) is 25 mW. (From Johnston and Kolb,[428] by courtesy of the IEEE.)

and 0.3250-μm lines in a cw HCD 6.35 mm in diameter and 85 cm long. Power outputs of 10 mW/line were estimated for 0.4416- and 0.5378-μm lines at 4 A discharge current. Fukuda and Miya[430] have developed a segmented HCD of metal–ceramic construction, using nine cathodes each 4.2 mm in diameter and 5 cm long. They report outputs for various

---

[430] S. Fukuda and M. Miya, *Jpn. J. Appl. Phys.* **13**, 667 (1974).

TABLE XVI. Scaling Relations for ^{4}He–Cd$^+$ Lasers of $1 < D < 1.6$ mm^a

Quantity		Value	Units
Helium pressure	$p_{He}D$	1.00	torr-cm
Cadmium pressure	$p_{Cd}D$	2.5×10^{-4}	torr-cm
Electronic field	ED	7.9	V
Current density	i/D	0.43	A/cm
Power output	P_{out}/LD	4.3	mW/cm^2
Gain	gD	1.4×10^{-4}	—
Efficiency^b	η/D	1.4×10^{-3}	cm^{-1}
Pressure sensitivity^c	$\Delta p_{He}D_{0.5}$	0.93	torr-cm
Noise^d	—	4	%

^{a} T. F. Johnston, Jr. and W. P. Kolb, *IEEE J. Quantum Electron.* **qe-12**, 482 (1976).
^{b} $\eta = P_{out}/iEL$.
^{c} Change in $p_{He}D$ to decrease P_{out} to 0.5 its maximum value.
^{d} Measured over 10-Hz to 1-MHz bandwidth at maximum P_{out}.

groups of lines oscillating simultaneously: (a) 12.5 mW at 0.8878 μm, 2.5 mW at 0.8531 μm, and 5.2 mW at 0.8067 μm at an optimum pressure of 8 torr helium and a current of 1.3 A., (b) 4.5 mW at 0.7284 μm and 1.1 mW at 0.7237 μm, and (c) 10 mW at 0.6360 μm, 7 mW at 0.5378 μm, and 2.5 mW at 0.5337 μm at 10 torr helium. Fukuda and Miya report that the 0.4416-μm line also oscillates in their HCD with about the same output as in positive column discharges.

Oscillation of the 0.4416-μm line in an rf-excited discharge has been reported by Vernyi et al.[431] They obtained 10 mW output from a 6-mm-diameter tube 50 cm long, excited at 40 MHz. The optimum helium pressure was 2.5 torr.

Toyoda et al.[432] report oscillation on the 0.5378-μm line in a laser-produced plasma. A 10-mW TEA–CO$_2$ laser was directed into a vacuum chamber, evaporating atoms from the surface of a cadmium target and ionizing them to form a plasma. The ion density was estimated to be 1.5×10^{16} cm^{-3} over a path length of about 1 cm. The laser oscillation persisted for 1 μsec, peaking 0.2 μsec after the 80-nsec CO$_2$ pulse. Since no helium was present in the laser chamber a process other than (2.4.10) plus cascade must be responsible for the inversion, probably direct electron excitation of the 4f ^{2}F$^o_{7/2}$ level. It should not be too surprising that

[431] E. A. Vernyi, L. D. Marsh, and B. M. Rabkin, *J. Appl. Spectrosc. (Engl. Transl.)* **16**, 121 (1972).

[432] K. Toyoda, M. Kobiyama, and S. Namba, *Jpn. J. Appl. Phys.* **15**, 2033 (1976).

more than one process can populate this level, since pulsed oscillation on the 0.5378-μm line has previously been reported in a *neon*–cadmium discharge by Fowles and Silfvast.[386]

2.4.4. Zinc Ion Laser

The He–Zn$^+$ laser closely parallels the He–Cd$^+$ laser in its early history and development. Indeed, most of the early references report work done on both lasers. However, the principal wavelengths of the He–Zn$^+$ laser are not as uniquely useful as the blue and uv laser lines in Cd$^+$, so that the corresponding development into a commercial product has not occurred for the He–Zn$^+$ laser. Figure 31 is a simplified Grotrian diagram of Zn$^+$, indicating all of the Zn II laser transitions that are known to date. The similarity of the diagram to Fig. 26 for cadmium is striking. Again, both Penning and charge-exchange collisions with helium populate the different zinc laser levels:

$$\text{He}^*(2s\ ^3S_1) + \text{Zn} \longrightarrow \text{He} + \text{Zn}^+(4s^2\ ^2D) + e + \text{K.E.}, \quad (2.4.11)$$

$$\text{He}^+(^2S_{1/2}) + \text{Zn} \longrightarrow \text{He} + \text{Zn}^+(6s, 5d) + \text{K.E.} \quad (2.4.12)$$

2.4.4.1. Spectroscopy of the Zinc Ion Laser.
Fowles and Silfvast[386] reported the first observation of oscillation in ionized zinc at 0.4924 and 0.7758 μm in a pulsed helium–zinc discharge. The operating parameters were essentially identical to those observed for cadmium. The tube wall temperatures were correspondingly higher, 300–500°C to yield vapor pressures in the same range as cadmium, 1 mtorr to 1 torr. Silfvast[390] later obtained cw operation on the 0.7479- and 0.7588-μm lines, proposing the Penning process (48) for the excitation of the 0.7479-μm line. Jensen *et al.*[433,434] reported pulsed operation on the 0.4912- and 0.5894-μm lines and cw operation on the 0.5894-, 0.7479-, and 0.7588-μm lines, proposing the charge-exchange process (2.4.12) for those lasers originating from the 6s, 5d, and 4f states, based on measurements of spontaneous emission decay rates. An additional line at 0.6021 μm was reported by Jensen *et al.*[435] Continuous oscillation was reported for the 0.4924-μm line by Bloom and Goldsborough.[392] Continuous oscillation in a hollow-cathode discharge was reported by Sugawara *et al.*[395,396] for the 0.4912-, 0.4924-, 0.5894-, 0.6103-, 0.7479-, and 0.7588-μm lines. An additional cw line at 0.7732 μm was obtained in a HCD by Iijima and Sugawara.[436] The weak-

---

[433] R. C. Jensen and W. R. Bennett, Jr., *IEEE J. Quantum Electron.* **qe-4**, 356 (1968).
[434] R. C. Jensen, G. J. Collins, and W. R. Bennett, Jr., *Phys. Rev. Lett.* **23**, 363 (1971).
[435] R. C. Jensen, G. J. Collins, and W. R. Bennett, Jr., *Appl. Phys. Lett.* **18**, 50 (1971).
[436] T. Iijima and Y. Sugawara, *J. Appl. Phys.* **45**, 5091 (1974).

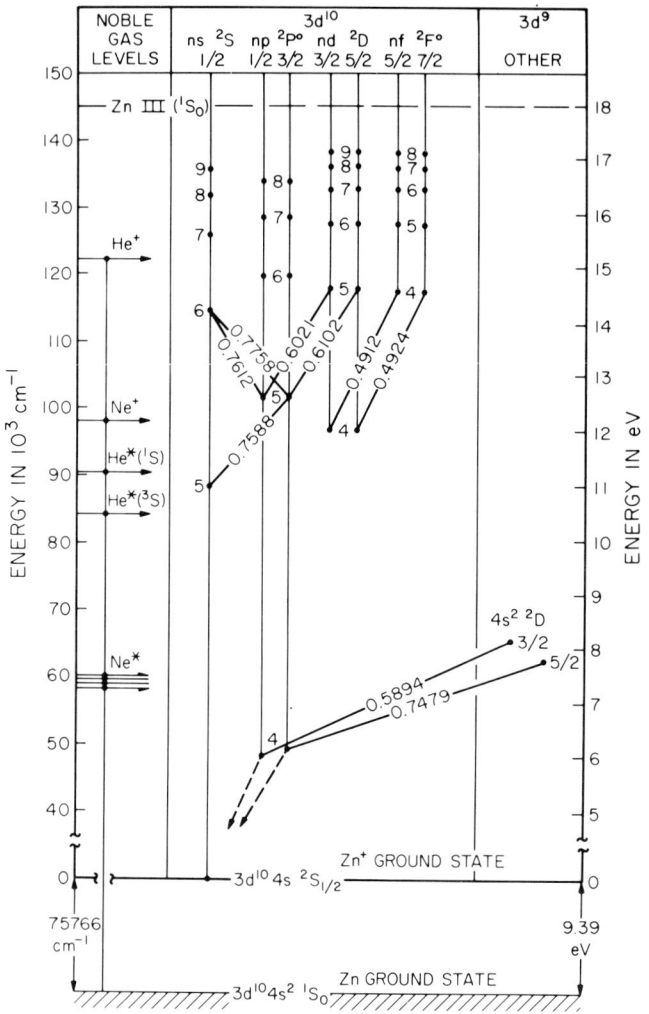

FIG. 31. Energy level diagram for singly ionized zinc, showing all the Zn II laser lines observed to date. The energies of the helium and neon metastables and ground-state ions are also shown with respect to the ground state of the Zn neutral atom.

est member of the lines originating from the $4s^2$ ^2D levels at 0.6214 μm was obtained by Piper and Gill[437] also in a hollow-cathode discharge.

2.4.4.2. Excitation Mechanisms in the He–Zn⁺ Laser. As stated above, Silfvast[390] was the first to propose the Penning process (2.4.11) for the excitation of the upper level of the 0.7479-μm laser line. Jensen et

[437] J. A. Piper and P. Gill, *J. Phys. D* **8**, 127 (1975).

al.[434] originally considered the helium molecular ion He_2^+ rather than the $He^*(2s\ ^3S_1)$ metastable as the species responsible for the 0.5894- and 0.7479-μm lines on the basis of an observed difference in decay rate variation with zinc vapor pressure for the atomic metastables and the laser lines. However, later measurements of decay rates by Riseberg and Schearer[438] gave identical decay rate variations for the $He^*(2s\ ^3S_1)$, $Zn(4s^2\ ^2D_{3/2})$, and $Zn(4s^2\ ^2D_{5/2})$ populations. Additional measurements by Collins et al.[439] also showed this to be the case. The flowing afterglow measurements of Webb et al.[400,401] make it clear that the Penning process (2.4.11) is responsible for the 0.5894- and 0.7479-μm laser lines.

Jensen et al.[433,439] originally proposed the charge-exchange reaction (2.4.12) as the populating mechanism for the 6s, 5d, and 4f levels (and also the 5p level by cascade), based on their measured identical decay rate variations with zinc vapor pressure and an estimate of the decay rate of He^+. The flowing afterglow measurements of Webb et al.[400,401] also indicated that these levels were populated by charge exchange. However, later flowing afterglow measurements by the same group[440] showed that only the 6p and 5d levels are populated primarily by the direct charge exchange process (2.4.12), and that the charge-exchange contribution to the 4f levels and the 6s levels is very small in comparison with the rate to the 5d levels. Thus some other mechanism must be invoked to explain the observed strong laser oscillation in the 0.4912- and 0.4924-μm lines. Green et al.[440] propose that the 4f levels receive their population by electron-collision-induced transfer from the 5d levels. This proposal is based on observations by Piper and Gill[437] of a faster-than-linear increase in laser output with discharge current, and improved performance at higher helium pressures and in smaller-diameter HCDs (higher electron density). A recent study by Gill and Webb[441] of the radial profiles of the various excited states in a Zn hollow cathode discharge is also consistent with the excitation mechanisms cited above.

2.4.4.3. Performance of the He–Zn$^+$ Laser.
A relatively small amount of performance data exist for the He–Zn$^+$ laser. Jensen et al.[434] give some early results for a cw positive-column laser 4 mm in diameter by 90 cm in length. They obtained 12 mW output at 0.7479 μm, 7 mW at 0.5894 μm, and 1 mW at 7588 μm. The optimum discharge conditions for the Penning lines were sharply peaked at 4 torr helium pressure and 80–100 mA discharge current, while those for the 0.7588-μm line exhib-

---

[438] L. A. Riseberg and L. D. Schearer, *IEEE J. Quantum Electron.* **qe-7**, 40 (1971).

[439] G. J. Collins, R. C. Jensen, and W. R. Bennett, Jr., *Appl. Phys. Lett.* **18**, 282 (1971); erratum **19**, 122 (1971).

[440] J. M. Green, G. J. Collins, and C. E. Webb, *J. Phys. B* **6**, 1445 (1973).

[441] P. Gill and C. E. Webb, *J. Phys. D* **11**, 245 (1978).

ited broad optima around 6 torr helium and 150 mA discharge current. The dependence on the zinc source temperature was relatively critical for all three lines, exhibiting sharp optima in the 320–340°C range with half-widths of 15–20°C.

Piper and Webb[429] have reported the performance of the zinc ion lines in a cw hollow-cathode discharge 6.35 mm in diameter by 85 cm long. They obtained 30 mW output total on the two blue lines at 16 torr helium pressure, 425°C zinc source temperature, and 5 A discharge current. An output of 4 mW at 0.5894 μm was obtained under the same conditions but with a clear optimum in current at 3 A. Higher output powers were obtained by Piper and Gill[437] with a smaller-diameter HCD, 3 mm by 50 cm operated in a quasi-cw fashion with 1.2-msec pulses. They observed 95 mW at 0.4924 μm, 50 mW at 0.4912 μm, and 25 mW at 0.7588 μm at the maximum available current of 6 A; at this current the output for all three lines was still increasing. They obtained 65 mW at 0.7479 μm and 15 mW at 0.5894 μm from the same tube with an optimum current of 4 A. The optimum helium pressure for the blue lines was about 450°C for the Penning lines and 470°C for the blue lines. These correspond to zinc vapor pressures an order of magnitude higher than the positive column He–Zn$^+$ laser.

2.4.5. Iodine Ion Laser

Iodine was the second element after mercury to exhibit ion laser oscillation on lines excited by collisions with noble gases. Fowles and Jensen[442] observed pulsed oscillation on two lines at 0.5761 and 0.6127 μm in a helium–iodine discharge in 1964. They reported four additional lines at 0.5407, 0.5678, 0.6585, and 0.7033 μm a few months later.[443] Fowles and Jensen proposed charge-exchange collisions with helium ground-state ions as the process responsible for the upper level population,[442] the first time this process had been proposed for any laser. They also noted that the Wigner spin rule[443] was obeyed for this reaction, since only triplet upper levels were observed to oscillate:

$$\text{He}^+(^2S_{1/2}) + \text{I}(^2P_{3/2}) \longrightarrow \text{He}(^1S_0) + \text{I}^+(6p'\ ^3X). \quad (2.4.13)$$

Additional iodine lines in the blue through infrared were reported by Jensen and Fowles,[444] Willett,[445] Willett and Heavens,[446] Piper,[447] and

---

[442] G. R. Fowles and R. C. Jensen, *Proc. IEEE* **52**, 851 (1964).
[443] G. R. Fowles and R. C. Jensen, *Appl. Opt.* **3**, 1191 (1964).
[444] R. C. Jensen and G. R. Fowles, *Proc. IEEE* **52**, 1350 (1964).
[445] C. S. Willett, *IEEE J. Quantum Electron.* **qe-3**, 33 (1967).
[446] C. S. Willett and O. S. Heavens, *Opt. Acta* **14**, 195 (1967).
[447] J. A. Piper, *J. Phys. D* **7**, 323 (1974).

Piper and Webb,[448] for a total of 27 identified lines. All but three of these lines originated from the 6p' $^3D_{1,2}$ and $^3F_{2,3}$ levels, which lie less than 0.5 eV below the He$^+$ ground state. The lower levels are primarily the 5d' and 6s' levels and are presumed to be depopulated by radiative decay to the iodine ion ground state.

Collins speculated in his thesis[399a] that continuous oscillation should be obtainable in a He–I$^+$ laser, and such an operation was first obtained by Collins et al.[449] with a positive-column discharge containing a mixture of helium and CdI_2. Only the 0.6127-μm iodine line would oscillate; attempts to produce simultaneous oscillation on the Cd II 0.4416-μm line were unsuccessful. Cw operation on 25 of the 27 known lines was later demonstrated by Piper et al.[447,448,450] in hollow cathode discharges. Continuous oscillation on additional lines in positive-column discharges has been obtained by Pugnin et al.,[451] Hattori et al.,[452] and Kano et al.[453]

The excitation mechanisms responsible for the He–I$^+$ laser have received comparatively little study, probably because the observed lines all seem to fit the charge-exchange model so well. Energy defects less than 0.5 eV occur for all but one line, and spin is conserved on 25 of the 27 known lines. Shay et al.[454] have measured the variation of the upper level decay rates with iodine vapor pressure for the four triplet levels responsible for 24 of the 27 lines; they show that the variations are identical for all four levels and that the zero pressure intercept agrees with the estimated value for the helium ion, and not the helium metastables. They obtained a value of 1.5×10^{-15} cm^2 for the charge-exchange cross section summed over the four levels, about three times larger than the value obtained by Tolmachev[455] summed over the same four lines. Kano et al.[456] have measured the electron temperature and density in a He–CdI_2 discharge; they conclude that the Penning electrons formed by process (2.4.9) decreases the electron temperature so much that the higher vapor pressures optimum for process (2.4.13) cannot be employed. CdI_2 does not seem to be a good source of iodine for the He–I$^+$ laser.

[448] J. A. Piper and C. Webb, IEEE J. Quantum Electron. qe-12, 21 (1976).

[449] G. J. Collins, H. Kano, S. Hattori, K. Tokutome, M. Ishikawa, and N. Kamiide, IEEE J. Quantum Electron. qe-8, 679 (1972).

[450] J. A. Piper, G. J. Collins, and C. E. Webb, Appl. Phys. Lett. 21, 203 (1972).

[451] V. I. Pugnin, S. A. Rudelev, and A. F. Stepanov, J. Appl. Spectrosc. (Engl. Transl.) 18, 667 (1973).

[452] S. Hattori, H. Kano, K. Tokutome, G. J. Collins, and T. Goto, IEEE J. Quantum Electron. qe-10, 530 (1974).

[453] H. Kano, T. Goto, and S. Hattori, J. Phys. Soc. Jpn. 38, 586 (1975).

[454] T. Shay, H. Kano, and G. J. Collins, Appl. Phys. Lett. 26, 531 (1975); Erratum 29, 221 (1976).

[455] Yu. A. Tolmachev, Opt. Spectrosc. (Engl. Transl.) 33, 653 (1972).

[456] H. Kano, T. Goto, and S. Hattori, IEEE J. Quantum Electron. qe-9, 776 (1973).

Goto et al.[457] have recently described the operating characteristics of a cw positive-column He–I$^+$ laser. A simple discharge tube 2.5 mm in diameter by 100 cm long was used with a cataphoretic flow of iodine. Instead of the usual oven and condenser arrangement employed with He–Cd$^+$ tubes, molecular sieve material (Union Carbide Co. type 5A) was used to contain the iodine both in the source (which was heated by a small oven) and the sink (which was cooled by airflow from a fan); the molecular sieve container at the cathode sink was located so that it prevented the iodine vapor from reaching the hot cathode. The optimum operating conditions were estimated to be an iodine pressure of 20 mtorr at a source temperature of 130°C, a helium pressure of 6 torr, and a discharge current of 280 mA. The laser output was 20–30 mW total on the five visible lines at 0.5407, 0.5678, 0.5761, 0.6127, and 0.6585 μm. The maximum output with an optimum value of output coupling was estimated to be 40–50 mW. The measured noise modulation was 10–15% summed over the 1–100-kHz range and exhibited a $1/f$ variation. With infrared mirrors the laser oscillated at 0.7033, 0.7618, 0.8170, and 0.8804 μm. The operation of this sealed-off tube was monitored for 1000 hours, with only a few percent decrease in performance.

Piper and Webb[448] have obtained the best performance for iodine in a hollow-cathode discharge. They employed a HCD 2.5 mm in diameter and 75 cm long with individually ballasted anodes in side-arms disposed every 1 cm along the cathode. A helium flow of 100 atm-cm^3 min^{-1} at pressures of 5–50 torr was introduced into the center of the cathode and extracted at the ends. The typical iodine vapor pressure was 0.1 torr. Current pulses of 1.6 msec duration and amplitudes up to 25 A were superposed on a 100-mA dc "keep-alive" current. This quasi-cw excitation was chosen to avoid thermal dissipation difficulties in the apparatus; however, the long pulse lengths assured that the discharge conditions were the same as those for a truly continuous discharge. Power outputs of 100 mW at 0.5761, 70 mW at 0.6127, and 55 mW at 0.5678 μm were obtained at 20 A current, 10 torr helium and 0.3 torr iodine. The output power variation with discharge current was essentially linear up to the maximum current for these three strongest lines. Power outputs of tens of mW were typical for the other strong lines at 0.5405, 0.6585, 0.7033, and 0.8804 μm. A total power of almost 1 W summed over all the wavelengths was observed. Competition effects were observed among lines sharing common lower levels: 0.5761 and 0.6585 μm; 0.8804 and

---

[457] T. Goto, H. Kano, N. Yoshino, J. K. Mizeraczyk, and S. Hattori, *J. Phys. E* **10**, 292 (1977).

0.8170 μm; 0.4987, 0.5216, and 0.5593 μm, thus indicating the limits in the radiative depopulation rates for lower levels.

2.4.6. Selenium Ion Laser

Oscillation in ionized selenium was first observed by Bell et al.[458] in a pulsed rf-excited ring discharge with an active region 5 mm in diameter and 60 cm long. Neon at 0.1 torr gave the best performance of various buffer gases employed. Continuous oscillation was subsequently observed in a He–Se discharge by Silfvast and Klein[459] on 24 wavelengths in the range 0.4604–0.6536 μm. The six strongest lines were 0.4976, 0.4993, 0.5069, 0.5176, 0.5228, and 0.5305 μm. The line at 0.5228 μm had the highest measured gain, 5.4%/m and was one of the two lines observed by Bell et al.[458] Output powers up to 30 mW summed over the stronger lines were observed in a 4-mm by 100-cm discharge tube at a current of 200 mA and a helium pressure of 8 torr. The output peaked at a selenium source temperature of 265°C, corresponding to a vapor pressure of 5 mtorr. Silfvast and Klein proposed that charge exchange was the primary excitation mechanism since all 24 lines originate from the 5p levels of Se II lying close to the He$^+$ ground-state energy. The 5s and 4d lower levels are assumed to be depopulated by radiative decay. Competition among lines sharing common lower levels was noted by Silfvast and Klein, indicating that the radiative decay may not be too rapid.

Keidan et al.[460] independently observed both pulsed and continuous oscillation on 17 of the 24 lines reported by Silfvast and Klein.[459] The strengths of the various lines observed by Keidan et al. are remarkably like those reported by Silfvast and Klein; they obtained 20–30 mW output summed over the strongest 7–10 wavelengths at 4.5 torr helium pressure, 7 mtorr selenium pressure and 200 mA discharge current in a tube 2 mm in diameter by 100 cm long. Both groups used cataphoretic flow[416-418] to introduce selenium into the discharge. Keidan et al. also proposed charge exchange as the population mechanism.

Additional wavelengths were later reported by Klein and Silfvast[461] for a total of 46 known lines. Of these 22 additional lines, two were classified as strong, 1.0409 and 1.2588 μm. They also reported improved operation

---

[458] W. E. Bell, A. L. Bloom, and J. P. Goldsborough, *IEEE J. Quantum Electron.* **qe-1**, 400 (1965).

[459] W. T. Silfvast and M. B. Klein, *Appl. Phys. Lett.* **17**, 400 (1970).

[460] V. F. Keidan, V. S. Mikhalevskii, and M. F. Sém, *J. Appl. Spectrosc. (Engl. Transl.)* **15**, 1089 (1971).

[461] M. B. Klein and W. T. Silfvast, *Appl. Phys. Lett.* **18**, 482 (1971).

TABLE XVII. Operating Parameters and Performance of He–Se$^+$ Ion Lasers

Ref.	Discharge dimensions		Total wavelengths observed	Power output		Optimum values			Comment
	Diameter (mm)	Length (cm)		mW	Wavelengths	I (mA)	He (torr)	Se (mtorr)	
a	3	50	—	—	—	500	6	5	
a	4	100	27	30	6	>200[e]	8	5	
b	2	100	17	20–30	7–10	200	4.5	7	
c	3	200	46	250	?	400	7	5	
d	1	10	5	—	—	50	20	—	$V_{\text{TUBE}} = 850$ V
d	1.5	50	17	28	6	115	20	—	$V_{\text{TUBE}} = 2750$ V
d	2	90	24[f]	—	—	175	15	—	$V_{\text{TUBE}} = 3600$ V

[a] W. T. Silfvast and M. B. Klein, *Appl. Phys. Lett.* **17**, 400 (1970).
[b] V. F. Keidan, V. S. Mikhalevskii, and M. F. Sém, *J. Appl. Spectros. (Engl. Transl.)* **15**, 1089 (1971).
[c] M. B. Klein and W. T. Silfvast, *Appl. Phys. Lett.* **18**, 482 (1971).
[d] K. G. Hernqvist and D. C. Pultorak, *Rev. Sci. Instrum.* **43**, 290 (1972).
[e] Not optimum current; maximum value limited by glass tube walls.
[f] Some wavelengths not confirmed by other references.

on the strong visible lines previously observed, 50 mW each on 0.5069, 0.5176, and 0.5228 μm, and 250 mW total in combination with other lines. The laser parameters are described in Table XVII. In all, 13 of the Se II 5p levels are responsible for the 46 laser lines. These levels are distributed from 0.2 eV above to 0.8 eV below the He$^+$ ground state energy. Klein and Silfvast[461] proposed that mixing collisions with electrons or helium atoms may play a role in populating some of the 5p levels, especially those lying above the He$^+$ level. Under the discharge conditions typical of He–Se$^+$ lasers the Se II 5p levels are spaced only about kT in energy.

Hernqvist and Pultorak[462] described the performance of three different He–Se$^+$ lasers, all utilizing the cataphoretic flow technique; tube parameters and performance are compared in Table XVII. They also reported competition among the weaker lines sufficiently severe that an intracavity prism was required to make the red lines oscillate. Hernqvist and Pultorak observed noise fluctuations on the light output of about 10% peak to peak.

Piper and Webb[429] obtained cw oscillation in a hollow-cathode discharge on 13 of the lines marked *strong* or *moderate* by Silfvast and Klein.[459] Curiously, they did not observe oscillation on the 0.5305-μm line marked *strong* by Silfvast and Klein. Their HCD was 6.35 mm in diameter by 85 cm long and optimized at 7 torr helium, 80 mtorr selenium, and 2 A discharge current. No output powers were given.

Only a little work has appeared on He–Se$^+$ laser mechanisms. From their flowing-helium afterglow measurements, Turner-Smith et al.[401] confirm that charge exchange is indeed the dominant process in the He–Se$^+$ laser. Electron temperature and density have been measured as a function of selenium vapor pressure by Goto et al[463]; both quantities are relatively constant over the pressure range corresponding to oscillation in the He–Se$^+$ laser, indicating that Penning reactions play little or no role in the discharge. McKenzie[464] has measured the radial profiles of the stronger upper laser levels. His measurements show that a relative deficiency develops on the discharge axis at higher currents, and the maximum emission originates from an annular region of diameter equal to about 60–70% of the discharge diameter.

2.4.7. Other Metal Vapor Ion Lasers

In addition to those ion lasers discussed in previous sections, there are several other ions that exhibit oscillation from levels excited by Penning

---

[462] K. G. Hernqvist and D. C. Pultorak, *Rev. Sci. Instrum.* **43**, 290 (1972).
[463] T. Goto, H. Kano, and S. Hattori, *J. Appl. Phys.* **43**, 5064 (1972).
[464] A. L. McKenzie, *J. Phys. B* **7**, L141 (1974).

or charge-exchange collisions, but that have received less attention to date. Thermally produced vapors of Te, As, Mg, Sn, Tl, and Be have exhibited oscillation when excited in helium or neon discharges. Additional lines have been predicted for ionized gallium, germanium, and calcium by Green and Webb[355,356] based on their flowing afterglow studies of the He–Ga$^+$, Ne–Ge$^+$, and Ar–Ca$^+$ systems.

2.4.7.1. **Tellurium Ion Laser.** As in the case of selenium, the first oscillation on tellurium ion lines was obtained by Bell et al.[465] in a pulsed rf-excited ring discharge. The tube used had a 3-mm diameter and a 60-cm active length, and optimized at 0.2 torr neon and 1–2 mtorr tellurium vapor pressure. Oscillation occurred at 0.5576, 0.5708, and 0.6350 μm. Webb[466] later observed the two shorter wavelengths and an additional five lines in pulsed discharges 6 and 9 mm in diameter and 120 cm long. Neon was used as the buffer gas at an optimum pressure of 0.1–0.25 torr for one group of lines that oscillated during the excitation pulse and in the early afterglow, and at 0.5–1.0 torr optimum pressure for a second group of lines that oscillated only in the late afterglow, 35–50 μsec after the 5-μsec excitation pulse.

Continuous operation was obtained by Watanabe et al.[467] on the 0.5708-μm line in a 4-mm by 140-cm discharge with 2–8 torr of *helium* and 200 mA discharge current. A nonoptimized output of 0.15 mW was reported. Continuous oscillation on 31 Te II transitions was reported by Silfvast and Klein,[468] spanning the wavelength range 0.4843–0.9378 μm. They used a 3-mm by 200-cm tube similar to that used for selenium.[461] Of the 31 lines observed, 15 were obtained only with neon as the buffer gas, 11 only with helium as the buffer gas, and 5 with either neon or helium. Several mW output was obtained on the stronger lines. The 0.5708-μm line exhibited the highest gain, 4.3%/m with helium (but only 0.5%/m with neon). The optimum tellurium vapor source temperature was 420°C, corresponding to 50 mtorr tellurium vapor pressure. The optimum discharge current was about 400 mA for either He or Ne gas fills. For the neon-excited lines the optimum buffer gas pressure was about 2 torr and for the helium-excited lines about 5 torr. Silfvast and Klein have identified 24 of the 31 lines as far as the spectroscopic literature allows: J values and level parity. From the location of the known Te II upper levels they concluded that charge exchange with the He$^+$ and Ne$^+$ ground-state ions is responsible for the two sets of lines. Silfvast and

[465] W. E. Bell, A. L. Bloom, and J. P. Goldsborough, *IEEE J. Quantum Electron.* **qe-2**, 154 (1966).
[466] C. E. Webb, *IEEE J. Quantum Electron.* **qe-4**, 426 (1968).
[467] S. Watanabe, M. Chihara, and I. Ogura, *Jpn. J. Appl. Phys.* **11**, 600 (1972).
[468] W. T. Silfvast and M. B. Klein, *Appl. Phys. Lett.* **20**, 501 (1972).

Klein point out that at the discharge operating temperature the neutral Te metastables also have significant thermal populations, so that the process

$$\text{Ne}^+ + \text{Te}^M \longrightarrow \text{Ne} + \text{Te}^+(**) \tag{2.4.14}$$

cannot be ruled out for some laser lines. The overall performance of the He–Te$^+$ and Ne–Te$^+$ lasers seems to be inferior to He–Se$^+$ lasers of similar dimensions.

Three of the lines observed in Bell *et al.*[465] and Webb[466] with low buffer gas pressures were not seen by Silfvast and Klein. They are also not identified as Te II lines. It is reasonable to assume that these lines originate from higher Te II levels than those excited by He$^+$ or from higher ionization states of Te, and are probably excited by electron collisions.

2.4.7.2. **Arsenic Ion Laser.** Laser oscillation in ionized arsenic was first observed by Bell *et al.*[458] in the same pulsed rf-excited ring discharge that was used for selenium. Four lines at 0.5498, 0.5558, 0.5651, and 0.6170 μm were observed with 0.1 torr neon as a buffer gas. (It is possible that the 0.5498-μm line was actually two lines.) Because of the low buffer gas pressure it is likely that the electron collisions rather than collisions with noble-gas atoms or ions were responsible for oscillation on these lines.

Continuous oscillation was later observed by Piper and Webb[469] in a flowing-helium hollow-cathode discharge tube 3 mm in diameter by 50 cm in length. Three of the four lines observed by Bell *et al.* plus five additional lines were obtained. Piper and Webb classified the eight cw lines into three categories according to the experimentally observed optimum helium pressure:

(1) The four lines at 0.5385, 0.5497, 0.5838, and 0.6512 μm optimized at 55 torr helium. These lines originate from the 6s $^1P_1$ and 6s $^3P_2$ levels, which lie less than 2000 cm^{-1} below the ground-state He$^+$ energy and are presumed to be populated directly by charge-exchange collisions.

(2) The two lines 0.5498 and 0.7103 μm optimized at around 40 torr helium pressure, which are populated largely by radiative cascade from the 6s levels excited by charge exchange.

(3) The two lines 0.5558 and 0.6170 μm optimized at about 25 torr helium pressure, which are populated by radiative cascade from the 5d levels, which are even closer to the He$^+$ energy than the 6s levels.

The optimum arsenic vapor pressure was estimated to be 0.8 torr, corresponding to an arsenic source temperature of 360°C. The output power increased approximately linearly with discharge current up to the maximum

[469] J. A. Piper and C. E. Webb, *J. Phys. B* **6**, L116 (1973).

value of 2.6 A. The highest output power observed was 12.5 mW for the 0.6512-μm line.

A recent study by Piper et al.[469a] using the Clarendon Laboratory flowing afterglow apparatus[401] indicates that the lines in group (1) are probably pumped by a somewhat different route than that originally proposed in Ref. (469). The results of the flowing afterglow studies suggest that the 6s states of As$^+$ are populated by electron collisional "knockdown" from higher lying As$^+$ states that have been populated by charge exchange collisions between ground state helium ions and neutral arsenic *metastables*.

2.4.7.3. Magnesium Ion Laser. Oscillation in ionized magnesium was first obtained by Hodges[470] in a cw hollow-cathode discharge 2 mm in diameter and 60 cm in length. Four lines at 0.9218, 0.9244, 1.0915, and 1.0952 μm were observed with helium pressures in the 0.5–5 torr range and magnesium vapor at a pressure corresponding to a temperature of 395°C. The Penning excitation process

$$\mathrm{He}^*(2s\ ^3S_1) + \mathrm{Mg}(3s^2\ ^1S_0) \longrightarrow \mathrm{He}(^1S_0) + \mathrm{Mg}^+(4p\ ^2P_{1/2,3/2}) + e + \mathrm{K.E.} \quad (2.4.15)$$

is assumed to be responsible for the upper laser level excitation. Rapid radiative decay depopulates the lower laser levels. An output power of 2 mW was observed on the two longer wavelength lines at a discharge current of 0.8 A and a helium pressure of 1 torr.

Turner-Smith et al.[401] have measured the relative charge-exchange excitation of Mg II lines by ground-state Ne$^+$ in their flowing neon afterglow apparatus. Collins[357] has listed nine possible candidates for oscillation on ultraviolet transitions of Mg$^+$, but as yet no oscillation has been reported. The lines listed by Collins are exactly the same lines shown by Manley and Duffendack in 1935[354a] to be greatly enhanced by charge-exchange collisions in a Ne–Mg$^+$ discharge.

2.4.7.4. Tin Ion Laser. Oscillation in ionized tin was first observed by Silfvast et al.[387] on three lines at 0.5799, 0.6453, and 0.6488 μm in a pulsed He–Sn discharge. The two longer wavelength lines originate from the 6p $^2P^o_{1/2}$ and 6p $^2P^o_{3/2}$ levels in Sn II, which lie about 3.5 eV below the He*(2s $^3S_1$) metastable. Silfvast[390] later obtained continuous oscillation on these two lines in a He–Sn discharge and attributed their excitation to Penning collisions. He used a discharge tube 5 mm in diameter and 75 cm long and found an optimum helium pressure of 8 torr. The 0.6453-μm line was the stronger of the two, and its output increased with Sn vapor pressure

---

[469a] J. A. Piper, I. M. Littlewood, and C. E. Webb, *J. Phys. B*, **11**, 3731 (1978).
[470] D. T. Hodges, *Appl. Phys. Lett.* **18**, 454 (1971).

from threshold at a 1020°C source temperature to 1100°C, the maximum temperature limited by devitrification of the fused-silica discharge tube envelope. The highest source temperature corresponds to 1.5 mtorr Sn vapor pressure. The output also increased linearly with discharge current from threshold at 30–100 mA, the highest current used. The maximum observed gain was 2–3% in the 75 cm effective tube length.

No further observations have been reported for these three Sn II lines. The 0.5799-μm line originates from the Sn II 4f $^2F^o_{7/2}$ level, which lies 1.4 eV below the He*(2s $^3S_1$) metastable level. However, since this line was originally observed in both He–Sn and Ne–Sn discharges[387] it is assumed to be excited by electron collisions.

2.4.7.5. Lead Ion Laser. Pulsed oscillation in ionized lead was first observed by Silfvast et al.[387] on a single line in at 0.5372 μm originating from the 5f $^2F^o_{7/2}$ level. Excitation of this level may occur by electron collision even though it lies 0.93 eV below the He*(2s $^3S_1$) level; it is also isoelectronic with the 4f $^2F^o_{7/2}$ level in Sn II responsible for the pulsed 0.5799-μm laser. Silfvast[471] later obtained continuous oscillation on the Pb II lines at 0.5609 and 0.6660 μm, which originate from the 7p $^2P^o_{1/2}$ and 7p $^2P^o_{3/2}$ levels lying 2.8 and 3.2 eV *below* the He*(2s $^3S_1$) metastable. These two transitions are isoelectronic with the two red cw lines in Sn II. Silfvast attributes the excitation of these lines to Penning collisions. He observed an optimum pressure of 6 torr and a Pb vapor pressure of a few mtorr in a 3.5-mm-diameter discharge tube.

Green and Webb[356] have studied the Ne–Pb$^+$ system in their flowing afterglow apparatus and report data for four levels that are excited by charge exchange. Collins[357] lists seven ultraviolet transitions as possible candidates for oscillation in the Ne–Pb$^+$ system, but no oscillation has been reported to date. The lines listed by Collins are identical with the seven lines showing enhancements greater than about 5 in the study of 31 Pb$^+$ lines by Duffendack and Gran in 1937.[354b] (Enhancement was defined as the ratio of line strengths in a discharge with a few torr of neon to the line strength in a low-pressure helium discharge where electron excitation dominates.)

2.4.7.6. Thallium Ion Laser. Ivanov and Sém[472] have reported pulsed oscillation on nine lines of ionized thallium in He–Tl and Ne–Tl discharges 8 mm in diameter and 120 cm long. Thallium vapor pressures of 0.5–5 mtorr and pulsed currents of 5–100 A were employed. Three lines at 0.5949, 0.6951, and 0.9350 μm were observed in Ne–Tl discharges

---

[471] W. T. Silfvast, private communication to C. S. Willett, as reported in Willett,[2] Appendix, Table 27, p. 433.

[472] I. G. Ivanov and M. F. Sém, *J. Appl. Spectrosc.* (*Engl. Transl.*) **19**, 1092 (1973).

only. The two shorter wavelengths originate from the 7p $^3P_2^o$ and 7p $^1P_1^o$ levels, which lie about 0.3 eV below the Ne⁺ ground state. Oscillation was observed to peak 10–50 μsec into the discharge afterglow. The optimum neon pressure was 10 torr and the observed gains were 15, 13, and 5%, respectively, for the three lines. Ivanov and Sém attribute the excitation of these lines to charge-exchange collisions with Ne⁺ ground states and find a cross section of 3×10^{-15} cm² for this process.

Six additional lines at 0.4737, 0.4981, 0.5079, 0.5152, 1.1350, and 1.1750 μm were observed only in He–Tl discharges. The four shorter wavelengths originate from the 5f $^1F_3^o$, 5f $^3F_2^o$, 5f $^3F_3^o$, and 5f $^3F_4^o$ levels, respectively. The 1.1350-μm line is unidentified, and the 1.1750-μm line originates from the 6f $^3F_3^o$ level. The observed gains were 5, 8, 12, and 15% for the blue–green lines, and 4 and 5% for the infrared lines. Ivanov and Sém observe a more complex output waveform for these lines than for the neon-excited transitions; two laser output maxima occur in the discharge afterglow 5 and 20 μsec following the excitation pulse. They attribute the first maximum to excitation of the upper laser levels by recombination of Tl²⁺ ions; they attribute the second output maximum to charge-exchange excitation of the Tl 5g^{1,3}G levels, which lie only 0.25 eV from the He⁺ ground state, followed by radiative decay to the 5f upper laser levels. They find a charge-exchange cross section for the 5g levels of 1.5×10^{-15} cm² and speculate that cw operation of He–Tl⁺ and Ne–Tl⁺ lasers may be possible.

Green and Webb[355,356] have studied the He–Tl⁺ system in their flowing afterglow apparatus and have determined that all of the excitation of the 5g levels occurs directly by charge-exchange collisions with He⁺. They predict that cw laser oscillation should be obtainable on four of the 5g → 5f transitions in the 0.92-μm region. Several other charge-exchange excited lasers are also predicted for the He–Tl⁺ system.[355]

2.4.7.7. Beryllium Ion Laser. Pulsed-ion laser oscillation has recently been observed by Zhukov et al.[473] in He–Be and Ne–Be discharges. The 4f $^2F_{7/2}^o$ → 3d $^2D_{5/2}$ transition at 0.4675 μm and the 4s $^2S_{1/2}$ → 3p $^2P_{3/2}^o$ transition at 0.5272 μm oscillate with helium, and the 3p $^2P_{3/2}^o$ → 3s $^2S_{1/2}$ transition at 1.2096 μm oscillates with neon. Both the position of the upper levels with respect to He⁺ and Ne⁺ ion ground states and the occurrence of the pulsed outputs in the discharge afterglow suggest charge-exchange collisions as the upper level excitation process. Oscillation on these three lines was obtained in an oven-heated BeO ceramic discharge tube at about 1300°C, with an optimum Be vapor pressure of about 10 mtorr and

---

[473] V. V. Zhukov, V. G. Il'yushko, E. L. Latush, and M. F. Sém, *Sov. J. Quantum Electron. (Engl. Transl.)* **5**, 757 (1975).

an optimum buffer gas pressure of several torr. The blue and infrared lines exhibited low threshold and saturation currents (e.g., 5 A for the 0.4675-μm line). Further increases in current only delayed the output pulse farther into the afterglow. In contrast, the 0.5272-μm line exhibited high threshold currents, greater than 30 A, and continued to increase in output with increasing current. This behavior is in agreement with theoretical predictions by Zhukov *et al.* that the inversion on the 4s $^2S_{1/2} \to$ 3p $^2P^o_{3/2}$ transition requires depopulation of the 3p lower level by electron collision, and that sufficient electron density, greater than 10^{15} cm^{-3}, occurs only at high currents.

Operation in a pulsed hollow-cathode discharge 6 mm in diameter by 40 cm in length was also obtained by Zhukov *et al.* on the 0.4675- and 1.2096-μm lines, but not on 0.5272 μm. The optimum Be vapor pressure was an order of magnitude higher, 100 mtorr.

2.2.4.8. Sputtered Metal Vapor Ion Lasers

Metal atoms can be introduced into the active region of a gas discharge by sputtering them from a nearby solid metal surface. Indeed, such sputtering can be a major cause of wall erosion in high-power, segmented-bore ion lasers.[132,163,175] The sputtering is particularly efficient in a hollow-cathode configuration, where the incident ions accelerated through the full cathode fall in potential. Hollow-cathode discharges have been used for decades as sources for metal spectra; commercial versions are available for virtually every metal in the periodic chart.

Heavens and Willett[474] were the first to propose such a configuration for a laser. In 1965, they proposed to sputter aluminum with neon atoms in an HCD and excite the Al II 3p^2 ^{1}D level by collisions with neon metastables, much as in the He-Ne laser. Laser oscillation on the Al II 3p^2 ^{1}D $\to$ 3p ^{1}P line at 0.39 μm was anticipated. Evidently the experimental work was unsuccessful, but all of the essential ingredients were present in their 1965 publication. On the basis of his work with the He-Zn$^+$ laser and Duffendack's observations,[354] Collins speculated in his thesis[399a] that a cw Ne-Cu$^+$ laser might be possible.

The first successful laser using a hollow cathode configuration to sputter the active atoms was reported by Karabut *et al.*[397] in 1970; they obtained several He-Cd$^+$ and He-Zn$^+$ laser lines from a HCD without heating the active metal cathodes. This demonstration did not stimulate additional work, no doubt because vapor pressures of Cd and Zn sufficient for laser operation can be obtained easily in oven-heated or self-heated discharges. The next sputtered metal vapor laser was reported in

---

[474] O. S. Heavens and C. S. Willett, *Z. Angew. Math. Phys.* **16**, 87 (1965).

1974 by Csillag et al.,[4] who accidentally obtained oscillation on a single line in singly ionized copper while studying the He–Kr$^+$ laser system in a tube containing a hollow copper cathode. They obtained a maximum output of 0.5 mW on the Cu II 0.7808-μm 6s $^3D_3 \rightarrow$ 5p $^3F_4^o$ transition. The optimum conditions were about 20 torr helium with 0.1 torr argon or krypton added to improve the sputtering rate for Cu atoms. The output continued to increase with discharge current up to the maximum of 1.5 A used in the 1.65-mm diameter by 50-cm-long HCD.

Since 1975 there have been several publications, primarily from the groups at the Colorado State University and the National Bureau of Standards in Boulder, Colorado, and the Central Research Institute for Physics, Budapest. These groups have reported oscillation on many lines in Cu, Ag, Au, and Al in noble gas HCDs.

2.4.8.1. Copper Ion Laser. About one year after the report of a single Cu II line by Csillag et al.,[4] McNeil et al.[475] reported oscillation on 23 Cu II lines in a He–Cu$^+$ hollow cathode discharge. The transitions observed are listed in Table XVIII; they fall into four groups as illustrated in Fig. 32: six near-infrared lines originating from the 6s levels, eight blue and blue–green lines originating from the 4f levels, four near-infrared lines from the 5p levels, and three blue-green lines from the 4s^2 levels. McNeil et al. identify three different processes for these lines; they attribute the excitation of the 6s and 4f levels to direct charge-exchange collisions with He$^+$ ground-state ions, the excitation of the 5p levels to radiative cascade from the 6s levels, and the excitation of the 4s^2 levels to Penning collisions with the He(2s $^3S_1$) metastables. Note that the Wigner sum rule is obeyed for the Penning excited lines but violated for all others. Four additional laser lines in the 0.25-μm ultraviolet region originating from the 5s levels were later reported by McNeil et al.[476,476a] and Hernqvist[477] in a neon–copper discharge. These five lines are attributed to charge-exchange collisions with Ne$^+$ ground-state ions.

McNeil et al.[475] investigated two different hollow-cathode configurations; one employed a 2-mm-wide by 6-mm-deep slot cut in one side of a 16-mm-diameter solid copper bar 50 cm long, with an anode positioned 5 mm from the slot entrance. Quasi-cw oscillation was obtained on all lines with 250-μsec current pulses up to 15 A superposed on a 1-A cw keep-alive discharge. Thresholds varied from 1.3 A for the strongest

---

[475] J. R. McNeil, G. J. Collins, K. B. Persson, and D. L. Franzen, *Appl. Phys. Lett.* **27**, 595 (1975).

[476] J. R. McNeil, G. J. Collins, K. B. Persson, and D. L. Franzen, *Appl. Phys. Lett.* **28**, 207 (1976).

[476a] J. R. McNeil and G. J. Collins, *IEEE J. Quantum Electron.* **qe-12**, 371 (1976).

[477] K. G. Hernqvist, *IEEE J. Quantum Electron.* **qe-13**, 929 (1977).

TABLE XVIII. Singly Ionized Copper Laser Lines

Excitation	Upper level		Lower level	Wavelength (μm)	Threshold[g] (A)
He[+]	4f $^3H_6^o$	→	4d $^3G_5$	0.4910	8[a]
	$^3H_5^o$	→	$^3D_3$	0.4932	6[a]
	$^3G_5^o$	→	$^3G_5$	0.4855	11[a]
	$^3G_5^o$	→	$^3F_4$	0.5052	3[a]
	$^3G_4^o$	→	$^3F_3$	0.5013	10[a]
	$^3D_3^o$	→	$^3D_3$	0.5021	8[a]
	$^3P_1^o$	→	$^3S_1$	0.4682	10[a]
	$^3P_0^o$	→	$^3S_1$	0.4673	4[a]
He[+]	6s $^1D_2$	→	5p $^1F_3^o$	0.7845	7[a]
	$^3D_3$	→	$^3P_2^o$	0.7404	3.1[a], 1.3[d], 1.95[f]
	$^3D_3$	→	$^3F_4^o$	0.7808	1.3[a], 0.6[d], 0.75[f]
	$^3D_3$	→	4p″$^1F_3^o$	0.7988	8.7[a]
	$^3D_2$	→	5p $^3F_3^o$	0.7665	5.2[a], 1.3[d], 1.9[f]
	$^3D_2$	→	4p′ $^1D_2^o$	0.7779	9.3[a]
	$^3D_2$	→	5p $^3P_1^o$	0.7805	6.5[a], 2.25[f]
	$^3D_1$	→	$^3F_2^o$	0.7739	5.3[a], 1.6[d]
He[+] + cascade	5p $^3F_4^o$	→	5s $^3D_3$	0.7826	2.4[a], 1.1[d], 1.25[f]
	$^1F_3^o$	→	$^1D_2$	0.7903	9[a]
	$^3F_2^o$	→	$^3D_1$	0.7896	5.7[a], 1.5[d]
	$^3P_1^o$	→	$^3D_2$	0.7944	9[a]
Ne[+]	5s $^3D_2$	→	4p $^3F_3^o$	0.2506	40[b]
	$^3D_2$	→	$^3F_2^o$	0.2599	7[b], 16[c]
	$^3D_1$	→	$^3F_2^o$	0.2486	14[b,e]
	$^3D_1$	→	$^3D_2^o$	0.2591	6[b,e], 10[c]
	$^3D_1$	→	$^3D_1^o$	0.2703	8[c], 32[e]
He[+]	4s² $^3P_2$	→	4p $^3P_2^o$	0.4556	9[a]
	$^3P_1$	→	$^3P_2^o$	0.4506	8[a]
	$^3P_1$	→	$^3P_0^o$	0.5060	10[a]

[a] J. R. McNeil, G. J. Collins, K. B. Persson, and D. L. Franzen, *Appl. Phys. Lett.* **27**, 595 (1975).

[b] J. R. McNeil, G. J. Collins, K. B. Persson, and D. L. Franzen, *Appl. Phys. Lett.* **28**, 207 (1976).

[c] K. G. Hernqvist, *IEEE J. Quantum Electron.* **qe-13**, 929 (1977).

[d] K. Rózsa, M. Jánossy, L. Csillag, and J. Bergou, *Opt. Commun.* **23**, 162 (1977).

[e] J. R. McNeil and G. J. Collins, *IEEE J. Quantum Electron.* **qe-12**, 371 (1976).

[f] J. A. Piper and D. E. Neely, *Appl. Phys. Lett.* **33**, 621 (1978).

[g] Threshold values for a, b, e refer to a 2 mm × 6 mm × 50 cm hollow cathode; those for c refer to a 2 mm × 4.5 mm × 70 cm hollow cathode; those for d refer to a hollow-cathode-anode configuration 3.4 mm in diameter by 19 cm long; those for f refer to a hollow cathode discharge 50 cm long by 3 mm in diameter, with CuCl as the source of copper vapor.

line, 0.7808 μm, to 11 A for the weakest lines, as listed in Table XVIII. The Penning-excited lines and all of the 4f lines except those two originating from the lowest-lying 4f ^{3}H levels were obtained only in this quasi-cw fashion. The lines originating from the 6s, 5p, and 4f ^{3}H levels were also obtained in a true cw discharge. The optimum helium pressure

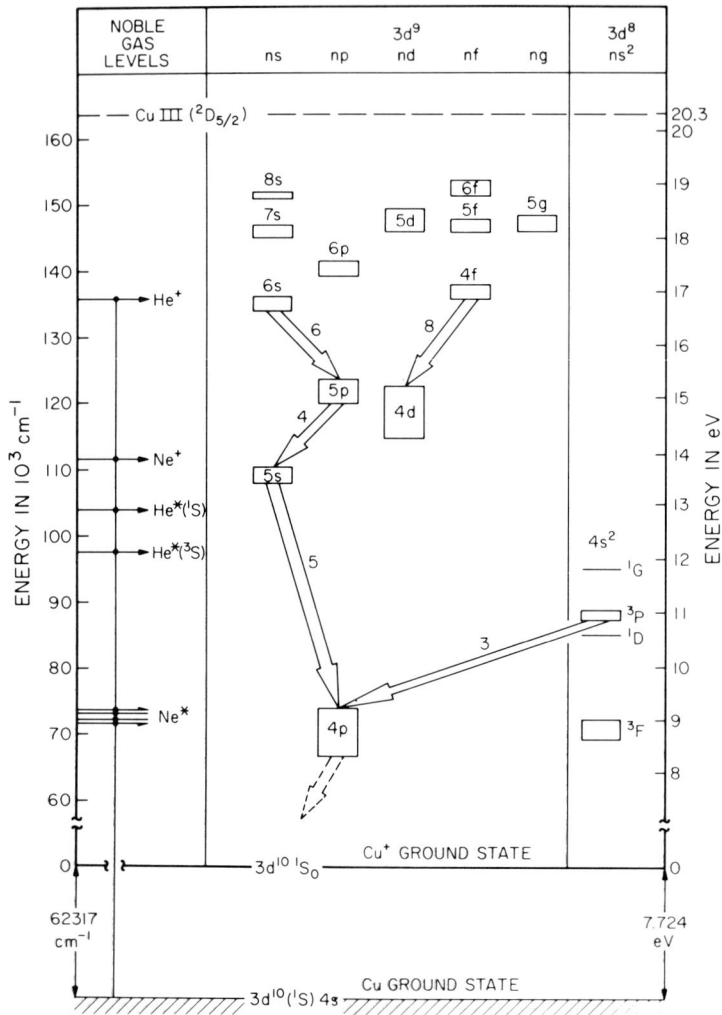

FIG. 32. Simplified energy level diagram for singly ionized copper showing the number of laser lines in each supermultiplet. The energies of the helium and neon metastables and ground-state ions are also shown with respect to the ground state of the Cu neutral atom.

was about 15 torr, but oscillation was observed from 8 to 23 torr. Mixtures of Helium and Ne, Ar, or Xe in the ratio of 8:1 were also used to lower the oscillation threshold currents by increasing the number of copper atoms sputtered by the discharge. The lowest threshold currents occurred in He:Xe and the highest in He:Ar.

A second HCD configuration investigated by McNeil et al.[475] consisted of a stack of copper disks totaling 80 cm in length. Each disk had a 20-mm-diameter hole for the laser bore; groups of adjacent disks were connected together as cathodes, with disks at either end serving as anodes. The strongest Cu II lines were obtained in cw oscillation; 150 mW output was observed on the 0.7808-μm line at 48 A discharge current with 6 torr He and 0.4 torr Ar.

In a more recent publication by the same group,[478] a gain of 1%/cm on the 0.7808-μm line is reported for a 6-mm-diameter HCD, based on observation of oscillation in an active length of 1 cm at a threshold of 0.2 A. Warner et al.[478] also report a cw output of 1 W in a HCD with a 4-mm by 12-mm rectangular cross section. Twelve individually ballasted cathodes 10 cm long were employed for a total discharge length of 120 cm. The discharge potential (cathode fall) was 400 V at 40 A current. A 20:1 mixture of He:Ar was used at 8 torr total pressure. Continuous operation for 50 hours with less than 10% variation in output power was observed.

Warner et al.[478] also report 0.5 W continuous laser output for a self-heated discharge in which the copper was thermally vaporized in a slotted molybdenum hollow cathode 9.5 mm in diameter and 13 cm in length. This discharge ran at 3.8 A and 440 V, giving five times higher efficiency than the sputtered copper laser. Pure helium at 25 torr was used. The cathode temperature was estimated to be between 1350 and 1500°C, giving a copper vapor pressure of 0.1 torr.

Rózsa et al.[479] have reported oscillation on six of the strongest Cu II lines 0.7404, 0.7665, 0.7739, 0.7808, 0.7826, and 0.7896 μm in an unusual HCD. They employ a hollow copper cathode 4.5 mm in diameter and 19 cm long; however, the anode is located coaxially *inside* the cathode and consists of a perforated thin wall tube 3.4 mm in diameter. The bright region of the discharge occurs inside the anode tube. (A similar discharge configuration had been used previously by the same group to obtain oscillation on argon,[480] krypton, and xenon[481] ion laser lines.) Threshold currents of 0.6–1.5 A were observed as listed in Table XVIII. The maximum multiline output of 30 mW was obtained with a He:Ne mixture and 10-msec, 2.4-A, 1600-V pulses.

The ultraviolet laser lines reported by McNeil et al.[476,476a] were observed in the same 2-mm by 6-mm by 50-cm rectangular slot HCD used for the blue-green and near-infrared lines.[475] Pure neon at 6–22 torr was used as the buffer gas. True cw oscillation was obtained on the 0.2591- and

---

[478] B. E. Warner, D. C. Gerstenberger, R. D. Reid, J. R. McNeil, R. Solanki, K. B. Persson, and G. J. Collins, *IEEE J. Quantum Electron.* **qe-14,** 568 (1978).
[479] K. Rózsa, M. Jánossy, L. Csillag, and J. Bergou, *Opt. Commun.* **23,** 162 (1977).
[480] M. Jánossy, L. Csillag, and K. Rózsa, *Phys. Lett. A* **63,** 84 (1977).
[481] K. Rózsa, M. Jánossy, J. Bergou, and L. Csillag, *Opt. Commun.* **23,** 15 (1977).

0.2599-μm lines, with a total of 7 mW output at 10 A and 270 V input. Peak output powers up to 210 mW were obtained with higher-current (50-A) pulses 200 μsec in duration. Although the lower laser levels are in near coincidence with the neon metastable levels, Green and Webb[355] have shown that Penning excitation between levels in such close coincidence is relatively unlikely.

Hernqvist[477] has obtained higher cw output powers on the Cu II uv lines in a HCD 2 mm by 4.5 mm by 70 cm: 34 mW at 0.2703 μm, 5 mW at 0.2599 μm, and 21 mW at 0.2591 μm, all at 20 A, 290 V input. The optimum neon pressure was 12 torr.

De Hoog et al.[482] have measured the copper vapor density by optical absorption in neon-sputtered copper HCDs similar to those employed by McNeil et al.[475,476] They find a linear dependence on discharge current above a current density of 0.1 A/cm (amperes per centimeter of discharge length along a 2-mm by 6-mm slot), with maximum values approaching 10^{14} atoms/cm^3 at 0.2 A/cm. Their observations of the variation of spontaneous emission with current for the Cu I, II and Ne I, II lines are consistent with their model of the processes in the Ne–Cu discharge.

Piper and Neely[482a] have recently obtained oscillation on five of the strongest Cu II laser lines of the 0.7-μm group in a hollow cathode discharge with thermally evaporated CuCl CuBr or CuI as the Cu source. They employed an HCD 3 mm in diameter and 50 cm in length. The operating temperature corresponded to 0.05–0.2 torr of the halide (typically 410°C for CuCl to 520°C for CuI). The optimum helium pressures were 35–50 torr, somewhat higher than those observed in the sputtered Cu laser case. The threshold currents listed in Table XVIII are lower than those observed by McNeil et al.[475] in a sputtered Cu laser of comparable dimensions, and are similar to the lowest threshold currents observed by Rózsa et al.[479] in the HAC discharge. Piper and Neely attribute this to their ability to optimize the Cu vapor pressure independently of the discharge current.

2.4.8.2. *Silver Ion Laser.* Oscillation in ionized silver was first reported by Karabut et al.[483] on two infrared lines at 0.8005 and 0.8403 μm originating from the Ag II 6p levels as listed in Table XIX. A ceramic discharge tube with a 40-cm active length was heated to about 1000°C to produce Ag vapor at a pressure of 1–10 mtorr. Helium was used as a buffer gas at about 7 torr. Lasing occured 20 μsec into the afterglow of the cur-

---

[482] F. J. de Hoog, J. R. McNeil, G. J. Collins, and K. B. Persson, *J. Appl. Phys.* **48**, 3701 (1977).

[482a] J. A. Piper and D. E. Neely, *Appl. Phys. Lett.* **33**, 621 (1978).

[483] É. K. Karabut, V. F. Kravchenko, and V. F. Papakin, *J. Appl. Spectrosc.* (*Engl. Transl.*) **19**, 938 (1973).

TABLE XIX. Singly Ionized Silver Laser Lines

Excitation	Upper level		Lower level	Wavelength (μm)	Threshold current (A)
He$^+$	5d	$^1S_0$ →	5p $^1P_1^o$	0.2243	7[a]
		$^3D_2$ →	$^3P_1^o$	0.2278	7[a]
	6p	$^3D_3^o$ →	6s $^3D_3$	0.8005	12[b], 3[c]
		$^3D_3^o$ →	$^3D_2$	0.8255	3[c], —[d]
		$^3D_2^o$ →	$^3D_2$	0.8325	—[d]
		$^3F_4^o$ →	$^3D_3$	0.8403	12[b], 1.2[c]
Ne$^+$	5s^2	$^1G_4$ →	5p $^3F_3^o$	0.3181	9.5[a]
		$^1G_4$ →	$^1F_3^o$	0.4086	35[e]
		$^1D_2$ →	$^1P_1^o$	0.4788	5.5[e]
		$^1D_2$ →	$^1D_2^o$	0.5027	14[e]

[a] J. R. McNeil, W. L. Johnson, G. J. Collins, and K. B. Persson, *Appl. Phys. Lett.* **29**, 172 (1976).

[b] É. K. Karabut, V. F. Kravchenko, and V. F. Papakin, *J. Appl. Spectrosc. (Engl. Transl.)* **19**, 938 (1973).

[c] B. E. Warner, D. C. Gerstenberger, R. D. Reid, J. R. McNeil, R. Solanki, K. B. Persson, and G. J. Collins, *IEEE J. Quantum Electron.* **qe-14**, 568 (1978).

[d] R. D. Reid, D. C. Gerstenberger, J. R. McNeil, and G. J. Collins, *J. Appl. Phys.* **48**, 3994 (1977).

[e] W. L. Johnson, J. R. McNeil, G. J. Collins, and K. B. Persson, *Appl. Phys. Lett.* **29**, 101 (1976).

rent pulse. A gain of about 5% was observed above the 12-A threshold current. Karabut *et al.* attributed the excitation of the upper laser levels to charge-exchange collisions with ground-state He$^+$ ions, and they measured a charge-exchange collision cross section of 1.5×10^{-16} cm^2.

Johnson *et al.*[484] used neon in a silver hollow-cathode discharge to produce several additional visible Ag II laser lines. In this case sputtering of the cathode material by incident neon ions produced the silver atoms. Continuous oscillation at 18 wavelengths was initially reported, but in a later publication[485] all but the three lines that could be identified with known Ag II levels were attributed to grating ghosts in the original measurements. The three classified lines originate from the 5s^2 singlet levels, which lie in close coincidence with the neon ion ground-state energy. Later work by the same group[486] reported additional lines at 0.3181 μm in a Ne–Ag$^+$ HCD and two uv lines at 0.2243 and 0.2278 μm in a He–Ag$^+$ HCD. The hollow cathodes used in these experiments consisted of a

[484] W. L. Johnson, J. R. McNeil, and G. J. Collins, *Appl. Phys. Lett.* **29**, 101 (1976).
[485] R. D. Reid, D. C. Gerstenberger, J. R. McNeil, and G. J. Collins, *J. Appl. Phys.* **48**, 3994 (1977).
[486] J. R. McNeil, W. L. Johnson, and G. J. Collins, *Appl. Phys. Lett.* **29**, 172 (1976).

2-mm-wide by 6-mm-deep slot milled in a solid silver bar 50 cm long. A stainless steel mesh anode extended the full length of the cathode. Currents up to 10 A were used to obtain continuous oscillation, and current pulses up to 50 A were reported for the three visible lines[484] and 350 mW for the 0.3181-μm lines.[486] Oscillation was observed with pure neon over the range 6–34 torr, with the lowest threshold currents at 12–14 torr and the highest output power at 26 torr. For the He–Ag$^+$ laser a 200:1 mixture of He:Ne or Ar was used to enhance the sputtering rate. The pressure behavior for He–Ag$^+$ laser was similar to that of the Ne–Ag$^+$ laser. Reid et al.[485] have listed four Ag II infrared lines as oscillating in a He–Ag$^+$ discharge, including the two previously reported by Karabut et al.,[483] but they give no details on pressures, thresholds, or powers. Warner et al.[478] later obtained 1 W peak power from three of these lines, 0.8005, 0.8255, and 0.8403 μm in the same 2-mm by 6-mm by 50-cm HCD as that used in references 484 and 486. They used 60-A, 200-μsec pulses superposed on a 1-A keep-alive current. An average output of 5 mW was obtained at 25 Hz repetition frequency. A 100:1 He:Xe mixture was used at 12 torr to enhance the sputtering efficiency.

2.4.8.3. Gold Ion Laser. Reid et al.[487] have also reported 20 transitions in ionized gold vapor in a He–Au HCD. Fifteen of these lines were obtained on a cw basis and five on a quasi-cw basis (200-μsec pulses). Nine of the wavelengths lie between 0.25 and 0.30 μm, six between 0.55 and 0.7 μm, and five between 0.75 and 0.77 μm. Because of the rather poor state of knowledge about the Au II spectrum, only eight of the lines could be associated with an identified upper or lower level, and none could be fully classified. Six of the partially classified lines originate from the 7s $^1D_2$ and $^3D_1$ levels; these lie less than 2000 cm^{-1} below the He$^+$ ion ground-state energy, so that charge-exchange excitation seems to be the mechanism responsible, as proposed by Reid et al. Similarly, two of the near-infrared lines originate from known but unclassified levels lying near the He$^+$ energy.

The same dimensions for the gold hollow cathode were employed by Reid et al. as in their previous work on Cu and Ag. Gold 0.1 mm thick was plated onto the base metal of the cathode. Oscillation was obtained with helium pressures of 8–35 torr, exhibiting a maximum at 25 torr. A 200:1 mixture of He:Ar was used to enhance the sputtering rate. A maximum of 125 mW peak output was obtained for the 0.25–0.29-μm lines with 50-A, 200-μsec pulses. Threshold currents varied from 1.3–35 A.

2.4.8.4. Aluminum Ion Laser. Heavens and Willett[474] were the first to propose a sputtered-metal, hollow-cathode discharge laser, a Ne–Al$^+$

[487] R. D. Reid, J. R. McNeil, and G. J. Collins, Appl. Phys. Lett. **29**, 666 (1976).

2.4. IONS EXCITED BY ATOMIC COLLISION

TABLE XX. Singly Ionized Aluminum Laser Lines

Excitation	Upper level		Lower level	Wavelength (μm)	Threshold current (A)
He$^+$	4f $^3F_4^o$	→	3d $^3D_3$	0.3587	24[a], 33[b]
	$^1F_3^o$	→	$^1D_2$	0.7471	5.4[a], 10[b], 5.2[c], 32[d]
	5s $^1S_0$	→	4p $^1P_1^o$	0.6920	4.8[a], 8[b], 5[c]
Ne$^+$	4p $^3P_2^o$	→	4s $^3S_1$	0.7042	3.5[a], 15[b], 2.8[c]
+ cascade	$^3P_1^o$	→	$^3S_1$	0.7057	11.5[a]

[a] D. C. Gerstenberger, R. D. Reid, and G. J. Collins, *Appl. Phys. Lett.* **30**, 466 (1977).
[b] W. K. Schuebel, *Appl. Phys. Lett.* **30**, 518 (1977).
[c] K. Rózsa, M. Jánossy, L. Csillag, and J. Bergou, *Phys. Lett.* A **63**, 231 (1977).
[d] Threshold with helium instead of neon, as reported in Gerstenberger *et al.*[a]

system in which the Al II 3p^2 ^{1}D level would be excited by Penning collisions with neon metastables. The first successful aluminum ion laser was demonstrated by Gerstenberger *et al.*,[488] who reported oscillation on the five Al II lines listed in Table XX, and independently by Schuebel[489] who observed four of the same five lines. Three of these lines originate from the 4f and 5s levels lying less than 3000 cm^{-1} below the Ne$^+$ ion ground-state energy. The remaining two originate from the 4p levels lying approximately 20,000 cm^{-1} below the Ne$^+$ level; they are likely to be populated by cascade from the charge-exchange excited states. Gerstenberger *et al.* were able to obtain only the 0.7471-μm line in a He–Al HCD, and then only at much higher threshold currents, indicating that charge exchange with Ne$^+$ is the dominant mechanism.

Gerstenberger *et al.* observed cw oscillation on the 0.6920-, 0.7042-, and 0.7471-μm lines with a maximum of 1 mW at 10 A current. A maximum of 1 mW was observed on the 0.3587-μm line with 50-A, 200-μsec pulses superposed on a 1-A cw keep-alive current, and 5 mW was observed on the infrared lines under the same conditions. Their HCD consisted of a 2-mm-wide by 6-mm-deep slot milled in the side of a 2.5-cm-diameter, 50-cm-long water-cooled aluminum alloy rod. The alloy used was 90% Al, 5% Cu, 2% Mg, and 3% other metals. The optimum neon pressure was 4 torr, substantially below the optimum values observed[476,486] for Ne–Cu$^+$ and Ne–Ag$^+$ lasers. Gerstenberger *et al.* argue that this lower optimum pressure results from the difficulty with which aluminum is sputtered compared to Cu or Ag, about 2.5 times lower yield for the same incident ion energy.[490] By reducing the pressure, the energy

[488] D. C. Gerstenberger, R. D. Reid, and G. J. Collins, *Appl. Phys. Lett.* **30**, 466 (1977).
[489] W. K. Schuebel, *Appl. Phys. Lett.* **30**, 516 (1977).
[490] N. Laegrid and G. K. Wehner, *J. Appl. Phys.* **32**, 365 (1961).

of the ions striking the cathode is increased, thereby increasing the number of Al atoms sputtered for a given discharge current. The oxide film that forms readily on aluminum also presents some difficulty in obtaining stable sputtering conditions. Gerstenberger et al. used both mechanical polishing and discharge cleaning with Ar–H_2 mixtures prior to laser operation to remove the oxide layers.

Schuebel[489] also employed an HCD with a 2-mm by 6-mm-deep slot milled in a 12-mm-diameter water-cooled aluminum bar 30 cm long. Gain values of 12, 2.2, and 12%/m were reported for the 0.6920-, 0.7402-, and 0.7471-μm lines, respectively. Schuebel used currents up to 48 A, pulse lengths up to 3 msec, and repetition rates up to 700 Hz with no indications of limitations that would prevent true cw operation. He observed a linear dependence of output power on discharge current above the threshold values listed in Table XX. A 20:1 mixture of Ne:H_2 was found to lower the threshold currents below their values for pure neon. Schuebel attributes the improvement with H_2 to an increased sputtering rate resulting from an increase in the cathode fall voltage and hence an increase in the energy of the neon ions striking the cathode. The optimum mixture pressures were about 3 torr.

Rózsa et al.[491] observed three of the Al II laser lines in their internal-anode HCD configuration, with thresholds as listed in Table XX. They used a 7-mm-diameter cathode 40 cm long with six 1-mm-diameter tungsten anode rods positioned 0.5 mm from the cathode surface. The bright discharge region occurred on the axis of the cathode cylinder. The discharge voltages were greater than 2 kV, much higher than the 300–700 V values typical of the slot HCDs.[489,490] The optimum neon pressures were 1–1.4 torr, with 0.02-torr Ar added to aid sputtering.

[491] K. Rózsa, M. Jánossy, L. Csillag, and J. Bergou, *Phys. Lett. A* **63**, 231 (1977).

3. SOLID STATE LASERS*

3.1. Introduction

Historically, the first laser was a solid state laser. Stimulated emission from synthetic ruby (Cr^{3+}–doped Al_2O_3), a material whose spectroscopic properties were well known, was reported in 1960.[1] This was quickly followed by the report of laser action from U^{3+} in single-crystal CaF_2.[2] The absorption and fluorescence properties of rare earth ions had been the subject of intense study in the preceding decade and therefore in the next year, 1961, laser action was reported from divalent samarium in CaF_2[3] and trivalent neodymium in glass[4] and in $CaWO_4$.[5] The remainder of the decade of the 1960s saw an intensive search for other solid state laser ions and hosts.[6-8] This effort has culminated in laser action from ions of several transition metal groups in more than 100 different crystals and innumerable glasses. In this chapter the general properties and operating characteristics of solid state lasers composed of paramagnetic ions in insulating solids are surveyed. Laser action in semiconducting materials and involving color centers is discussed in separate chapters.

The spectral range of solid state lasers is shown in Fig. 1 and extends from approximately 0.5 to 3.0 μm, a wavelength factor of six. While this is small compared to that for gas or semiconductor lasers, it is larger than that for organic dye and many other specific types of lasers. The operating wavelengths of some selected lasers are included in Fig. 1. Solid state lasers have been reported that operate at intervals of about every 0.1 μm or less throughout the range shown.

[1] T. H. Maiman, *Nature (London)* **187**, 493 (1960).
[2] P. P. Sorokin and M. J. Stevenson, *Phys. Rev. Lett.* **5**, 557 (1960).
[3] P. P. Sorokin and M. J. Stevenson, *IBM J. Res. Dev.* **5**, 56 (1961).
[4] E. Snitzer, *Phys. Rev. Lett.* **7**, 444 (1961).
[5] L. F. Johnson, G. Boyd, K. Nassau, and R. Soden, *Proc. IRE* **50**, 213 (1962).
[6] L. F. Johnson, *in* "Lasers" (A. K. Levine, ed.), Vol. 1, p. 137. Dekker, New York, 1966.
[7] E. Snitzer and G. Young *in* "Lasers" (A. K. Levine, ed.), Vol. 2, p. 191. Dekker, New York, 1966.
[8] R. B. Chesler and J. E. Geusic, *in* "Laser Handbook" (F. T. Arecchi and E. O. Schulz-DuBois, eds.), Vol. 1, p. 325. North-Holland Publ. Amsterdam, 1972.

* Part 3 is by **M. J. Weber**.

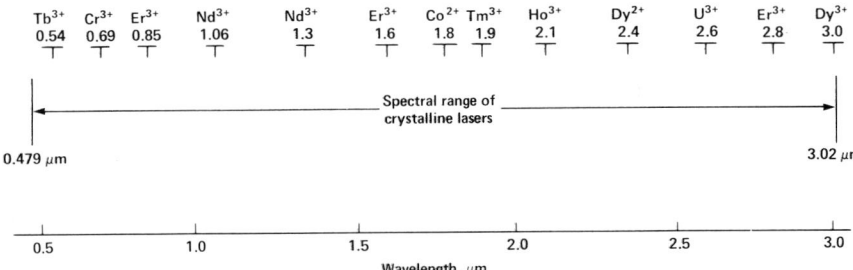

FIG. 1. Spectral range and wavelength of selected solid state lasers.

While the number of solid state lasers investigated is large, the number of technologically important solid state lasers is small. Pulsed ruby lasers were developed extensively in the early 1960s.[9] As the applications and requirements for lasers became more explicit, desirable properties became better defined and the choice of laser materials more selective. In the mid-1960s, lasers with neodymium in yttrium aluminum garnet ($Y_3Al_5O_{12}$ or YAG) appeared.[10] The properties of this crystal were so favorable that Nd:YAG lasers quickly became the dominant solid-state source of continuous and pulsed coherent radiation at moderate powers. Neodymium glass lasers with their good energy storage capabilities provide the highest peak powers and shortest pulse durations.[7] With the advent of integrated optics, miniature lasers using stoichiometric materials[11] have received increasing attention. Today, commercial solid state lasers are limited principally to Nd:YAG, Nd:glass, and ruby.

A solid state laser consists of a crystalline or amorphous material, usually in the form of a cylindrical rod, containing a paramagnetic lasing ion. The optical resonance cavity is formed either by depositing mirrors on the ends of the rod or by placing the rod between two external mirrors. The lasing material is excited by optical pumping with continuous or

[9] V. Evtuhov and J. K. Neeland, in "Lasers" (A. K. Levine, ed.), Vol. 1, p. 1. Dekker, New York, 1966.
[10] J. E. Geusic, H. Marcos, and L. Van Uitert, Appl. Phys. Lett. **4**, 182 (1964).
[11] H. G. Danielmeyer, "Stoichiometric laser materials," in Festkörperprobleme Adv. Solid State Phys. **15**, 253 (1975).

3.1. INTRODUCTION

pulsed lamps or, in special cases, with another laser. The most commonly used pumping configuration is a reflector in the form of an elliptical cylinder with the laser rod at one focus and the lamp at the other. Auxiliary equipment includes power supplies for the pumping sources, optical switches and modulators, controls, and, depending upon the repetition rate, air or water cooling of the laser rod and lamps. Engineering developments with respect to sources and techniques for optical pumping, cavity design, Q-switching, mode locking, and stabilization have resulted in rugged, reliable solid-state lasers for myriad applications.[12] These include laboratory use and research (optical spectroscopy, nonlinear optics, holography, laser-plasma interactions, inertial confinement fusion, medical), materials processing (cutting, scribing, drilling, welding), communications (integrated optics, high-data-rate transmission, satellite communication systems), and military (rangefinders, target designators).

The materials and operating conditions of solid state lasers differ in many respects from those of other types of lasers. General characteristics include:

Laser ions: Principally trivalent rare earths; divalent rare earth, iron group, and actinide ions have also been used.

Hosts: Oxide and fluoride crystals; silicate, phosphate, and other glasses.

Size: Volumes of the lasing media range from $\sim 10^{-8}$ cm³ in miniature stoichiometric lasers to $\sim 10^4$ cm³ for individual Nd:glass disks in large amplifiers.

Excitation: Optical pumping by flashlamps, continuous lamps, or other lasers.

Output: Continuous powers to several hundred watts; pulses as short as picoseconds with peak powers exceeding gigawatts.

The large lasing ion densities and long fluorescence lifetimes of many solid state lasers combine to provide high energy storage. Using Q-switching, mode-locking, and cavity-dumping techniques, this stored energy can be released in varying time intervals and can produce extremely high peak powers. Other general properties that can be contrasted with corresponding values for other types of lasers are summarized in Table I. Since properties of the crystalline or glass host, ion type, and mode of operation vary widely, the table only indicates the realm of demonstrated values.

Given the above characteristics, the physical processes specific to

---

[12] W. Koechner, "Solid State Laser Engineering." Springer-Verlag, Berlin and New York, 1976.

Table I. General Properties of Solid State Lasers

Wavelength range	0.48–3.0 μm
Stimulated emission cross section	10^{-18}–10^{-20} cm²
Inversion for 1% gain/cm	10^{16}–10^{18} ions cm^{-3}
Fluorescence lifetime (storage time)	10^{-5}–10^{-2} sec
Laser ion density	10^{18}–10^{22} cm^{-3}
Energy storage	≤1 J/cm³
Efficiency	≤3% overall

solid-state lasers that dictate their limits are examined in Chapter 3.2. These include the general energy-level schemes and radiative and nonradiative transition probabilities of paramagnetic ions in solids. These, in turn, depend upon the particular laser ion and host. The lasers considered here are insulating solids, either ordered or disordered materials, in which the active ion is present either as an impurity or, in stoichiometric materials, as a component of the host. Laser materials—both the active ions used and the hosts employed—are surveyed in Chapter 3.3. This is followed in Chapter 3.4 by a discussion of the principal lasers in use today, their operating properties and characteristics, and comparative features. Other lasers and materials available are also surveyed. The chapter concludes with a brief review of hazards associated with the operation of solid-state lasers.

In this part many essential features and properties of solid state lasers are not covered in detail. References to review articles on specific topics are given throughout as the topics appear. A good, up-to-date survey of the design, operating techniques and applications of solid state lasers is presented by Koechner.[12] This book is concerned mainly with engineering aspects but also includes background information.

3.2. Physical Processes

The lasing schemes of solid state lasers obey the same general principles of other lasers, but the physical processes determining the energy levels and transition probabilities are different. The gain of the media is given by a product of the stimulated emission cross section and the population inversion. For solid state lasers, population inversion is achieved by optical pumping and is dependent upon the absorption spectrum of the laser ion in the host, the spectral match of this spectrum with the spectrum of the pumping source, the lifetime of the upper laser level, which determines the pumping rate required, and the quantum efficiency. The last quantity includes the fluorescence conversion efficiency (the number of ions excited to the fluorescing level per incident pump photon absorbed) and the quantum efficiency of the fluorescing level (the fractional number of photons emitted per excited ion). For low-threshold laser operation it is desirable to have

(1) large stimulated emission cross section (narrow fluorescence linewidth),
(2) broad (or numerous) absorption bands for optical pumping
(3) high fluorescence conversion efficiency,
(4) high fluorescence quantum efficiency,
(5) four-level lasing scheme (terminal laser level at energy $\gg kT$), and
(6) no excited-state absorption from the upper laser level.

The spectroscopic properties of ions in solids and the ability to satisfy the above criteria are affected by the interaction of the ion with its local environment. To illustrate this and the essential features of three- and four-level solid state lasers, we shall use Cr^{3+} in Al_2O_3 and Nd^{3+} in $Y_3Al_5O_{12}$ as paradigms.

3.2.1. Energy Levels

The electronic energy levels of a free ion are determined by the electrostatic interaction of the electrons V_{el} and the spin–orbit coupling V_{so}. When the ion is introduced into a solid, it is subject to an additional electric field due to the neighboring ions. This crystalline Stark field V_{cf} reduces the degeneracy of the free-ion states and causes further energy level splittings.

3.2.1.1. Iron Group Ions.
Iron transition group ions have a $1s^22s^22p^63s^23p^63d^n$ electronic configuration. For these ions the outer 3d electrons are strongly affected by the local crystal field and the relative strengths of the interactions are $V_{cf} > V_{el} \gg V_{so}$.[13] When an ion such as trivalent chromium ($3d^3$) is substituted into a solid such as Al_2O_3, the local field splits the ^4F and ^2G free-ion states and produces the energy level scheme shown in Fig. 2. The actual field at the Cr^{3+} site has trigonal symmetry that causes further small splittings of the ^2E state into $2\overline{A}$ and $\overline{E}$ states (not shown).

Lattice vibrations cause time-dependent fluctuations of the crystal field. The resulting energy level variations appear as line broadening. This accounts for the width of the ^4T levels in Fig. 2. In the strong-field limit, the separations of certain levels such as $^4A_2$, ^2E, $^2T_1$, and $^2T_2$ are approximately independent of the crystal field. The linewidths of transi-

---

[13] J. S. Griffith, "The Theory of Transition Metal Ions." Cambridge Univ. Press, London and New York, 1961.

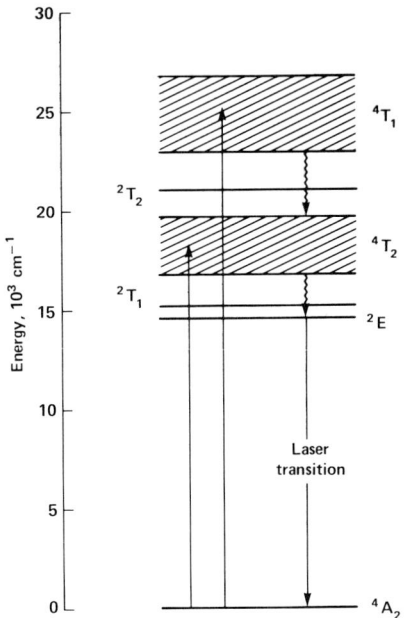

FIG. 2. Energy level diagram of Cr^{3+} in Al_2O_3 illustrating a three-level laser scheme. Wavy lines denote nonradiative transitions.

tions between these levels are narrow. The absorption spectrum of $Cr^{3+}:Al_2O_3$ is shown in Fig. 3. The broad bands centered at about 550 and 400 nm correspond to the $^4A_2 \to {}^4T_2$ and $^4A_2 \to {}^4T_1$ transitions, respectively; the weak, sharp lines at 470 and 694 nm are the $^4A_2 \to {}^2T_2$ and $^4A_2 \to {}^2E$ transitions.

The energy level diagram of Cr^{3+} provides an ideal three-level laser scheme. Optical pumping is via the broad $^4T_2$ and $^4T_1$ bands in the visible and higher-lying absorption bands. Ions excited into these bands rapidly decay nonradiatively to the ^2E level (<1 nsec). Since electric-dipole $^2E \to {}^4A_2$ transitions are both parity and spin forbidden, the ^2E state is metastable with a long lifetime (≈ 3 msec). Sharp emission lines (called R lines) are observed from ^2E to $^4A_2$ and are used for laser action. Characteristic of a three-level laser scheme, to achieve threshold for stimulated emission approximately one-half of the ions in $^4A_2$ must be pumped into the ^2E state. Absorption at the lasing wavelength from ^2E to higher energy levels is possible and may reduce the net gain or prevent oscillation. In ruby the effect of excited-state absorption is very weak.

In addition to purely electronic energy states and transitions, the ion interacts with lattice vibrations to produce electronic-vibrational product states. Thus, associated with each electronic level in Fig. 2 is a series of

3.2. PHYSICAL PROCESSES

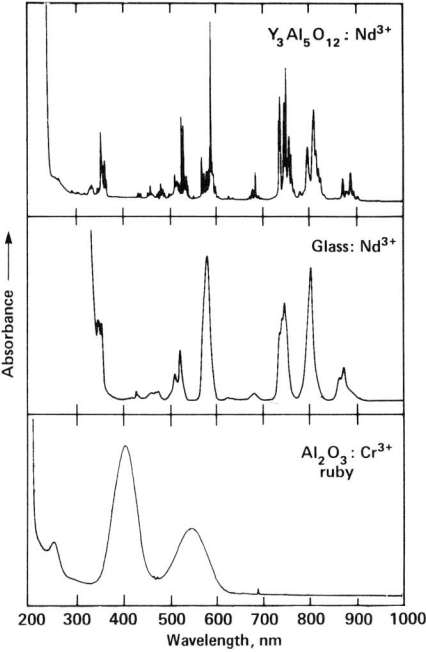

FIG. 3. Comparison of the absorption spectra of solid state laser materials: Top, Nd^{3+} in $Y_3Al_5O_{12}$ (YAG); middle, Nd^{3+} in phosphate glass (LHG-7); bottom, Cr^{3+} in Al_2O_3 (ruby). All spectra recorded at 300 K.

vibronic levels. The transition from the lowest level of ^2E to the lowest level of $^4A_2$ is the zero-phonon line; transitions terminating on vibronic levels of $^4A_2$ appear as sidebands of the zero-phonon line. The intensity of these sidebands depends on the strength of the ion–phonon coupling and varies with host. The threshold for laser action involving vibronic transitions depends on the transition probability and the thermal Boltzman population in the terminal level. Phonon-terminated laser action is observed for several iron group ions.

3.2.1.2. **Rare Earth Ions.** Lanthanide series ions have an electronic configuration $1s^2 2s^2 2p^6 3s^2 3p^6 3d^{10} 4s^2 4p^6 4f^n 5s^2 5p^6$. Optical transitions occur between states of the 4f electrons. These electrons are partially shielded from the crystal field by the outer $5s^2 5p^6$ shells. Therefore, in contrast to other transition group ions, the Stark splitting of the free-ion levels is small ($\sim 10^2$ cm^{-1}), $V_{el} \gg V_{so} \gg V_{cf}$, and J is a good quantum number.[14] The positions of the free-ion states of trivalent rare earths are

---

[14] B. G. Wybourne, "Spectroscopic Properties of Rare Earths." Wiley (Interscience), New York, 1965.

determined by electrostatic and spin–orbit interactions and are well established up to $\approx 40,000$ cm^{-1}.[15] Because the crystal-field interaction is small, the centers of gravity of the J states exhibit only small variations (a few hundred cm^{-1}) from host to host; the details of the Stark splitting, however, are dependent on the strength and symmetry of the local field. Energy levels associated with the excited $4f^{n-1}5d$ configuration occur as higher energies. In many hosts these levels are near or above the fundamental absorption edge of the host and therefore not useful for optical pumping (exceptions are wide-bandgap hosts, such as fluoride crystals).

Trivalent rare earths offer many possibilities for stimulated emission because they have a large number of energy levels in the optical region for pumping, metastable states with high quantum efficiency, and terminal levels with energies sufficiently high about the ground state to have negligible population at ambient temperatures. Neodymium epitomizes these properties.

The lower energy levels of Nd^{3+} in $Y_3Al_5O_{12}$ are shown in Fig. 4. The extent of the Stark splitting is indicated by the width of the bands; explicit examples of the Stark structure for the $^4I_{9/2}$, $^4I_{11/2}$, and $^4F_{3/2}$ states are shown at the right. The natural linewidths of transitions between Stark levels are small, ≤ 10 cm^{-1},[16] and are resolved in the absorption spectrum of $Nd:Y_3Al_5O_{12}$ in Fig. 3. Approximately one-half of the light in the 400–900-nm region is absorbed by Nd^{3+} ions at concentrations typically used in laser materials. Optically excited Nd^{3+} ions rapidly relax nonradiatively to the $^4F_{3/2}$ level from which fluorescence occurs to levels of the ^4I multiplet. The most intense fluorescence is to the $^4I_{11/2}$ manifold; this spectrum for $Nd^{3+}:Y_3Al_5O_{12}$ is included in Fig. 5. The strongest line at 1064 nm arises from a superposition of the two transitions shown in Fig. 4. The $^4I_{11/2}$ state is sufficiently far removed from the ground state that its room temperature population is very small. The threshold for 4-level laser action is very low. Transitions to $^4I_{13/2}$, $^4I_{15/2}$, and high-lying Stark levels of $^4I_{9/2}$ can also be used for laser action. Relaxation of the terminal laser level to complete the four-level lasing scheme occurs nonradiatively by phonon processes.

For divalent rare earths, the spin–orbit constants are smaller and the separations of the free-ion states of $4f^n$ are reduced.[15] In addition, the $4f^{n-1}5d$ bands occur at lower energies and, for many ions, extend into the visible. Since these are broad, parity-allowed electric-dipole transitions, they provide strong absorption bands for optical pumping. Lasing transi-

---

[15] G. H. Dieke, "Spectra and Energy Levels of Rare-Earth Ions in Crystals." Wiley, New York, 1968.

[16] See, for example, T. Kushida, *Phys. Rev.* **185,** 500 (1969).

3.2. PHYSICAL PROCESSES

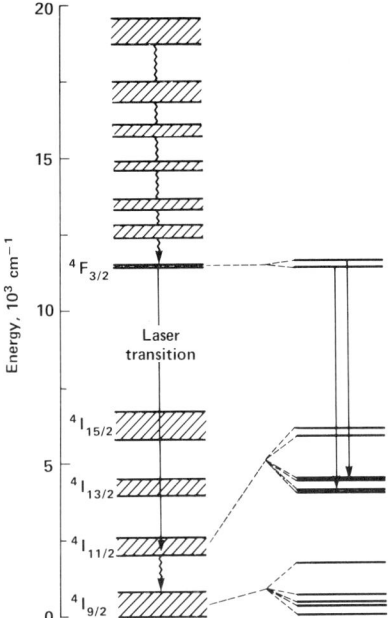

FIG. 4. Energy level diagram of Nd^{3+} in $Y_3Al_5O_{12}$ illustrating a four-level laser scheme. Wavy lines denote nonradiative transitions.

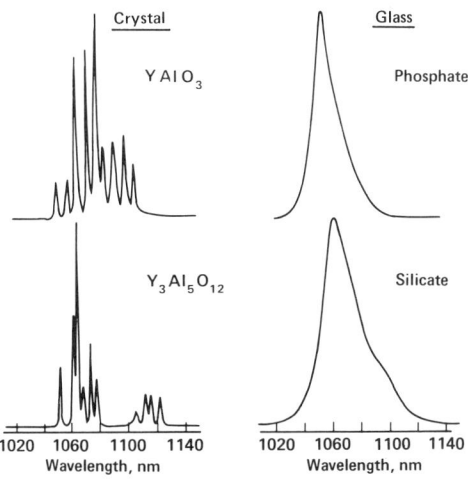

FIG. 5. Comparison of the $^4F_{3/2} \rightarrow {}^4I_{11/2}$ fluorescence spectra of Nd^{3+} in different oxide crystals and glasses at 300 K.

tions between Stark levels of $4f^n$ occur as described for trivalent rare earths.

The energy levels of paramagnetic ions in amorphous materials are essentially the same as in crystals. However, laser ions enter glasses as network modifier ions and the number, type, and spatial distribution of neighboring ions differ from site to site. As a consequence, the Stark splittings vary. An example of the dependence of the Stark splitting and transition probabilities on lattice site is illustrated in Fig. 5 by the $4F_{3/2} \rightarrow {}^4I_{11/2}$ fluorescence spectra of Nd^{3+} in two different yttrium aluminum oxide crystals. Neodymium ions substitute for Y^{3+} and have 8-fold nearest-neighbor oxygen coordination in both crystals. It is the difference in the site symmetry—D_2 in $Y_3Al_5O_{12}$ and C_{1h} in $YAlO_3$—and the more distant neighbor distances that cause the spectral changes. In a glass the number of physically inequivalent sites is very large. This results in inhomogeneous broadening of optical transitions. In Figs. 3 and 5 the Stark structure in the Nd:glass spectra is poorly resolved. The large inhomogeneous linewidths in glass increase the spectral coverage for optical pumping, but the effective cross sections for absorption and emission are correspondingly smaller. Note in Fig. 5 that the extent of the Stark splitting, as reflected in the emission bandwidth, and the wavelengths of transitions between Stark manifolds vary with the host glass.

Inhomogeneous line broadening in crystals due to strains or other imperfections is generally small, ≤ 1 cm^{-1}.

3.2.2. Transition Probabilities

3.2.2.1. Radiative Transitions.

Optical transitions between electronic states of laser ions in solids, either those used for pumping or for stimulated emission, are predominantly of electric-dipole nature. Although transitions between states of the same electronic configuration are parity-forbidden by Laporte's rule, if the paramagnetic ion is located in a noncentrosymmetric site, odd-order terms in the expansion of the static or dynamic crystal field admix states of higher, opposite-parity configurations (4p for Cr^{3+} and $4f^{n-1}5d$ for rare earths) into the ground configuration and electric-dipole transitions become allowed.[17] Spin forbiddenness of transitions is reduced by admixing of different spin states by the spin–orbit interaction. The resulting oscillator strengths of forced electric-dipole transitions are dependent on the degree of admixing by the spin–orbit and crystal field interactions which in turn are host-dependent. In the case of rare earth ions, the oscillator strengths of transitions

[17] D. S. McClure, "Electronic Spectra of Molecules and Ions in Crystals." Academic Press, New York, 1959.

between J states are small, $\sim 10^{-6}$. For iron group ions, the oscillator strengths are large for spin-allowed transitions and small for spin-forbidden transitions.

The spectra intensities of f–f transitions of rare earths have been treated using a phenomenological approach of Judd[18] and Ofelt.[19] The electric dipole line strength is expressed as a sum of products of experimentally derived intensity parameters and matrix elements of tensor operators connecting states of the $4f^n$ configuration. Intensity parameters determined for a given ion-host are used to calculate the probability of transitions between any $4f^n$ levels of interest for laser action. This includes absorption and fluorescence intensities, excited-state absorption, radiative lifetimes and branching ratios, and, combined with fluorescence spectra, stimulated emission cross sections. Since the Judd–Ofelt parameters do not differ greatly for adjacent ions in the lanthanide series, estimates can be made using extrapolated values. The validity and application of the Judd–Ofelt theory is reviewed in Peacock.[20]

Electric-dipole transitions between 4f and 5d configurations of rare earths are parity-allowed. The oscillator strengths for f–d transitions are, therefore, much larger than for f–f transitions and have magnitudes of 10^{-1} to 10^{-2}. Emission from 5d states, while not common, has been observed for several rare earths where there is a large energy gap from the lowest 5d state to lower-lying 4f states.[21,22]

Magnetic-dipole and electric-quadrupole transitions are allowed between states of the same configuration and can be calculated, given appropriate eigenstates.[14] Since magnetic-dipole transitions are subject to $|\Delta J| \leq 1$ and $\Delta S = \Delta L = 0$ selection rules, they are usually only significant for intramultiplet transitions. Electric-quadrupole transitions are much less probable than dipolar processes and are generally neglected.

In glasses, transition probabilities as well as energy levels vary from site to site.[23] Following broadband, nonselective excitation, the fluorescence decay consists of a sum of exponentials rather than a simple exponential time dependence.

3.2.2.2. Nonradiative Transitions. Relaxation of electronic states by phonon processes arises from the interaction of the ion with the fluctuating crystalline field. In these processes the electronic energy of the excited ion is transferred to vibrational energy of the host lattice. Nonra-

---

[18] B. R. Judd, *Phys. Rev.* **127**, 750 (1962).
[19] G. S. Ofelt, *J. Chem. Phys.* **37**, 511 (1962).
[20] R. D. Peacock, *Struct. and Bonding (Berlin)* **22**, 83 (1975).
[21] M. J. Weber, *Solid State Commun.* **12**, 741 (1973).
[22] K. H. Yang and J. A. DeLuca, *Phys. Rev. B* **17**, 4246 (1978).
[23] C. Brecher and L. A. Riseberg, *Phys. Rev. B* **13**, 81 (1976).

diative relaxation between levels separated by energies up to a few hundred cm^{-1} occur by one- and two-phonon processes. These direct and Raman processes can be very fast, $\leq 10^{-11}$ sec, and are the origin of natural line broadening. The linewidth of transitions is determined by broadening of both initial and final states. At low temperatures, only spontaneous phonon processes are active; as the host temperature is increased, stimulated phonon processes contribute. In ruby, the homogeneous linewidth of the R lines is caused by fast phonon processes, which limit the lifetimes of ions in the $2\overline{A}$ and $\overline{E}$ states. For rare earth ions, Stark level splittings are typically ≤ 100 cm^{-1}. Since this is within the range of the single-phonon spectrum, nonradiative transitions between Stark levels are very fast and result in lifetime broadening.[16]

Nonradiative relaxation between levels separated by several hundreds to thousands of cm^{-1} requires the simultaneous emission of many phonons to conserve energy and hence proceed at much slower rates. Multiphonon relaxation processes determine three important laser properties: pump conversion efficiency, radiative quantum efficiency, and for four-level lasing schemes, the lifetime of the terminal laser level. These transitions are indicated by wavy lines in Figs. 2 and 4.

The ion–phonon interaction causing transitions between 4f levels of rare earths is characteristic of weak coupling. In higher-order processes, the detailed properties of the individual electronic states and phonon modes involved are averaged out, and the most important factor influencing the rate is the energy gap to the next-lower level. Experimental studies have shown that rare earth decay rates exhibit an approximate exponential dependence on energy gap ΔE to the next-lower level of the form $W(T)\exp(-a\,\Delta E)$, where the constants $W(T)$ and a are dependent on the host and the strength of the ion–lattice coupling but not, with few exceptions, on the specific rare earth ions or electronic state.[24] Because higher-energy vibrations conserve the transition energy in a lower-order process, multiphonon relaxation is stronger in materials with high frequency vibrations. In glasses, multiphonon relaxation is site-dependent and is due predominantly to vibrational modes associated with molecular groups of the glass network former (e.g., for silicate glasses the SiO_4^{4-} tetrahedron). Depending on the glass, these modes have frequencies ranging from ~ 700 to 1400 cm^{-1}.[25] In both crystalline and amorphous hosts, materials having low vibrational frequencies generally have more fluorescing levels because of the smaller multiphonon emission rates and hence more possibilities for efficient laser action.

[24] See L. A. Riseberg and M. J. Weber, *in* "Progress in Optics" (E. Wolf, ed.), Vol. 14, p. 91. North-Holland Publ., Amsterdam, 1976, and references therein.
[25] C. B. Layne, W. H. Lowdermilk, and M. J. Weber, *Phys. Rev. B* **16**, 10 (1977).

3.2. PHYSICAL PROCESSES

Rare earth ions in 5d states interact with the local field much more strongly than ions in 4f states. Nonradiative decay from 5d levels is characteristic of intermediate-strength rather than weak coupling.[24] For most divalent and trivalent lanthanide ions, levels of the $4f^{n-1}5d$ configuration overlap those of the $4f^n$ configuration. Thus excitation cascades from 5d to 4f states and 5d bands can be used for optical pumping.

Transition metal ions interact more strongly with the crystal field than rare earths ions. Multiphonon processes are very probable and only states with very large energy gaps to the next-lower level ($\geq 10^4$ cm^{-1}) have sufficient radiative quantum efficiency to be readily observable in fluorescence.

Nonradiative processes are a major factor in determining the long-wavelength operating limit of solid state lasers.[26] The probability for radiative transitions is proportional to $(\Delta E)^3$ and decreases as the energy level separation becomes smaller; the probability for nonradiative transitions, on the other hand, increases. Eventually the radiative quantum efficiency becomes too small to be useful for practical lasers.

Because of the ion–lattice interaction, both the frequency and linewidth of optical transitions of paramagnetic ions in solids are temperature-dependent. For rare earths, the frequency shifts are typically ≤ 10 cm^{-1} between 0 and 300 K.[16] Line narrowing by cooling increases the stimulated emission cross sections and can result in lower-threshold operation. The rate of multiphonon relaxation is also temperature dependent and is a strong function of temperature for high-order processes.[24]

3.2.3. Ion–Ion Interactions

Nonradiative energy transfer between like and unlike paramagnetic ions occurs at concentrations where the average ion separations become small (≤ 1 nm) and interactions via electric multipolar or exchange forces are sufficiently strong to cause transitions with rates comparable with radiative rates. Depending upon the ions and transitions involved, this can lead to energy migration, fluorescence quenching, or an increase in the optical pumping efficiency by fluorescence sensitization. All of these processes are operative in various solid state lasers.

To illustrate the effects of ion–ion energy transfer on rare earth laser action, consider an excited Nd^{3+} ion in the $^4F_{3/2}$ state. Energy levels, radiative transitions, and pairs of energy-conserving transitions for two nearby ions are shown in Fig. 6. The process in the center leads to spatial

---

[26] L. F. Johnson and H. J. Guggenheim, *Appl. Phys. Lett.* **23**, 96 (1973).

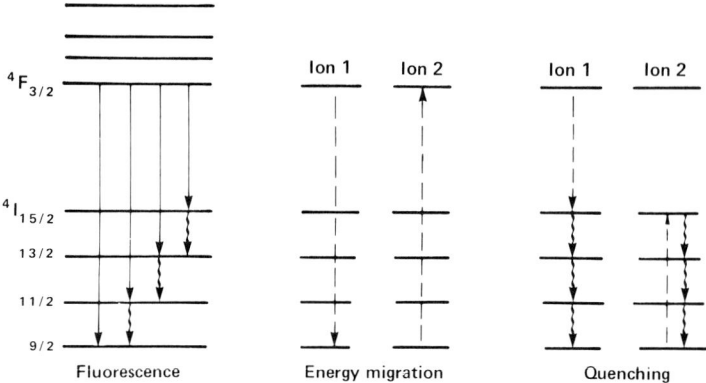

FIG. 6. Radiative transitions (solid lines) of Nd^{3+} and pairs of energy conserving transitions (dashed lines) resulting from ion–ion interactions. Wavy lines denote nonradiative transitions.

energy migration but no net relaxation. Migration has two effects. One, rapid energy diffusion, leads to a spatial equilibrium of excitation. Since standing waves in an optical resonator cause spatial hole-burning in the laser gain, this is ameliorated by energy migration.[27] Second, if other ions or imperfections are present that act as quenching centers, relaxation may occur either via a direct one-step process[28] or a multistep process involving energy migration and transfer.[29] The rate and time-dependence of the donor ion decay are dependent upon whether the process is characteristic of fast-diffusion or diffusion-limited relaxation.[30] Energy transfer in materials exhibiting inhomogeneous broadening is more complicated because of the energy mismatch and possible differences in the transition probabilities for ions in physically different sites.[31]

An important consideration for rare earths, and Nd^{3+} in particular, is the existence of pairs of transitions that cause self-quenching. This is illustrated at the right in Fig. 6. Self-quenching occurs for many fluorescing states and causes serious reduction in the fluorescence lifetime and quantum efficiency at ion concentrations $\geqslant 1\%$. When present, concentration quenching limits the improvements in optical pumping possible by increasing the laser ion concentration, that is, increased absorption of the pump radiation is nullified by reduced quantum efficiency.

[27] H. G. Danielmeyer, *J. Appl. Phys.* **42**, 3125 (1971).
[28] D. L. Dexter, *J. Chem. Phys.* **21**, 836 (1953).
[29] D. L. Dexter and J. H. Schulman, *J. Chem. Phys.* **22**, 1063 (1954).
[30] M. J. Weber, *Phys. Rev. B* **4**, 2932 (1971).
[31] N. Motegi and S. Shionoya, *J. Luminescence* **8**, 1 (1973).

The rate of self-quenching is a function of the oscillator strengths and the spectral match of the two transitions and the separation of the two ions.[28] If the transitions are sufficiently nonresonant or the ions well separated, self-quenching is reduced and materials with rare earth concentrations up to 100% can be used for lasers; these stoichiometric laser materials are discussed later.

Ion–Ion energy transfer may occur even when the energy differences between pairs of transitions become large, the energy difference being conserved by phonon creation or annihilation. The rate of phonon-assisted transfer between like and unlike ions depends on the available phonon energies and exhibits an approximately exponential dependence on energy mismatch similar to that found for multiphonon emission.[32,33]

Another case of self-quenching not included in Fig. 6 is that of two excited Nd^{3+} ions in $^4F_{3/2}$ states. The process involves de-excitation of one ion accompanied by excitation of the other ion to a higher excited state followed by subsequent nonradiative relaxation back to $^4F_{3/2}$. The rate of this process becomes significant only when the excited ion density is large.

Energy transfer transitions of the types illustrated above also occur radiatively wherein a photon emitted by one ion is absorbed by a second ion. This leads to radiation trapping and migration but not to lifetime shortening.

3.3. Laser Materials

A laser material is a combination of an active ion and a host. The principal types of paramagnetic ions and host materials that have been used for solid state lasers are surveyed below. For a comprehensive listing of solid state laser materials, the tables in Kaminskii and Osiko,[34] Weber,[35] and Kaminskii[36] should be consulted. These tables contain references to the original work.

3.3.1 Laser Ions

Paramagnetic ions from several transition groups have been used for solid state lasers and are listed in Table II. The laser ion is usually intro-

---

[32] T. Miyakawa and D. L. Dexter, *Phys. Rev. B* **1**, 2961 (1970).
[33] N. S. Yamada, S. Shionoya, and T. Kushida, *J. Phys. Soc. Jp.* **32**, 1577 (1972).
[34] A. Kaminskii and V. Osiko, *Izv. Akad. Nauk. SSSR Neorg. Mater.* **6**, 629 (1970).
[35] M. J. Weber, *in* "Handbook of Lasers" (R. J. Pressley, ed.), p. 371. CRC Press, Cleveland, Ohio, 1971.
[36] A. A. Kaminskii, "Laser Crystals" (in Russian) Nauka, Moscow, 1975.

TABLE II. Paramagnetic Ions Used for Solid State Lasers[a]

Group	Ions
Transition metals	$V^{2+}(3d^3)$, $Cr^{3+}(3d^3)$, $Co^{2+}(3d^7)$, $Ni^{2+}(3d^8)$
Divalent rare earths	$Sm^{2+}(4f^6)$, $Dy^{2+}(4f^{10})$, $Tm^{2+}(4f^{13})$
Trivalent rare earths	$Pr^{3+}(4f^2)$, $Nd^{3+}(4f^3)$, $Eu^{3+}(4f^6)$, $Tb^{3+}(4f^8)$, $Dy^{3+}(4f^9)$, $Ho^{3+}(4f^{10})$, $Er(4f^{11})$, $Tm^{3+}(4f^{12})$, $Yb^{3+}(4f^{13})$
Actinides	$U^{3+}(5f^3)$

[a] Electronic configurations are given in parentheses.

duced into the host as a substitutional impurity with fractional concentrations in the range from 0.1 to several atomic percent, the concentration being selected to achieve uniform, efficient optical pumping. In some materials the laser ion is present as a stoichiometric component of the host.

3.3.1.1. Rare Earth Ions. Rare earths are the most widely used ions in solid state lasers.[37,38] The wide applicability and versatility of rare earths arise from several attractive spectroscopic properties favorable for achieving low-threshold, efficient laser action:

(1) The electronic states of the ground $4f^n$ configuration provide complex and varied optical energy level structures, thus, there are many possible three- and four-level lasing schemes.

(2) There are a large number of excited states suitable for optical pumping.

(3) These excited states decay nonradiatively to metastable states having high radiative quantum efficiencies and narrow 4f–4f emission lines.

Because the energy levels do not change greatly with host, if a given ion is demonstrated to lase in one host, there are usually many other host possibilities.

Figure 7 summarizes the energy levels, transitions, and approximate wavelengths of trivalent lanthanide-ion lasers. For several ions, stimulated emission has been observed between more than one pair of J states and in cascade laser schemes. To accomplish this, a frequency-selective element (for example, a prism, grating, filter, or narrow-band mirror) is added to the resonator cavity. In cases where the lasing transition terminates on Stark levels of the ground J-state manifold, intense pumping or

[37] M. J. Weber, in "Handbook on the Physics and Chemistry of Rare Earths," (K. A. Gschneidner, Jr. and L. Eyring), Vol. 4. North-Holland Publ., Amsterdam, 1979.

[38] R. Reisfeld and C. K. Jørgensen, "Lasers and Excited States of Rare Earths." Springer-Verlag, Berlin, Heidelberg, New York, 1977.

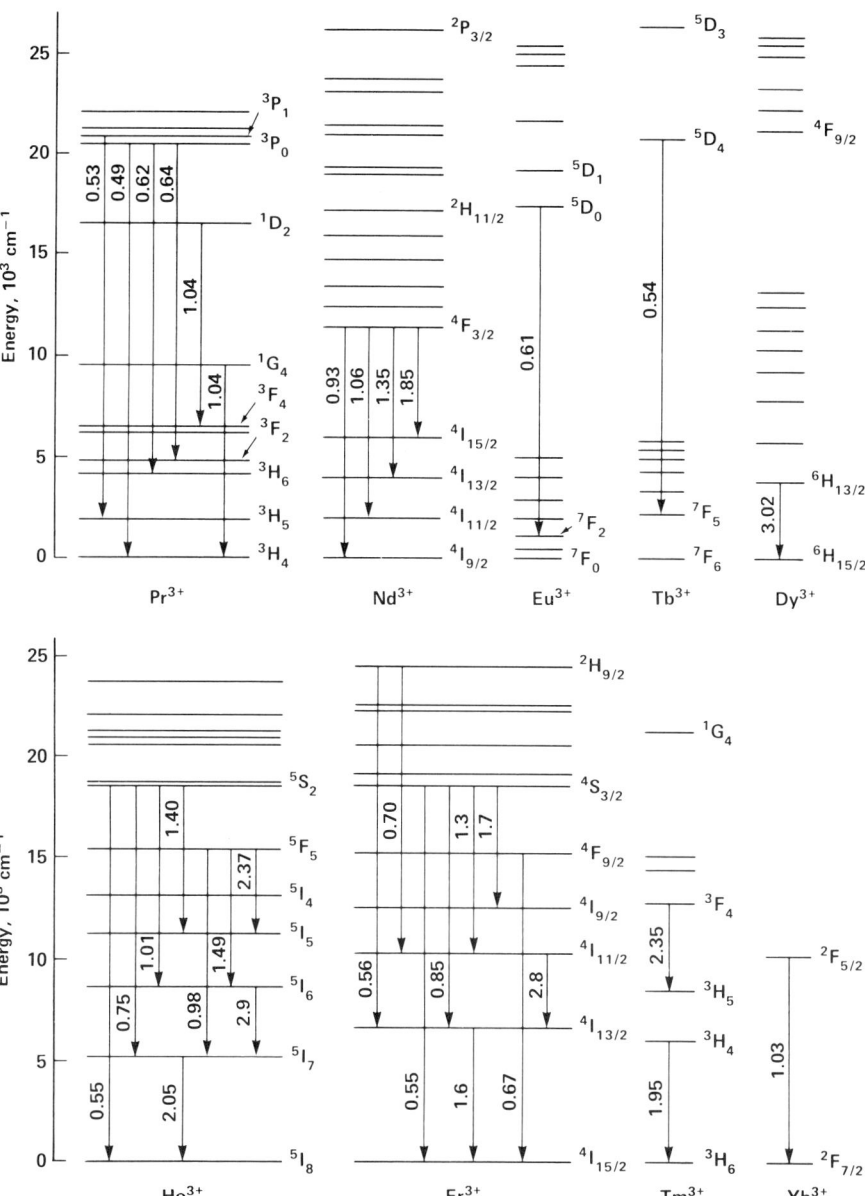

Fig. 7. Energy levels and laser transitions of trivalent rare earth ions. Wavelengths of transitions are in μm (Weber[37]).

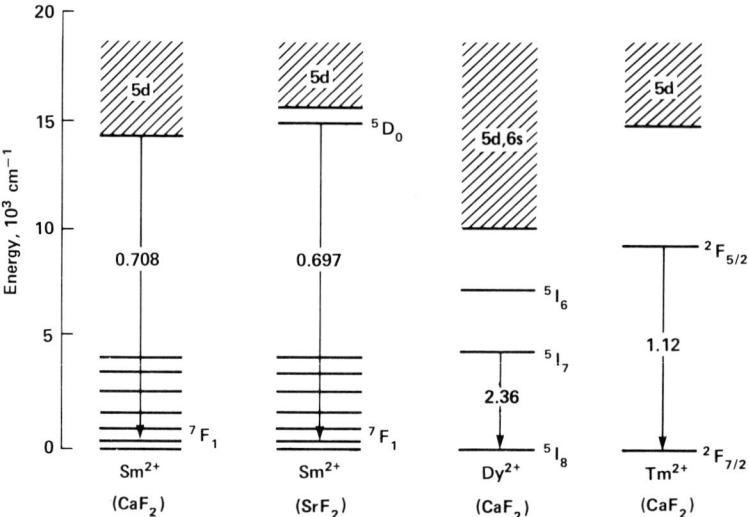

FIG. 8. Energy levels and laser transitions of divalent rare earth ions. Wavelengths of transitions are in μm and are for the host crystals noted in parentheses.

cooling of the laser material is required to achieve threshold. Operating characteristics of Nd^{3+} and other trivalent rare earths are presented in Chapter 3.4.

A summary of the energy levels, transitions, and wavelengths of divalent rare earth ion lasers is given in Fig. 8. Stimulated emission involves purely electronic $4f \rightarrow 4f$ transitions for Sm^{2+}, Dy^{2+}, and Tm^{2+}; phonon-terminated transitions for Sm^{2+} in CaF_2, and $5d \rightarrow 4f$ transitions for Sm^{2+} in SrF_2 at or below liquid-nitrogen temperatures. Dy^{2+} lasers have been operated cw using tungsten filament lamps, high-pressure xenon arc lamps, and sunlight as the optical excitation sources. The $Tm^{2+}:CaF_2$ laser involves a magnetic-dipole transition and has been operated in both pulsed and cw modes.

Other rare earth laser schemes are possible in addition to those in Figs. 7 and 8. There are 1639 free-ion energy levels associated with the $4f^n$ configurations of the 13 trivalent lanthanide ions. Yet, of the 192,177 transitions between pairs of levels, only 34 have been used for lasers. Given suitable pump sources and materials, stimulated emission involving many more transitions undoubtedly can be obtained. This is particularly true with the increasing availability of lasers at new wavelengths for pump sources and of tunable lasers for selective excitation of levels.

3.3.1.2. Iron Transition Metal Ions. The energy levels and laser transitions of iron-group ions are shown in Fig. 9. Laser wavelengths range

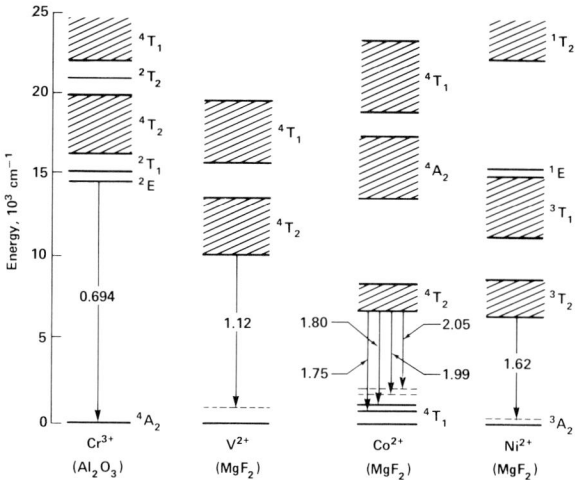

FIG. 9. Energy levels and laser transitions of transition metal ions. Wavelengths of transitions are in μm and are for the host crystals noted in parentheses. Vibronic levels are dashed. Vibronic laser action of Cr^{3+} in $BeAl_2O_4$ has recently been reported.[40]

from approximately 0.7 to 2.0 μm. The only purely electronic laser transitions are those of Cr^{3+} and Co^{2+}. Room temperature operation is readily obtainable for $Cr^{3+}:Al_2O_3$ (ruby) lasers, which have been operated both pulsed and cw.[9] In heavily doped ruby ($\approx 0.5\%$ Cr), additional lasing transitions involving exchange-coupled pairs of Cr^{3+} ions are observed at 700.9 and 704.1 nm.[39] Trivalent chromium has also been lased in $Y_3Al_5O_{12}$ and $BeAl_2O_4$. In the latter crystal laser action involved vibronic transitions and was tunable from approximately 700 to 800 nm at room temperature.[40]

Laser action of the other iron-group ions involves phonon-assisted transitions.[41] Cooling the laser material is generally required to reduce the population in the terminal vibronic level. Vibronic sidebands extend over a range of 1000–2000 cm^{-1}. Since the lowest threshold transition changes with temperature, phonon-terminated lasers are tunable over a frequency range of $\geqslant 5$–10%.[42]

Divalent vanadium is isoelectronic with Cr^{3+}. In MgF_2, the local field causes the minimum of the $^4T_2$ band to occur below the ^2E state, and

---

[39] A. L. Schawlow and G. E. Devlin, *Phys. Rev. Lett.* **6**, 96 (1961).

[40] J. C. Walling, H. P. Jenssen, R. C. Morris, E. W. O'Dell, and O. G. Peterson, to be published.

[41] L. F. Johnson, H. J. Guggenheim, and R. A. Thomas, *Phys. Rev.* **149**, 179 (1966).

[42] D. E. McCumber, *Phys. Rev. A* **136**, 954 (1964).

broadband, Stokes-shifted $^4T_2 \to {}^4A_2$ emission is obtained instead of the $^2E \to {}^4A_2$ emission of ruby. (Emission from $^4T_2$ also occurs for Cr^{3+} in most glasses and in other crystals, particularly at elevated temperatures.[43]) The peak stimulated emission cross section is small, $\sim 8 \times 10^{-21}$ cm^2, and the threshold for oscillation is high. V^{2+} and Co^{2+} ions have only been operated as pulsed lasers. Divalent nickel lases cw in MgF_2 and MnF_2 hosts.[44]

3.3.1.3. Other Transitions and Ions. To date, solid state lasers have been based upon only the transition groups and ions listed in Table II. Emission with high quantum efficiency from other ions in these groups and from other transition and posttransition groups is known, however. Whereas narrow emission lines are not common for iron group ions, ions such as $Mn^{2+}(3d^5)$ and $Cu^+(3d^{10})$ exhibit broadband emission in the visible from metastable states. Trivalent molybdenum, which has a $4d^3$ electronic configuration and an energy level scheme similar to that of Cr^{3+}, exhibits $^2E \to {}^4A_2$ fluorescence in crystals and glasses and has been discussed as a laser candidate.[45]

Optical pumping and laser action of rare earths have utilized 4f–4f transitions. The only exception is the 5d $\to$ 4f lasing transition of Sm^{2+}. Broad, intense, Stokes-shifted 5d $\to$ 4f fluorescence bands of high quantum efficiency are observed for several divalent and trivalent rare earths. With either optical or electron-beam pumping rates sufficient to compensate for the faster decay rates of the 5d states and in the absence of deleterious excited-state absorption, it should be possible to obtain laser action using these transitions.[46] The emission wavelengths extend into the vacuum ultraviolet and because of the large bandwidth, such lasers would be tunable similar to organic dye lasers.[47] Candidate ions for 5d $\to$ 4f lasing include trivalent Ce, Pr, Nd, Er, and Tm and divalent Sm, Eu, Yb. Attempts to lase Ce^{3+} in $Y_3Al_5O_{12}$ crystals, however, have been unsuccessful because excited-state absorption is more probable than stimulated emission.[48] This process prevents oscillation of several potential laser ions.

There are many qualitative similarities between the energy levels and spectral features of the $4f^n$ configuration of lanthanide ions and the $5f^n$ configuration of actinide ions. Hence, many of the comments and discus-

[43] P. Kisliuk and C. A. Moore, *Phys. Rev.* **160**, 307 (1967).
[44] P. F. Moulton, A. Mooradian, and T. B. Reed, *Opt. Lett.* **3**, 164 (1978).
[45] S. Parke, S. Gomolka, and J. N. Sandoe, *J. Non-Cryst. Solids* **20**, 1 (1976).
[46] M. J. Weber, *Proc. 11th Rare Earth Research Conf., 1974* Vol. I, p. 361 (1974).
[47] K. H. Yang and J. A. DeLuca, *Appl. Phys. Lett.* **29**, 499 (1976); *Appl. Phys. Lett.* **31**, 594 (1977).
[48] R. R. Jacobs, W. F. Krupke, and M. J. Weber, *Appl. Phys. Lett.* **33**, 410 (1978).

sion of rare earth lasers are also apropos to possible actinide lasers. Thus far, only one actinide ion—uranium—has lased, although oscillation of other actinide ions has been attempted.[49] (Laser action involved a $^4I_{11/2} \rightarrow {}^4I_{9/2}$ transition of U^{3+} in alkaline earth fluoride crystals at low temperatures.) The radioactivity of most of the actinides limits the selection of ions. In comparison to the lanthanides, data on energy levels and transition probabilities are limited; but general spectroscopic features are known. The 6d bands of the actinides occur at lower energies than the 5d bands of the lanthanides. Since 5f electrons of the latter are less well shielded, the oscillator strengths of electric-dipole transitions are more intense due to greater admixing of opposite-parity states into $5f^n$. The ion–host interaction should also be stronger. Therefore, the homogeneous linewidths arising from transitions between nearby Stark levels are larger and, because of fast multiphoton decay processes, there are fewer emitting states with high radiative quantum efficiencies. These properties restrict potential candidates for actinide laser action.

Posttransition group ions such as $Ge^{2+}(4s^2)$; Sn^{2+}, $Sb^{3+}(5s^2)$; and Tl^+, Pb^{2+}, $Bi^{3+}(6s^2)$ are other laser possibilities. Allowed absorption and emission transitions include $^1S_0 - {}^1P_0$, $^3P_1$. The bands are broad ($>10^3$ cm^{-1}) and exhibit large Stokes shifts. The overall quantum efficiency and possibility of excited-state absorption are not well known. Thus far laser action has not been demonstrated.

3.3.2. Host Materials

Solid-state laser hosts include both ordered (crystalline) and disordered (glassy) materials. In the former, the spectroscopic properties of the laser ions are usually identical and homogeneous; in the latter, the sites are different and the material exhibits inhomogeneous spectroscopic properties. Each has relative advantages and disadvantages for laser action.

Generally desirable properties of the host material are high optical quality (no defects, index inhomogeneities, or strain birefringence), transparency to the excitation and lasing wavelengths, chemical durability, hardness sufficient for good optical finishing, and in the case of high-repetition-rate or continuous operation, good thermal characteristics to minimize induced aberrations. Intense optical fields in the laser beam can cause damage in transparent dielectric materials by laser-induced electric breakdown.[50] This phenomenon arises from electron avalanche or multiphoton ionization and is a performance-limiting property in both

---

[49] H. A. Friedman and J. T. Ball, *J. Inorg. Nucl. Chem.* **34**, 3928 (1972).
[50] N. Bloembergen, *IEEE J. Quantum Electron.* **qe-10**, 375 (1974).

crystals and glasses.[51] Damage thresholds are dependent on surface roughness, material purity, and laser wavelength and pulse duration. Even before catastrophic damage occurs, an optical beam of intensity I induces a refractive index change given by γI, where γ is the nonlinear refractive index coefficient. These changes lead to whole-beam and small-scale self-focusing[52] and a loss of focusable energy. For short-pulse, high-power lasers used for fusion experiments, materials having small refractive index nonlinearities are of paramount importance.[53] Fluoride crystals and glasses have the smallest index nonlinearities.[54]

The desirability of a host material for a given application therefore depends on a combination of spectroscopic, optical, thermal, mechanical, and chemical properties. Comparisons of the physical properties of solid state laser materials are given in Koechner[12] and Thornton et al.[55]

3.3.2.1. Crystals. Crystals are the most widely used hosts for laser ions. Crystals with cations having ionic radii and valence states compatible with that of the laser ion provide substitutional sites and homogeneous doping distribution. Cations having sizes suitable for trivalent rare earth ion substitution include Y^{3+}, La^{3+}, Bi^{3+}, and the alkaline earths Ca^{2+}, Sr^{2+}, Ba^{2+}. When the substitutional site for trivalent laser ions is divalent, such as in the alkaline earth fluorides or $CaWO_4$, excess fluorine or other charge-compensating ions are added to maintain charge neutrality.

The number of different crystalline hosts used for lasers presently totals approximately 140. Compositions include simple and mixed oxides, fluorides, and more complex formulations. Crystal structures include cubic (fluorite, garnet, perovskite), uniaxial (scheelite, tysonite), and biaxial (distorted perovskite, illmenite). The point group summetry at the laser ion site affects the pattern of the crystal-field splittings and radiative transition probabilities. Representative examples of crystalline laser hosts are given in Table III; extensive tabulation and references to the original papers are given in Kaminskii et al.[34,36] The growth and chemistry of laser crystals are discussed in detail by Nassau.[56]

Of the large number of crystals investigated, the garnet $Y_3Al_5O_{12}$ (YAG)

---

[51] W. L. Smith, *Opt. Eng.* **17**, 489 (1978).

[52] E. S. Bliss, J. T. Hunt, P. A. Renard, G. E. Sommargren, and H. J. Weaver, *IEEE J. Quantum Electron.* **qe-12**, 402 (1976).

[53] M. J. Weber, *in* "Critical Materials Problems in Energy Production" (C. Stein, ed.), p. 261. Academic Press, New York, 1976.

[54] M. J. Weber, D. Milam, and W. L. Smith, *Opt. Eng.* **47**, 463 (1978).

[55] J. Thornton, W. Fountain, G. Flint, and T. Crow, *Appl. Opt.* **8**, 1 (1969).

[56] K. Nassau, *in* Applied Solid State Science (R. Wolfe, ed.), Vol. 2, p. 173. Academic Press, New York, 1971.

TABLE III. Representative Types of Crystals
Used as Hosts for Laser Ions[a]

Oxides	
Simple	$MgO, Al_2O_3, Y_2O_3, Gd_2O_3$
Complex	$YAlO_3, Y_3Al_5O_{12}, BeAl_2O_4, La_2Be_2O_5$
Germanate	$Bi_4Ge_3O_{12}, Ba_2MgGe_2O_7$
Molybdate	$CaMoO_4, Gd_2(MoO_4)_3, NaLa(MoO_4)_2$
Niobate	$LiNbO_3, LaNbO_4, Ca(NbO_3)_2$
Tungstate	$CaWO_4, NaLa(WO_4)_2$
Vanadate	$YVO_4, Ca_3(VO_4)_2$
Fluorides	
Simple	$MgF_2, MnF_2, CaF_2, SrF_2, LaF_3, CeF_3$
Complex	$LiYF_4, KMgF_3, BaY_2F_8$
Solid solutions	$CaF_2 + YF_3, 5NaF \cdot 9YF_3$
Others	
Oxysulfide	La_2O_2S
Fluorapatite	$Ca_5(PO_4)_3F$
Oxyapatite	$CaY_4(SiO_4)_3O$

[a] See Table X for stoichiometric laser crystals.

has a particularly favorable combination of properties of being a very hard, optically isotropic crystal with sites suitable for trivalent rare earth ion substitution (Y^{3+}) or for trivalent iron-group ions (Al^{3+}) without charge compensation. Alkaline earth fluoride crystals have been the principal hosts for divalent laser ions. These are relatively soft, optically isotropic materials. (MgF_2 is an exception.)

POLYCRYSTALLINE MATERIALS. Polycrystalline materials have also been used as laser hosts. For crystals that are optically isotropic, a polycrystalline form is attractive because conventional ceramic processing techniques rather than single-crystal growth can be used. Stimulated emission has been reported for Dy^{2+} in hot-pressed CaF_2, and more recently for a solid solution of cubic 89% Y_2O_3, 10% ThO_2, and 1% Nd_2O_3.[57] At room temperature, the absorption and emission linewidths are intermediate to those measured in crystals and glasses. Polycrystalline materials to date have been plagued by residual porosity and undetermined submicroscopic scattering centers, which cause significant optical losses and reduced laser performance.

MIXED CRYSTALS—SOLID SOLUTIONS. Garnet materials with lower stimulated emission cross sections and gain coefficients have been ob-

[57] C. Greskovich and J. P. Chernock, *J. Appl. Phys.* **44**, 4599 (1973).

tained using a mixed-crystal approach.[58,59] It was noted that (1) the positions of the Nd^{3+} fluorescence lines in $Y_3Al_5O_{12}$ and $Y_3Ga_5O_{12}$ are slightly different and (2) crystals of $Y_3(Al_{1-x}Ga_x)_5O_{12}$ can be prepared that are isotropic and that maintain the garnet structure over the entire range of compositions $0 \leq x \leq 1$. Therefore, because of the different possible distributions of Ga^{3+} ions on Al^{3+} sites, the crystal field at the Nd^{3+} sites in the mixed crystal varies between the two extremes. Spectrally, this appears as inhomogeneous line broadening and results in lower cross sections.

Another large class of crystalline materials that has been utilized for rare earth lasers is solid solutions of fluorides such as the yttrofluorite $CaF_2 + YF_3$ and ternary systems such as $CaF_2 + NaYF_4 + YF_3$. These are not definite compositions but form solid solutions with no apparent ordering of the cations. Many of these materials are characterized by large optical linewidths and in homogenous spectral properties approaching those in glasses. Extensive listings of these laser materials are given in Kaminskii and Osiko[34] and Nassau.[56]

STOICHIOMETRIC MATERIALS. Stoichiometric materials are pure chemical compounds of rare earths; thus, there is no question of ion size, charge, or coordination to be considered as in other laser materials. Also, since the laser ion is not a dilute substitutional impurity, the inhomogeneous broadening in crystals is very small and there is no statistical distribution of ion–ion separations. Other spectroscopic properties, including intensities and cross sections, are similar to those in other crystalline hosts. Danielmeyer[11] has discussed the physical processes pertinent to stoichiometric materials.

At the high active ion concentrations present in stoichiometric materials the major question is to prevent severe quenching by identifying ions, transitions, and compounds for which the desired fluorescence is not severely quenched by ion–ion interactions. Resonant absorption losses in high rare earth concentration materials can also affect the threshold for laser oscillator and the slope efficiency.[60] The fluorescence quenching observed for materials in which Nd^{3+} ions are separated by complexes such as PO_4, BO_3, or WO_4 is similar to but not as serious as for other crystals. In $La_{1-x}Nd_xP_5O_{14}$, the fluorescence lifetime was only reduced from 310 to 120 μsec as x increased to unity. The model for the concentration quenching in Nd^{3+} stoichiometric materials is still incomplete.[61] In general, however, for high fluorescence quantum efficiency (i) there should be sufficient spectral mismatch of transitions to reduce cross relax-

---

[58] L. A. Riseberg, R. M. Brown, and W. C. Holton, *Appl. Phys. Lett.* **23**, 127 (1973).
[59] R. K. Watts and W. C. Holton, *J. Appl. Phys.* **45**, 873 (1974).
[60] K. Otsuka and T. Yamada, *Opt. Commun.* **17**, 24 (1976).
[61] A. Lempicki, *Opt. Commun.* **23**, 376 (1977).

ation, (ii) the minimum separation between laser ions should be large, (iii) these should be weak ion–lattice interaction to reduce phonon-assisted transfer, and (iv) imperfections that can serve as quenching centers via energy migration should be absent. Stoichiometric hosts, laser ions, and transitions are presented in Section 3.4.4.

3.3.2.2. Glasses. Most of the conditions required for obtaining stimulated emission in crystals also apply to glasses. Glasses possess excellent optical quality, are optically isotropic, have sufficient hardness and environmental durability to retain good optical surfaces, and can usually be doped homogeneously with rare earths to high concentrations. In addition, glass can be cast into various sizes and shapes and clad with specially doped or index-matching glass layers to control parasitic oscillations.[62a] Compared to crystals, the thermal conductivity of glass is lower and temperature gradients lead to thermally induced birefringence and optical distortion. In general, crystal hosts are better for cw and high-repetition-rate operation; glasses are better for low-repetition-rate, Q-switched, and large energy storage operation.

The inhomogeneous broadening in glass has several effects on laser performance. The large linewidths are favorable for efficient optical pumping. In addition, the effective cross sections are small. This increases the threshold for oscillation but reduces losses by amplified spontaneous emission; therefore, high-energy storage is possible, ~ 1 J/cm^3. Because of site-to-site differences in cross sections, there is a distribution of stimulated emission rates in glass. Whereas this is not important for small-signal gain, under saturated gain conditions it affects the rate of energy extraction and the gain profile. The extent of spectral hole burning in the gain profile is dependent upon the relative magnitudes of the homogeneous and inhomogeneous line broadening in the laser glass.[62b]

Laser action in glasses has been obtained only from trivalent rare earths. Because of the smaller number of metastable fluorescing states, the number of rare earth ions and transitions lased and the spectral range covered in Fig. 1 are substantially less for glasses than for crystals. Laser ions and transitions are listed in Table IV. Where a range of laser wavelengths is given, it arises from the use of different host glasses or temperatures. All glass laser transitions are included among those shown in Fig. 7. References to the original papers are given in various review articles.[63–65]

[62a] D. C. Brown, S. D. Jacobs, and N. Lee, *Appl. Opt.* **17,** 211 (1978).
[62b] A. Y. Cabezas and R. P. Treat. *J. Appl. Phys.* **37,** 3556 (1966).
[63] K. Patek, "Glass Lasers." Butterworth and Co., Ltd., London, 1968.
[64] C. G. Young, *Proc. IEEE* **57,** 1267 (1969).
[65] E. Snitzer, *Bull. Am. Ceram. Soc.* **52,** 516 (1973).

TABLE IV. Ions and Glasses Used for Glass Lasers

Ion	Transition	Wavelength (μm)	Sensitizer(s)	Glass(es)
Nd^{3+}	$^4F_{3/2} \to {}^4I_{9/2}$	0.918–0.921		Borate, silicate (77 K)
	$^4F_{3/2} \to {}^4I_{11/2}$	1.04–1.08	Mn^{2+}, UO_2^{2+}	Borate, silicate, phosphate, germanate, tellurite, aluminate, fluorophosphate, fluoroberyllate
	$^4F_{3/2} \to {}^4I_{13/2}$	1.32–1.40		Borate, silicate, phosphate
Ho^{3+}	$^5I_7 \to {}^5I_8$	1.95–2.08	Yb^{3+}, Er^{3+}	Silicate
Er^{3+}	$^4I_{13/2} \to {}^4I_{15/2}$	1.54–1.55	Yb^{3+}	Silicate, phosphate
Tm^{3+}	$^3H_4 \to {}^3H_6$	1.85–2.02	Yb^3, Er^{3+}	Silicate
Yb^{3+}	$^2F_{5/2} \to {}^2F_{7/2}$	1.01–1.06	Nd^{3+}	Borate, silicate

Inorganic glass formers range from elements (S, Se, . . .), simple (SiO_2, GeO_2, B_2O_3, P_2O_5, . . .) and conditional (TeO_2, Al_2O_3, WO_3, . . .) glass-forming oxides, halides (BeF_2, AlF_3, ZrF_4, $ZnCl_2$, . . .), chalcogenides (As_2S_3, As_2Se_3, . . .), fused salts, metal alloys, and hydrogen-bonded compounds.[66] Of these, only a few have been used as laser hosts—principally oxide glasses, one fluoride glass, and mixed anion glasses such as fluorophosphates. Neodymium laser action has been reported in simple fused silica[67] and in glass ceramics.[68] In the latter, Nd^{3+} ions are in the glassy phase. Many glasses are ruled out because of poor physical properties and inadequate optical transmission in the wavelength region of interest.

The above glass-forming ion groups form the random glass network; other compounds play the role of network modifiers. Examples of modifier ions are alkali, alkaline earth, and other higher-valence-state cations. These ions are added to improve glass-forming and other physical/chemical properties. Both the glass network former and network modifier ions affect spectroscopic and laser properties. Systematic changes with composition have been achieved in cross sections, fluorescence lifetimes, peak wavelength, and linewidths.[69] Therefore, laser glasses can be tailored, within limits, to achieve large stimulated emission cross sections or high energy storage. Other compositional changes may be made to minimize thermally induced distortion or refractive index nonlinearities.

Laser ions are assumed to be distributed randomly in most laser glasses, although clustering is possible. Ions that exhibit fluorescence

[66] H. Rawson, "Inorganic Glass-Forming Systems." Academic Press, London and New York, 1967.
[67] J. Stone and C. A. Burrus, *Appl. Phys. Lett.* **23**, 388 (1973).
[68] G. Müller and N. Neuroth, *J. Appl. Phys.* **44**, 2315 (1973).
[69] R. R. Jacobs and M. J. Weber, *IEEE J. Quantum Electron.* **12**, 102 (1976).

quenching by ion–ion interaction in crystals exhibit similar behavior in glasses. Nonradiative decay by multiphonon emission is very probable in glass hosts and has been studied as a function of glass composition.[25,70] Borate glasses, with their high vibrational frequencies, are unfavorable hosts for many ions, such as Nd^{3+}, because strong multiphonon emission from the $^4F_{3/2}$ state reduces the quantum efficiency.

The index of refraction of glasses changes with composition and ranges from ≈ 1.3 for beryllium-fluoride-based glasses to >2 for tellurite glasses and glasses containing large polarizable ions of high atomic number such as lead or tantalum. With increasing amounts of fluorine, the linear and nonlinear indices of fluorophosphate glasses have been reduced to approximately one-half that of phosphate glasses.[71] Pure fluoride glasses with BeF_2 as the network former have the lowest known refractive indices.[72] The toxicity of beryllium compounds necessitates extra precautions in melting and handling these glasses.

Undesirable absorption bands due to transient and stable color centers are frequently induced in glasses and crystals by exposure to ultraviolet radiation from the flashlamps.[73] When this occurs, either the deleterious wavelengths must be removed by spectral filtering or antisolarizing ions such as cerium, antimony, or molybdenum added to the glass to inhibit the formation of color centers. The presence of platinum particles or other absorbing impurities in a glass can lead to damage in the intense optical fields present in high-power lasers.[74]

3.3.3. Fluorescence Sensitization

The optical pumping efficiency and output power of many rare earth lasers have been increased by codoping the medium with other ions that absorb pump radiation and effectively transfer the excitation to the upper laser level. This transfer may be either radiative or nonradiative. Sensitization schemes used for phosphors and other luminescence phenomena[75] are also generally applicable to lasers. Requirements for the sensitizer ion include (i) no ground- or excited-state absorption at the laser wavelength, (ii) absorption bands that complement rather than compete with absorption bands of the laser ion, since the fluorescence conversion

---

[70] R. Reisfeld, *Struct. Bonding* (Berlin) **22**, 123 (1975).
[71] O. Deutschbein, M. Faulstich, W. John, G. Krolla, and N. Neuroth, *Appl. Opt.* **17**, 2228 (1978).
[72] M. J. Weber, C. F. Cline, W. L. Smith, D. Milam, D. Heiman, and R. W. Hellwarth, *Appl. Phys. Lett.* **32**, 402 (1978).
[73] R. J. Landry, E. Snitzer, and R. H. Bartram, *J. Appl. Phys.* **42**, 3827 (1971).
[74] R. W. Hopper and D. R. Uhlmann, *J. Appl. Phys.* **41**, 4023 (1970).
[75] L. G. Van Uitert, in "Luminescence of Inorganic Solids" (P. Goldberg, ed.), p. 465. Academic Press, New York, 1966.

efficiency is generally smaller for the former, (iii) one or more metastable energy levels above the upper laser level, and (iv) no other pairs of levels that can quench the activator fluorescence. In addition, for efficient transfer the concentration of sensitizer ions must be sufficiently high to provide significant transfer within the fluorescence lifetime of the activator.

Of possible sensitizer ions for solid state lasers, rare earth ions have found the widest and most varied use. They have been present both as impurities and as a component of the host crystal. Ions from other transition metal groups have also been employed because they frequently have broad absorption bands and therefore are attractive when broadband pumping sources are used. Of the many sensitization schemes reported, some offer dramatic improvement in pumping efficiency and reduced threshold while others offer only marginal improvement. The most efficient optically pumped laser is the "alphabet" holmium laser—Ho^{3+} sensitized by Er^{3+}, Tm^{3+}, and Yb^{3+}.[76] The absorption bands of these ions combine to form a quasi-continuous spectrum. Via a complex cascade, energy absorbed by the various ions is eventually transferred to the $^5I_7$ lasing level of Ho^{3+}. This sensitization scheme has been used in both crystals and glasses.

A list of laser transitions and sensitizer ions is given in Table V. Sensitizers include paramagnetic ions (dopants and stoichiometric components), molecular groups such as WO_4^{3-}, VO_4^{3-}, UO_2^{2+}, and, in one instance, color centers. Figures of the energy levels and transitions of the sensitizer and activator ions and original references are given in Weber[35] and Kaminskii.[36] Other fluorescence sensitization schemes are known (for example, $Cr^{3+} \rightarrow Yb^{3+}$, $Mo^{3+} \rightarrow Er^{3+}$), but only those actually used for lasers are included in Tables IV and V.

The concept of upconversion, which is well established in phosphors,[77] is one in which higher-lying states of an activator are excited by successive energy transfers from a less energetic sensitizer. This process has been applied to rare earth lasers.[78] The materials were codoped crystals of $BaY_{1.2}Yb_{0.75}Er_{0.05}F_8$ and $BaY_{1.4}Yb_{0.59}Ho_{0.1}F_8$. Pumping involved absorption by Yb^{3+} followed by, in the first case, two or more successive transfers to Er^{3+} with subsequent lasing from the $^4F_{9/2}$ state and, in the second case, two successive transfers to Ho^{3+}—first $^5I_8 \rightarrow {}^5I_6$ and then $^5I_6 \rightarrow {}^5S_2$, with lasing from $^5S_2$. The upconversion sensitization technique has the potential of providing excitation for short-wavelength laser action.

[76] L. F. Johnson, J. E. Geusic, and L. G. Van Uitert, *Appl. Phys. Lett.* **8**, 200 (1966).
[77] F. Auzel, *Proc. IEEE* **61**, 758 (1973).
[78] L. F. Johnson and H. J. Guggenheim, *Appl. Phys. Lett.* **19**, 44 (1971).

TABLE V. Ions Used as Sensitizers
for Optically Pumped Solid State Lasers

Laser ion	Laser transition	Sensitizer ion(s)
Ni^{2+}	$^3T_2 \rightarrow {}^3A_2$	Mn^{2+}
Nd^{3+}	$^4F_{3/2} \rightarrow {}^4I_{11/2}$	$Ce^{3+}, Cr^{3+}, Mn^{2+}, UO_2^{2+}, (VO_4)^{3-}$
Tb^{3+}	$^5D_4 \rightarrow {}^7F_5$	Gd^{3+}
Dy^{3+}	$^6H_{13/2} \rightarrow {}^6H_{15/2}$	Er^{3+}
Ho^{3+}	$^5I_7 \rightarrow {}^5I_8$	$Er^{3+}, Tm^{3+}, Yb^{3+}, Cr^{3+}, Fe^{3+}, Ni^{2+}$
	$^5S_2 \rightarrow {}^5I_8$	Yb^{3+a}
Er^{3+}	$^4I_{13/2} \rightarrow {}^4I_{15/2}$	Yb^{3+}, color center
	$^4F_{9/2} \rightarrow {}^4I_{15/2}$	Yb^{3+a}
	$^4S_{3/2} \rightarrow {}^4I_{13/2}$	Ho^{3+}
Tm^{3+}	$^3H_4 \rightarrow {}^3H_6$	$Er^{3+}, Yb^{3+}, Cr^{3+}$
	$^3F_4 \rightarrow {}^3H_5$	Cr^{3+}
Yb^{3+}	$^2F_{5/2} \rightarrow {}^2F_{7/2}$	Nd^{3+}

[a] Multistep upconversion process.

3.4. Properties and Comparison of Solid State Lasers

Stimulated emission has been observed from more than 200 ion–crystal combinations and numerous glasses, but only a very few of these lasers have received practical acceptance and, fewer still, successful commercialization. Manufacturer's catalogs and buyer's guides[79] list Nd:YAG, Nd:glass, ruby, plus a few miscellaneous lasers under the category of solid state lasers. The physical properties and operating characteristics of the principal solid state lasers in use today are compared in Table VI and discussed in individual sections below.

While the number of available solid state lasers is small, they provide coherent radiation at a wide range of wavelengths and offer a large variety of pulsed and cw operating modes and output powers. In long-pulse operation, oscillation occurs during the flashlamp pulse, typically ~ 1 msec, and the repetition rate is determined by the flashlamp. Repetition rates for Q-switched operation range up to ≈ 50 kHz; rates for mode-locked operation and cavity dumping are $\sim 10^8 - 10^9$ and $1 - 10^6$ Hz, respectively. Many solid state lasers also operate continuously. Pump sources have included xenon and krypton arc lamps, tungsten-halide lamps, and light-emitting diodes (LED). Table VII lists the laser ions, hosts, operating wavelengths, transitions, and temperatures of cw solid state lasers.

[79] See, for example, "Laser Focus Buyer's Guide," 13th ed. Laser Focus Magazine, Newton, Massachusetts, 1978. This book is updated annually.

TABLE VI. Comparison of Solid State Lasers[a]

	Ruby	Nd:YAG	ED-2 Glass	LHG-5 Glass
Host	Al_2O_3	$Y_3Al_5O_{12}$	silicate	phosphate
Dopant	Cr^{3+}	Nd^{3+}	Nd^{3+}	Nd^{3+}
Concentration (cm^{-3})	1.6×10^{19}	1.4×10^{20}	2.8×10^{20}	3.2×10^{20}
Physical Properties				
Density (g/cm^3)	3.99	4.56	2.55	2.68
Hardness (Knoop)	2100	1215	544	497
Refractive index n (at λ_L)	1.76	1.82	1.55	1.53
dn/dT ($10^{-6}\ K^{-1}$)	12.6	7.3	2.9	~0
Nonlinear index γ ($10^{-20}\ m^2/W$)	3.5	7.2	3.8	3.1
Loss coefficient (cm^{-1})	≈0.001	≈0.002	≈0.005	≈0.001
Thermal conductivity ($W\ cm^{-1}\ K^{-1}$)	0.42	0.13	0.013	0.012
Thermal expansion ($10^{-6}\ K^{-1}$)	5.8	7.7	10.3	9.8
Specific heat ($J\ g^{-1}\ K^{-1}$)	0.76	0.59	0.92	0.71
Laser Properties				
Stimulated emission cross section ($10^{-20}\ cm^2$)	2.5	50–80[b]	2.8	3.9
Wavelength λ_L (nm)	694(R_1)	1064	1061	1054
Fluorescence lifetime (μsec)	3000	230	300	290
Fluorescence linewidth (nm)	0.53	0.4	26	19
Inversion for 1% gain/cm (cm^{-3})	4.0×10^{17}	$1.2–2.0 \times 10^{16}$	3.6×10^{17}	2.6×10^{17}
Stored energy for 1% grain/cm (J/cm^3)	$2.18^c + 0.12$	0.002–0.004	0.07	0.05

[a] Values are for a temperature of 300 K.
[b] Range of reported stimulated emission cross sections.
[c] Required to equalize populations.

3.4.1. Nd:YAG Laser

Although Nd^{3+} has lased in approximately 100 crystalline hosts, Nd:YAG is by far the most widely used and important solid state laser. Nd:YAG lasers have higher efficiencies and higher repetition rates than ruby lasers. Their versatility is evident from applications ranging from laboratory use and spectroscopy, materials processing, and oscillators for Nd:glass fusion lasers to military uses as rangefinders and target illuminators and designators.

Laser action has been obtained for transitions from the $^4F_{3/2}$ state to all terminal J states and for transitions between many different pairs of Stark levels. Reported wavelength ranges and modes of operation are sum-

3.4. PROPERTIES AND COMPARISON OF SOLID STATE LASERS

TABLE VII. CW Solid State Lasers[a]

Active ion	Host(s)	Wavelength (μm)	Transition	Temperature (°K)
Cr^{3+}	Al_2O_3	0.694	$^2E \rightarrow {}^4A_2$	300
Ni^{2+}	MgF_2, MnF_2	1.67–1.93	$^3T_2 \rightarrow {}^3A_2$	82–192
Dy^{2+}	CaF_2	2.359	$^5I_7 \rightarrow {}^5I_8$	77
Tm^{2+}	CaF_2	1.116	$^2F_{5/2} \rightarrow {}^2F_{7/2}$	4–27
Nd^{3+}	$CaWO_4$, $Y_3Al_5O_{12}$, $YAlO_3$, $KY(WO_4)_2$, $Ca_5(PO_4)_3F$, La_2O_2S, $La_2Be_2O_5$, $CaMoO_4$, CeF_3, LaF_3, $Ca(NbO_3)_2$, glass	1.041–1.123	$^4F_{3/2} \rightarrow {}^4I_{11/2}$	300
	$Y_3Al_5O_{12}$, $YAlO_3$, $KY(WO_4)_2$	1.319–1.444	$^4F_{3/2} \rightarrow {}^4I_{13/2}$	300
Ho^{3+}	$Y_3Al_5O_{12}$, $ErY_3Al_5O_{12}$, Er_2O_3, $LiYF_4$, CaF_2–$(Er, Tm, Yb)F_3$	2.099, 2.122, 2.06–2.123	$^5I_7 \rightarrow {}^5I_8$	77
Tm^{3+}	$Y_3Al_5O_{12}$, $Er_{1.5}Y_{1.5}Al_5O_{12}$, Er_2O_3	2.013, 2.014, 1.934	$^3H_4 \rightarrow {}^3H_6$	77
U^{3+}	CaF_2	2.613	$^4I_{11/2} \rightarrow {}^4I_{9/2}$	77

[a] See Table X for cw stoichiometric laser materials.

marized in Table VIII. By using second and higher harmonics of the 1064- and 1319-nm transitions generated in KDP and ADP crystals and optical parametric oscillators, Nd:YAG lasers can readily provide coherent radiation at wavelengths ranging from 0.2 to 3.0 μm.

The properties and operation of Nd:YAG lasers have been thoroughly reviewed by Danielmeyer.[80] A number of physical and laser properties of Nd:YAG are included in Table VI. Although the stimulated emission cross sections and quantum efficiency have been the subjects of several

TABLE VIII. Wavelength Range of Lasing Transitions Reported for Nd:YAG

Transition	Wavelengths[a] (nm)	Operation[b] (temp., °K)	Reference
$^4F_{3/2} \rightarrow {}^4I_{9/2}$	890–946 (946)	P, ($\leq$295)	c
$^4F_{3/2} \rightarrow {}^4I_{11/2}$	1052–1123 (1064)	P, cw (300)	d,e
$^4F_{3/2} \rightarrow {}^4I_{13/2}$	1319–1444 (1319)	P, cw (300)	d,e
$^4F_{3/2} \rightarrow {}^4I_{15/2}$	1833 (1833)	P, ($\leq$293)	f

[a] The wavelength in parentheses has the lowest threshold.
[b] P, pulsed; cw, continuous.
[c] R. W. Wallace and S. E. Harris, *Appl. Phys. Lett.* **15**, 111 (1969).
[d] R. G. Smith, *IEEE J. Quantum Electron.* **qe-4**, 505 (1968).
[e] J. Marling, *IEEE J. Quantum Electron.* **qe-14**, 56 (1978).
[f] R. W. Wallace, *IEEE J. Quantum Electron.* **qe-7**, 203 (1971).

[80] H. G. Danielmeyer, *in* "Lasers" (A. K. Levine and A. J. DeMaria, eds.), Vol. 4, p. 1. Dekker, New York, 1976.

studies, the results differ,[81,82] which accounts for the range of values given in Table VI.

Pulsed Nd:YAG lasers have peak power outputs ranging from kilowatts to gigawatts and brightnesses up to 10^{15} W/cm^2/sr. Pulse durations are typically 0.1–1.0 msec long pulse, 10–100 nsec Q-switched, and $\leqslant 100$ psec mode-locked. Output energies of $\geqslant 1$ J/pulse can be obtained from a single oscillator rod. Continuous output powers ranging from fractions to hundreds of watts multimode and 20 W TEM$_{00}$ (fundamental transverse) mode have been obtained from Nd:YAG lasers. As an example, a krypton-lamp-pumped, 4-mm-diameter by 50-mm-long Nd:YAG rod can produce 4 W TEM$_{00}$ and 30 W multimode. The full-angle beam divergence is < 10 mrad. Output stabilities of $\pm 5\%$ pulsed and $\pm 3\%$ continuous have been achieved.

Krypton arc lamps provide efficient optical pumping because their emission spectrum is well matched to the strong near-infrared absorption bands of Nd:YAG in Fig. 3. In elliptical pumping cavities, cw laser action is obtainable with several percent overall efficiency. Potassium–mercury discharges provide even higher pumping efficiency. Nd:YAG lasers have also been pumped using other standard lamps as well as GaAsP light-emitting diodes, electron beams, and the sun.

For high repetition rate and continuous operation, cooling of the laser material is necessary. Tap-water cooling, air cooling, and self-contained closed-loop cooling systems have been used. An example of a compact cw Nd:YAG delivering ~ 100 mW at 1064 nm is shown in Fig. 10. Only power and cooling water are required; the remainder of the laser—pump cavity, resonator, and power supply—are all self-contained.

Crystals of Nd:YAG have been grown to 30 mm diameter and 150 mm long using top seeding and pulling (Czochralski) techniques. Because of the presence of a center core of different composition, useful diameters are more typically $\leqslant 10$ mm. Gallium garnets are to be grown core free.

Laser action has also been observed from epitaxially grown thin films[83] and single-crystal fibers[84] of Nd:YAG. Both pulsed[83] and cw[85] lasing have been reported.

---

[81] S. Singh, R. G. Smith, and L. G. Van Uitert, *Phys. Rev. B* **10**, 2566 (1974).

[82] E. M. Dianov, A. Ya. Kavasik, V. B. Neustraev, A. M. Prokhorov, and I. A. Scherbakov, *Sov. Phys.—Dokl.* (Engl. Transl.) **20**, 622 (1975).

[83] J. P. van der Ziel, W. A. Bonner, L. Kopf, S. Singh, and L. G. Van Uitert, *Appl. Phys. Lett.* **22**, 656 (1973).

[84] C. A. Burrus and J. Stone, *Appl. Phys. Lett.* **26**, 318 (1975).

[85] W. A. Bonner, *J. Electron. Mater.* **3**, 193 (1974).

3.4. PROPERTIES AND COMPARISON OF SOLID STATE LASERS

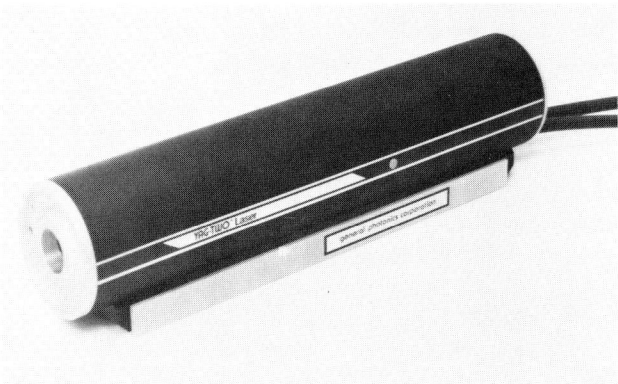

FIG. 10. Example of a commercial Nd:YAG laser (courtesy of General Photonics Corporation).

3.4.2. Nd:Glass Lasers

Neodymium glass lasers have the highest energy, power, and radiance and shortest pulse duration of solid state lasers. Applications include rangefinders, materials processing, picosecond spectroscopy, and amplifying media for large lasers used in inertial confinement fusion experiments. Nd:glass lasers have been operated at several wavelengths (see Table IV) and are tunable over a small range, ≈ 10–20 nm.

Xenon flashlamps are used for optical pumping of Nd:glass lasers. Because of the broad absorption linewidths, slope efficiency is very good. Output energies up to 100 J/pulse, multimode, are obtainable in long-pulse operation for rod diameters ≤ 10 mm; ~ 20 J/pulse in 15 nsec is obtainable Q-switched. Transform-limited, 5-psec, TEM_{00}-mode pulses have been generated from Nd:phosphate glass lasers.[86] Beam divergences are ≤ 5 mrad. Single transverse mode, diffraction-limited radiance of 2×10^{17} W/cm²-sr has been reported.[87] Because of the poor thermal properties of glasses, repetition rates are usually limited to ~ 1 pps or less unless special efforts such as forced air or liquid cooling are made. Glass lasers have been operated cw. High-gain glasses such as phosphates are more attractive for this purpose.

Current fusion laser designs require many parallel amplifier chains with beam diameters of 30 cm or more.[88] Whereas crystalline materials of

[86] J. R. Taylor, W. Sibbett, and A. J. Cormier, *Appl. Phys. Lett.* **31**, 184 (1977).
[87] W. F. Hagen, *IEEE J. Quantum Electron.* **4**, 361 (1968).
[88] J. A. Glaze, W. W. Simmons, and W. F. Hagen, *Proc. Soc. Photo-Opt. Instrum. Eng* **76**, 7 (1976).

these dimensions are presently impractical, high-optical-quality glass is available. In addition, for components of these sizes, amplified spontaneous emission and parasitic losses limit performance efficiency.[62a] Here the small effective stimulated emission cross section of glasses is an advantage. An example of a 4-cm-diameter, xenon-flashlamp-pumped rod amplifier is shown at the top of Fig. 11. Glass lasers have been built with rod diameters to ≈ 100 mm. To obtain a uniform spatial gain profile in large-diameter rod amplifiers, the Nd concentration must be decreased and the optical pumping and overall efficiency are reduced. For beam diameters $\geqslant 100$ mm, the glass is cut into slabs or disks for more uniform exposure to the pump radiation. A typical disk amplifier is shown at the bottom of Fig. 11. The elliptical-shaped glass disks are oriented at Brewster's angle with respect to the axis of the amplifier so that one polarization is passed without loss. The laser glass is heated during optical pumping, and as it cools thermal gradients and optical aberrations develop. In large Nd:glass amplifiers, up to 1 hour may be required to reestablish thermal equilibrium.

At the other extreme in size, thin-film waveguides of Nd silicate glass have been fabricated using an rf-sputtering technique.[89] The spectral properties of the bulk and thin-film glasses were the same. By optically pumping with a cw dye laser, large optical gains of ~ 1 cm^{-1} were obtained at 1058 nm and are potentially useful for integrated optics applications.

Until recently, the only laser glasses available commercially were silicates. Prompted by the requirements of fusion lasers, high-gain phosphate and low-refractive-index fluorophosphate glass compositions have been developed.[71] Stimulated emission cross sections for the $^4F_{3/2} \to {}^4I_{11/2}$ transition range from ≈ 1 to 5×10^{-20} cm^2.[69] Currently available glass types are listed in Table IX (as manufacturers refine the glass compositions, designations may change). Of the silicates, LSG-91H and ED-2 offer high gain; LG-650 has a lower cross section, 1.1×10^{-20} cm^2, and combined with a longer lifetime, ~ 700 μsec, provides good energy storage. The phosphate glasses, such as LHG-5 in Table VI, have narrower linewidths than silicate glasses and their cross sections and gain coefficients are higher by $\geqslant 50\%$. Phosphate glasses also have lower γ values and better properties with respect to thermally induced birefringence. Fluorophosphate glasses have the smallest available nonlinear refractive indices and gains comparable to those of silicate glasses.

---

[89] Bor-Ue Chen and C. L. Tang, *Appl. Phys. Lett.* **28**, 435 (1976).

3.4. PROPERTIES AND COMPARISON OF SOLID STATE LASERS 201

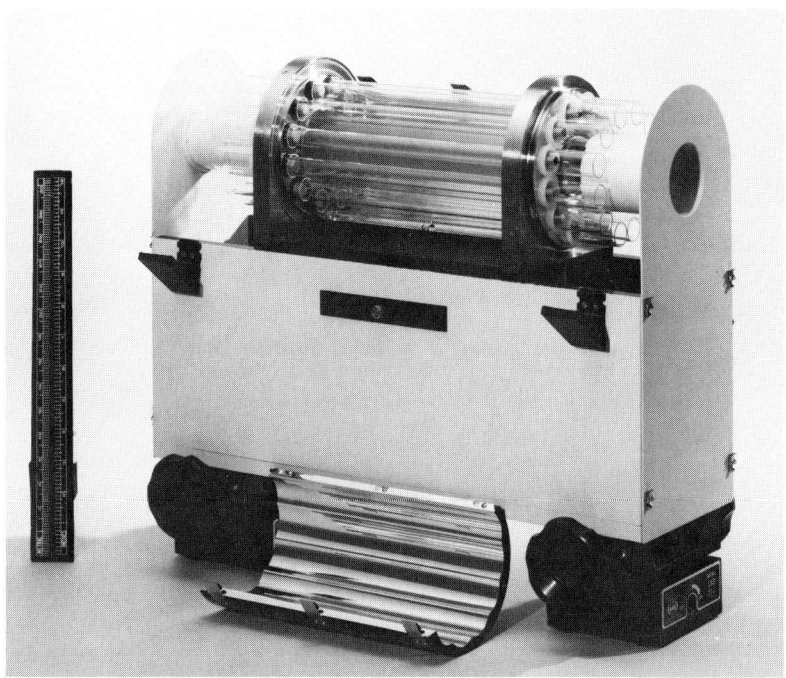

FIG. 11. Flashlamp-pumped Nd:glass rod (top) and disk (bottom) amplifiers used in lasers for fusion experiments (courtesy of Lawrence Livermore Laboratory). Laser beam diameters are 40 and 200 mm, respectively.

TABLE IX. Neodymium-Doped Laser Glasses

Glass type	Source and designation	Peak wavelength (nm)
Silicate	Hoya: LSG-91H	1061
	Owens-Illinois: ED-2	1061
	Schott: LG-650	1057
Phosphate	Hoya: LHG-5, -7, -8	1053
	Kigre: Q-88	1054
	Owens-Illinois: EV-2	1054
	Schott: LG-700, -710	1054
Fluorophosphate	Hoya: LHG-10	1051
	Owens-Illinois: EVF-1	1053
	Schott: LG-800	1052
	Schott: LG-810	1051

3.4.3. Ruby Lasers

The ruby laser, being the first, received considerable development in the 1960s.[9] Ruby is a very hard material with good thermal conductivity. These properties combined with large output powers and a visible wavelength make ruby lasers useful for many applications. A major use of ruby lasers is for rangefinders, where they provide a rugged power source for ambient temperature operation. Oscillation occurs at 694.3 nm (from the $\bar{E}$ state); the other R line lases at 629.9 nm. The second harmonic of the ruby laser is a good pulsed source of near-ultraviolet radiation.

Ruby lasers are pulsed, although cw operation is possible with appropriate pumping. Helical or linear xenon flashlamps or mercury lamps are used for optical pumping. Repetition rates are usually in the order of or less than a few pulses per second. Because $Cr^{3+}:Al_2O_3$ is a three-level laser and the stimulated emission cross sections are small, large excited-state populations are required to reach threshold. The overall efficiency is low, $\sim 0.1\%$. In long-pulse operation, output energies on the order of 10 mJ/pulse, TEM_{00}, and up to 80 J/pulse, multimode, are obtainable in a few milliseconds. In Q-switched and mode-locked operation, output powers of $\sim 10^8$ W (10^{-8} sec) and $> 10^9$ W (10^{-11} sec) have been reported. Oscillator-amplifier combinations provide increased power, energy, and radiance.

The output of ruby lasers normally exhibits irregular spiking.[90] This occurs because the stimulated emission rate increases rapidly and soon exceeds the pumping rate, which depletes the inversion and oscillation

---

[90] H. Statz and G. deMars, in "Quantum Electronics" (C. H. Townes, ed.) p. 530. Columbia Univ. Press, New York, 1960.

stops. Continued pumping reestablishes the population inversion, threshold is reached, and the cycle repeated. Spiking makes the peak power difficult to define.

Synthetic ruby crystals are grown to 25-mm diameter and 300-mm lengths. High-optical-quality ruby is essentially scatter-free; the distortion observed in an interferometer is less than 1 fringe/cm. The material is usually fabricated into right circular cylinders with plano-plano ends with multilayer dielectric reflecting or antireflection coatings. Brewster-angle rods are also commonly used. When the rod axis is cut perpendicular to the optic axis, the output is polarized with the electric field perpendicular to the optic axis. When the rod axis and optic axis are parallel, the output is unpolarized.

3.4.4. Stoichiometric Lasers

Stoichiometric laser materials are of interest for miniature oscillators and amplifiers for integrated optics technology. Because of the high density of active ions ($>10^{21}$ cm^{-3}), the absorption coefficients are large and high gain is achievable in small volumes. Efficient optical pumping has been obtained with semiconductor and dye lasers[91,92] and light-emitting diodes[93]; miniature xenon flashlamps have also been used.[94] Stoichiometric lasers offer small size, light weight, long lifetime, and high efficiency. Properties and applications of these lasers are surveyed by Danielmeyer[11] and Kaminskii.[36]

A list of stoichiometric lasers is given in Table X (references to the original work can be found in Danielmeyer[11] and Weber[37]). All materials are crystalline, although NdP_5O_{14} was also prepared and lased in the glassy phase.[95] Maximum dimensions of single crystals are generally only a few millimeters but NdP_5O_{14} has been grown to 30 mm.[94]

The active laser ions in stoichiometric materials have all been rare earths. Oscillation has been obtained at wavelengths ranging from 0.49 to 2.13 μm. Nd^{3+} has again received the most study. Reported stimulated emission cross sections ($^4F_{3/3} \rightarrow {}^4I_{11/2}$) range from 1.8×10^{-19} cm^2 for NdP_5O_{14} to 9×10^{-19} cm^2 for $LiNdP_4O_{12}$. The operating characteristics are dependent upon the pumping conditions. With cw laser excitation, thresholds for Nd^{3+} lasers are ≤ 1 mW at ambient temperatures. Output powers of several milliwatts are attainable with overall effi-

[91] S. R. Chinn, J. W. Pierce, and H. Heckscher, *Appl. Opt.* **15**, 1444 (1976).
[92] M. Saruwatari, T. Kiumira, and K. Otsuka, *Appl. Phys. Lett.* **29**, 291 (1976).
[93] M. Saruwatari, T. Kimura, T. Yamada, and J. Nakano, *Appl. Phys. Lett.* **27**, 682 (1975).
[94] S. R. Chinn and W. K. Zwicker, *Appl. Phys. Lett.* **31**, 178 (1977).
[95] H. P. Weber, T. C. Damen, H. G. Danielmeyer, and B. C. Topfield, *Appl. Phys. Lett.* **22**, 534 (1973).

TABLE X. Stoichiometric Solid State Lasers

Ion–crystal	Laser transition(s)	Wavelength (nm)	Operation[a] (temp., °K)
Pr^{3+}			
$PrCl_3$	$^3P_0 \to {^3H_4}, {^3H_6}, {^3F_2}$	489.2, 616.4, 645.2	P($\leq$65–300)
	$^3P_1 \to {^3H_5}$	529.8	P(35)
$PrBr_3$	$^3P_0 \to {^3F_2}$	640	P(300)
Nd^{3+}			
NdP_5O_{14}	$^4F_{3/2} \to {^4I_{11/2}}$	1051	cw(300)
	$^4F_{3/2} \to {^4I_{13/2}}$	1.323	P(300)
$MNdP_4O_{12}$ (M = Li, Na)	$^4F_{3/2} \to {^4I_{11/2}}$	1048, 1051	cw(300)
$M_3Nd(PO_4)_2$ (M = Na, K)	$^4F_{3/2} \to {^4I_{11/2}}$	1055	cw(300)
$KNdP_4O_{12}$	$^4F_{3/2} \to {^4I_{11/2}}$	1051	cw(300)
$NdAl_3(BO_3)_4$	$^4F_{3/2} \to {^4I_{11/2}}$	1065	cw(300)
$Na_5Nd(WO_4)_2$	$^4F_{3/2} \to {^4I_{11/2}}$	1063	cw(300)
Ho^{3+}			
HoF_3	$^5I_7 \to {^5I_8}$	2090	P(77)
$LiHoF_4$	$^5F_5 \to {^5I_5}, {^5I_6}, {^5I_7}$	2352, 1486, 979	P($\approx$90)
$Ho_3Al_5O_{12}$	$^5I_7 \to {^5I_8}$	2122, 2129	P(90)
Er^{3+}			
$LiErF_4$	$^4S_{3/2} \to {^4I_{9/2}}$	1732	P($\approx$90)

[a] P, pulsed; cw, continuous.

ciencies of ~15%. For fiber optics applications, the ≈ 1.32-μm wavelength of Nd^{3+} is of particular interest because of the low resonant losses and favorable dispersion of glass fibers at this wavelength. Because of the smaller stimulated emission cross section for the $^4F_{3/2} \to {^4I_{13/2}}$ transition, the thresholds are higher. Both the $^4F_{3/2} \to {^4I_{11/2}}$ and $^4F_{3/2} \to {^4I_{13/2}}$ lasing has been observed in several materials.

3.4.5. Other Lasers and Materials

In addition to the lasers above, a small number of other lasers are available commercially.[79] Several use $LiYF_4$ (YLF) as the host crystal. Nd:YLF lasers operate at 1047 nm in π polarization and 1053 nm in σ polarization. These wavelengths are well-matched to the peak gain of Nd^{3+}-doped fluoroberyllate and phosphate laser glasses; these lasers are attractive choices for oscillators in glass amplifier chains. Other available YLF lasers include Ho^{3+} sensitized with Er^{3+} and Tm^{3+}. In a cw/repetitively Q-switched mode, this laser delivered 5 W multimode at

2.06 μm. Flashlamp-pumped, moderate-repetition-rate Er:YLF lasers operate at 0.85, 1.22, and 1.73 μm using transitions illustrated in Fig. 7. The output at 0.85 μm is ~50 mJ/pulse. This wavelength is within the spectral range of high-quantum-efficiency photoemissive materials and is of interest in applications where good detectability is important.

Tunable lasers are convenient or essential for many applications. Recently efficient cw oscillation was obtained from a cooled Ni:MgF$_2$ crystal in the range 1.6–1.8 μm. Using a Nd:YAG laser for optical pumping, output powers up to 1.7 W were achieved at efficiencies up to 37%.[44] Another promising laser is Cr^{3+} in BeAl$_2$O$_4$ (alexandrite). Operating as a 4-level laser using vibronic transitions, a 100-nm continuous tuning range centered at 750 nm was observed.[40] Output pulse energies of 0.5 J in 200 μsec were easily obtainable at the 750-nm peak from a 6.3-mm diameter by 76-mm-long laser rod.

Pump cavities, optical resonators, and associated electronics of solid-state lasers are frequently adaptable for use with different laser materials. The number of laser materials commercially available is again limited and buyer's guides[79] should be consulted. Crystalline oxide materials currently available doped with Nd^{3+} (other rare earth dopants are generally also obtainable) include Y$_3$Al$_5$O$_{12}$ (YAG), YAlO$_3$ (YAP, YALO), La$_2$Be$_2$O$_5$ (BEL), and NdP$_5$O$_{14}$ (NPP); acronyms are given in parentheses. Fluoride host crystals for trivalent rare earths include LiYF$_4$ (YLF), LaF$_3$, and CeF$_3$; hosts for divalent rare earths are CaF$_2$, SrF$_2$, and BaF$_2$. Host crystals for iron group ions include Al$_2$O$_3$, BeAl$_2$O$_4$, MgAl$_2$O$_4$, MgF$_2$, MnF$_2$, and several of the above crystals.

Current laser glass types are listed in Table IX. These glasses can be melted with various rare earth and other transition group ions. The dopant levels, however, may be limited by the glass-forming properties.

3.5. Hazards

Hazards associated with the operation of solid-state lasers include radiation, which can cause eye damage and skin burns, and high-voltage electrical sources.[96] Ocular hazards at various wavelengths and their effects are summarized in Table XI. The most hazardous spectral region is 400–1100 nm. As seen from Fig. 1, this is the regime in which most solid state lasers operate and, in particular, Nd and ruby lasers operate. Lasers operating at wavelengths longer than 1400 nm, such as Ho^{3+} (2.1 μm) and Er^{3+} (1.6 μm) are of interest in part because of their greatly

---

[96] "Safe Use of Lasers," Ansi Standard Z-136.1 (1973), Am. Nat. Stand. Inst. Inc., New York, 1973.

TABLE XI. Optical Radiation Hazards

Wavelength (nm)	Hazard
<315	Absorbed by cornea; welder's flesh or photokeratitis; sunburn or erythema of skin
315–400	Absorbed by lens; may contribute to cataract; long-wave erythema
400–1400	Transmitted by ocular media and focused on retina; most hazardous region
>1400	Absorbed by ocular media
>3000	Absorbed by front surface of eye

reduced ocular hazard. Whereas no primary solid state lasers operate in the ultraviolet, the generation of harmonics can readily produce harmful levels of radiation in this spectral region.

With regard to potential ocular hazards, the Bureau of Radiological Health has established the four classifications of laser systems given below and requires laser manufacturers to label their products accordingly.[97]

Class I Under normal operating conditions cannot emit a hazardous level of optical radiation (exempt of labels and controls).

Class II Does not have enough optical power to injure accidentally but may produce retinal injury when stared at for a long period.

Class III Produces accidental injury if viewed directly or if radiation collected and directed into the eye.

Class IV Output sufficient to produce a hazardous diffuse reflected beam as well as a direct or specular reflected beam.

Most solid state lasers fall under categories III and IV.

Ocular hazards are minimized by enclosing the laser beam and target, although this is not always possible. In laboratory use, direct observation of beams and spectral reflection should be avoided. Wearing goggles that filter out the laser wavelength is a simple way to ensure eye protection. Adequate labels, warning lights, and controlled access to the laser area are also good practice.[97]

Electrical hazards are associated with the high-voltage power supplies and capacitors for flashlamps, electro-optic Q-switches, and modulators and can cause severe shock, burns, and be lethal. Safety precautions in-

[97] U. S. Department of Health, Education, and Welfare, Bureau of Radiological Health, *Fed. Regist.* **40**, No. 148, 32252-32265 (1975).

clude adequate enclosures, safety locks, insulation, connectors, and those operating practices normally prescribed for high-voltage equipment.

Acknowledgment

It is a pleasure to acknowledge, with thanks, the helpful and skillful assistance of Ms. Karen Wenzinger in the preparation of this manuscript.

4. SEMICONDUCTOR DIODE LASERS*

4.1. Introduction

Semiconductor lasers cover a spectral range extending from the far infrared (~ 33 μm) to the visible (~ 0.5 μm). However, because some interesting semiconductors cannot be doped both n- and p-type, only part of the range is attainable with p–n junction injection devices. This is the case, for example, for the wide-bandgap II–VI compounds CdS and ZnO. In such materials, the production of the high excess carrier population needed for stimulated emission requires optical or electron-beam pumping. Such lasers are of limited utility except as a means for exploring material properties.

The interest in semiconductor lasers centers on p–n junction devices, and more specifically on devices that incorporate heterojunctions. Such lasers have a recombination region bracketed by higher bandgap regions that provide efficient minority-carrier injection and confinement, as well as an internal waveguide for the stimulated radiation. Thus, the effective laser volume is accurately controlled. The incorporation of heterojunctions in injection lasers[1-3] provides, under favorable circumstances, device performance approaching the theoretical limits of the semiconductor. Of course, many restrictions must be considered in the construction of heterojunction lasers in order to obtain the desired objectives, and these are considered in this review.

Independent of the materials, however, the design of a semiconductor laser is guided by several basic principles:

(1) A recombination region is needed where population inversion is produced by the injection of electron–hole pairs. In general, the lower the temperature, the lower the pair density needed to produce sufficient gain to sustain oscillations in the cavity. For example, a pair density of about 2×10^{18} cm^{-3} is generally needed to initiate lasing at room temper-

---

[1] H. Kressel and H. Nelson, *RCA Rev.* **30**, 106 (1969).
[2] I. Hayashi, M. B. Panish, and P. W. Foy, *IEEE J. Quantum Electron.* **qe-5**, 211 (1969).
[3] Zh. I. Alferov, V. M. Andreev, E. L. Portnoi, and M. K. Trukan, *Fiz. Tekh. Poluprovodn.* **3**, 1328 (1969); *Sov. Phys.—Semicond.* (*Engl. Transl.*) **3**, 1107 (1970).

---

* Part 4 is by **Henry Kressel**.

ature in a GaAs heterojunction laser. The thinner the recombination region (i.e., the smaller the volume of material in which population inversion is needed), the lower the threshold current density—within limits, of course.

(2) An optical cavity is needed to produce feedback. A Fabry–Perot cavity is commonly used, which is produced by cleaving two parallel facets of the device a distance typically 300 μm apart. However, feedback can also be produced by periodic variations of the refractive index (in distributed-feedback lasers), thus eliminating the need for the cavity mirrors.

(3) An internal waveguide must be provided to confine, to a reasonable extent, the stimulated radiation to the region producing gain. The degree of radiation confinement to the recombination region depends on the dielectric steps bounding that region. A key feature that distinguishes homojunction from heterojunction lasers is that the dielectric steps of heterojunction structures provide this function in a controlled manner. The index steps need only be a few percent for essentially full confinement in a typical heterojunction laser. A much lesser degree of confinement exists in homojunction lasers.

The construction features and metallurgical quality of the structure are key features of useful lasers. Although many materials are candidates for heterojunction laser construction, few have so far been used because of the difficulty of producing certain semiconductors of sufficient quality by epitaxial growth techniques.

Epitaxial material synthesis lies at the heart of laser diode concentra-

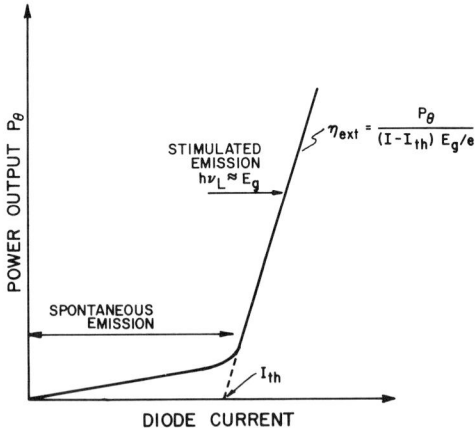

FIG. 1. Power output from laser diode as a function of current. The threshold current is I_{th} and the differential quantum efficiency is η_{ext}.

4.1. INTRODUCTION

tion because of the need to tailor the energy bandgap profile of the device. Liquid phase epitaxy has been the most popular technique, but vapor phase epitaxy and molecular beam epitaxy (i.e., evaporation) are used. Although the epitaxial growth technique should be irrelevant to the device performance, all factors being constant, subtle stoichiometric differences may impact the laser reliability and internal quantum efficiency.

Key injection laser parameters are the threshold current density, differential quantum efficiency, and power level that can be reliably produced. Figure 1 shows a typical curve of power emitted P_θ versus diode current, where we see the region of spontaneous emission below threshold and the steep increase in power emitted as the lasing threshold is traversed. The threshold current density (and threshold current) as well as the operating range of the laser depend on the diode topology and the internal geometry.

4.1.1. Laser Topology

Figure 2 shows the two basic laser diode configurations that differ in the way the junction area is defined with respect to laser operation. In both of the structures shown the Fabry–Perot resonator is formed by cleaving two parallel facets. A reflecting film is sometimes placed over one facet to increase the useful output at the opposite end. Figure 3 is the schematic of the basic heterojunction laser diode waveguide, which we will have occasion to refer to as we discuss the operation of these devices.

Waves propagating parallel to the cleaved facets must be suppressed by the introduction of high losses peculiar to them. For large-area (generally

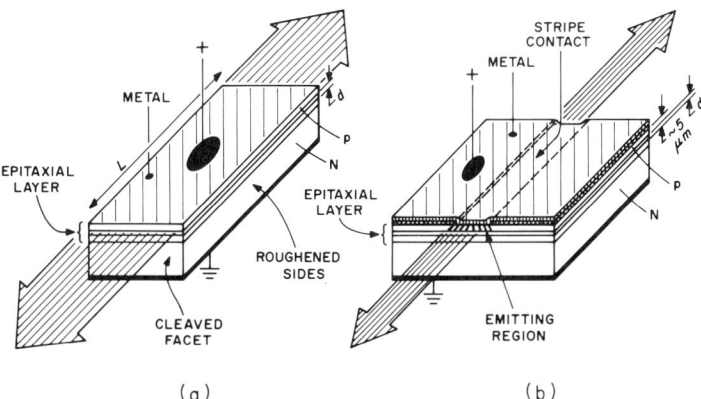

FIG. 2. (a) Broad-area laser formed by cleaving two facets and sawing the other two sides. (b) Stripe-contact laser formed by restricting the active diode area using selective metallization.

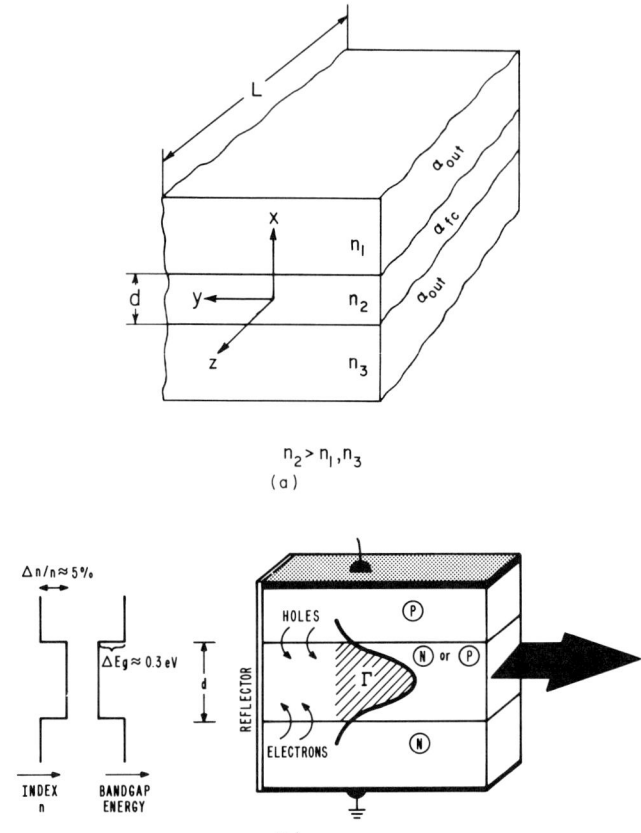

FIG. 3. (a) Model of the heterojunction laser diode as a three-region dielectric slab with refractive index n_1, n_2, and n_3 in the three regions, respectively. The absorption coefficient at the lasing wavelength is indicated for each of the regions. (b) Double-heterojunction laser diode showing a propagating wave with the fraction Γ within the recombination region. The bandgap energy steps and associated refractive index steps are shown on the left of the laser schematic.

denoted "broad-area") devices, sawing the sidewalls achieves this. In a stripe-contact diode the active junction area is restricted by other means. The simplest planar stripe-contact laser is the oxide-isolated one shown in Fig. 4a. Here, the current restriction produces a shallow maximum in the dielectric constant, which forms a dielectric waveguide and thus confines the optical field to a region below the stripe (although radiation spread beyond the striped region occurs). Other stripe-contact methods include methods for confining the current in the junction plane by increasing the lateral resistance by proton bombardment (Fig. 4e). The maximum lat-

4.1. INTRODUCTION

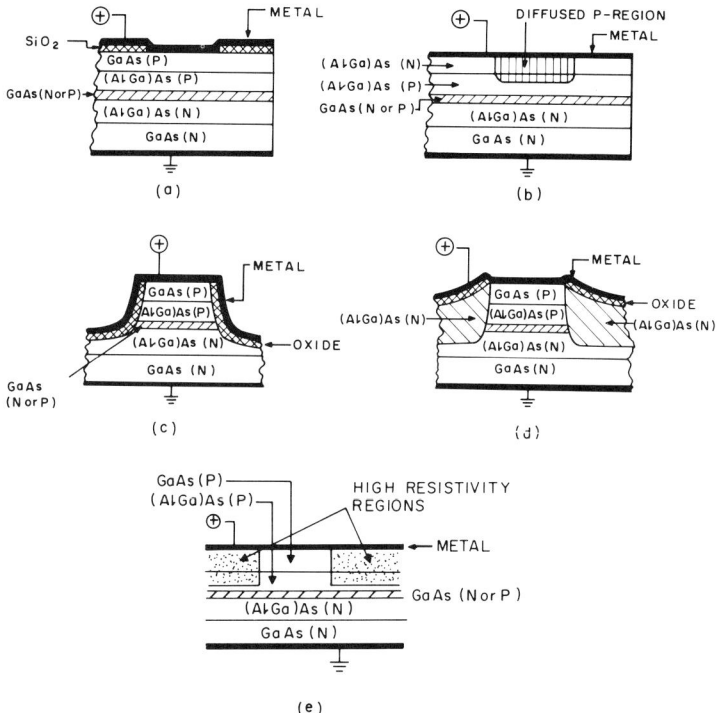

FIG. 4. Methods of forming stripe-contact lasers. (a) The metallization to the active area is restricted using an oxide layer (SiO_2); (b) selective diffusion is made into an n-type surface layer, resulting in a back-biased junction except in the active region; (c) narrow mesa formed by etching the sides of the active region; (d) the active area is defined by first etching a mesa and then regrowing high-bandgap (resistive) AlGaAs into the etched regions (this structure is denoted the "buried heterostructure"), (e) the resistivity on the two sides of the active region is increased by proton bombardment. Structures (a), (b), and (e) are denoted planar stripe structures.

eral confinement is obtained by introducing two heterojunctions, one on each side of the area to be defined as shown in Fig. 4d, or with a mesa, Fig. 4c.

4.1.2. Vertical Geometry

Figure 5 shows the schematic cross section of major classes of laser diodes using from one to four heterojunctions. For each structure, we show the energy diagram, the refractive index profile, distribution of the optical energy ($\propto E^2$) and the position of the recombination region. These structures (listed in historical stage of evolution) are of increasing complexity:

214　　　4. SEMICONDUCTOR DIODE LASERS

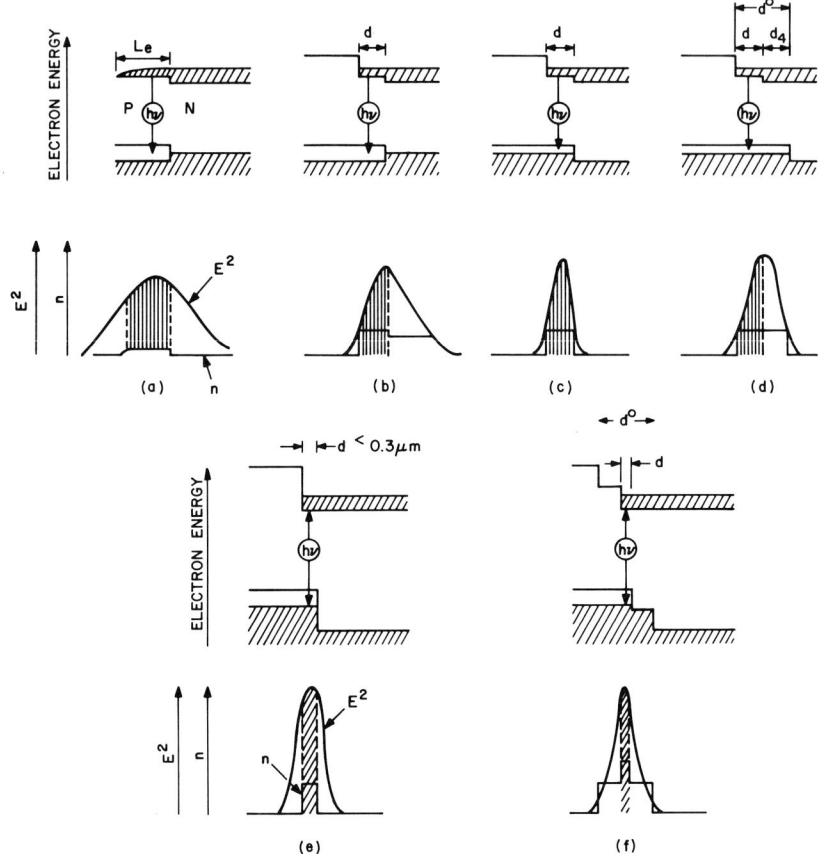

FIG. 5. Schematic cross section of 6 laser configurations. The shaded region in each case is the region of electron–hole recombination. The curve E^2 denotes the optical field intensity.

(1) In the homojunction laser[4-6] there are no abrupt refractive index steps for optical confinement or significant potential energy barriers for carrier confinement. The recombination region width is essentially set by the minority carrier diffusion length. The (small) radiation confinement is the result of refractive index gradients resulting from dopant concentra-

---

[4] R. N. Hall, G. E. Fenner, J. D. Kingsley, T. J. Soltys, and R. O. Carlson, *Phys. Rev. Lett.* **9**, 366 (1962).

[5] M. I. Nathan, W. P. Dumke, G. Burns, F. H. Dill, Jr., and G. Lasher, *Appl. Phys. Lett.* **1**, 63 (1962).

[6] T. M. Quist, R. H. Rediker, R. J. Keyes, W. E. Krag, B. Lax, A. L. McWhorter, and H. J. Ziegler, *Appl. Phys. Lett.* **1**, 91 (1962).

tion gradients and carrier concentration differences. Typically a p–p–n configuration is used, where the p–p interface provides a small potential barrier.

(2) In the single-heterojunction (close-confinement) diode, a p–p heterojunction forms one boundary of the waveguide as well as a potential barrier for carrier confinement within the p-type recombination region.[1,2] The refractive index step at the p–p heterojunction is much larger (typically a factor of 5) than that at the p–n homojunction. Thus, this is an asymmetrical waveguide. The threshold current densities are typically 1/4 to 1/5 of the homojunction values at room temperature ($\sim 10{,}000$ vs. $\sim 50{,}000$ A/cm²).

(3) In the double-heterojunction (DH) laser the recombination region is bounded by two high-bandgap regions to confine the carriers and the radiation. The device can be made either symmetrical or asymmetrical. The first reported DH laser had $J_{th} \approx 4000$ A/cm².[3] The reduction to a threshold current density at room temperature of about 2000 A/cm² was thereafter achieved with these devices with GaAs[7] and with (AlGa)As[8] in the recombination region. Values as low as 475 A/cm² have been achieved.[9]

(4) In the large-optical-cavity (LOC) laser, the waveguide region is wider than the recombination region, which occupies one side of the space between the two major heterojunctions.[10] The device was basically intended for efficient pulsed-power operation.

(5) The very narrowly spaced double-heterojunction laser is a subclass of the basic DH device, but it is designed with an extremely thin recombination region $d \approx 0.1$ adjusted to permit the wave to partially spread outside d but still provide full carrier confinement.[11] This is to minimize the beam divergence while keeping $J_{th} < 3000$ A/cm², a value desirable for room temperature cw operation.

(6) In the four-heterojunction (FH) laser, five regions are included.[12] A submicron-thick recombination region is bracketed by two heterojunctions, which are further enclosed within two outer heterojunctions. The basic concept of an extended waveguide region is similar to the thin DH concept. The recombination region is generally centered within the

---

[7] M. B. Panish, I. Hayashi, and S. Sumski, *Appl. Phys. Lett.* **16**, 326 (1970).

[8] H. Kressel and F. Z. Hawrylo, *Appl. Phys. Lett.* **17**, 169 (1970).

[9] H. Kressel and M. Ettenberg, *J. Appl. Phys.* **47**, 3533 (1976).

[10] H. F. Lockwood, H. Kressel, H. S. Sommers, Jr., and F. Z. Hawrylo, *Appl. Phys. Lett.* **17**, 499 (1970).

[11] H. Kressel, J. K. Butler, F. Z. Hawrylo, H. F. Lockwood, and M. Ettenberg, *RCA Rev.* **32**, 393 (1971).

[12] G. H. B. Thompson and P. A. Kirkby, *IEEE J. Quantum Electron.* **qe-9**, 311 (1973).

waveguide region. By adjusting the index step between the recombination region and the adjoining region, maximum design flexibility is achieved. High peak power operation is obtained by widening the optical field distribution. Alternatively, J_{th} values of 500–1000 A/cm² are obtained by restricting the heterojunction spacings.[13,14]

The heterojunction laser design flexibility, made possible by sophisticated epitaxial technologies, offers ample opportunity for specific device designs consistent with applications. Thus, lasers can be designed for high peak power, but low duty cycle operation. For example, peak power values in excess of 10 W are easily obtained from single-heterojunction lasers at duty cycles of about 0.1%, and much higher values can be achieved by stacking diodes, by the formation of arrays, or by widening the emitting region in the junction plane. On the other end of the power scale are lasers designed for cw operation at room temperature. Such lasers have small junction areas and are designed to emit continuous power levels of about 5–20 mW, although about 100 mW has been achieved with laboratory devices.[15]

Of course, the diode geometry varies with the intended device. Low threshold is not essential for low duty cycle pulsed lasers, but the ability to sustain high optical power levels without facet damage is required. For cw lasers, on the other hand, the lowest possible threshold current density is needed and this is generally achieved with narrow recombination region devices of the double-heterojunction class, although low threshold operation can also be obtained with LOC and four-heterojunction lasers suitably designed. The DH laser, however, is relatively simple to construct and is most widely used for the fabrication of cw lasers. The DH laser is also simple to model in terms of a three layer dielectric slab and is thus well suited to illustrate the principles of laser operation. We shall not be concerned in this review with specific laser applications, of which optical communications using fibers is one of the most important.

4.2. Injection

Heterojunctions are usually divided into two classes.[16] *Isotype heterojunctions* consist of two adjoining regions of different bandgap energy but

---

[13] G. H. B. Thompson and P. A. Kirkby, *Electron. Lett.* **9**, 295 (1973).

[14] M. B. Panish, H. C. Casey, Jr., S. Sumski, and P. W. Foy, *Appl. Phys. Lett.* **11**, 590 (1973).

[15] H. Kressel and I. Ladany, *RCA Rev.* **36**, 230 (1975).

[16] Extensive reviews are given by A. G. Milnes and D. L. Feucht, "Heterojunctions and Metal-Semiconductor Junctions." Academic Press, New York, 1972; B. L. Sharma and R. K. Purohit, "Semiconductor Heterojunctions." Pergamon, Oxford, 1974.

4.2. INJECTION

of the same conductivity type. *Anisotype heterojunctions* have different bandgaps and opposite conductivity types. In single-heterojunction lasers (Fig. 5b) an isotype heterojunction is placed within 2 μm of an injecting p–n homojunction. In the double-heterojunction structure (Fig. 5c), an isotype heterojunction and anisotype heterojunction form the borders of the recombination region and provide the carrier injection into that region.

To understand the operation of heterojunction lasers, we need not dwell on the many models proposed for the transport mechanism across heterojunctions. Complications are frequently introduced by interfacial defects (Chapter 4.7) and the current–voltage characteristics are controlled by a combination of thermal injection over the barrier, tunneling through the barrier, and space-charge region recombination. The relative importance of these factors depends on the applied voltage. For example, tunneling is dominant at very low bias values, whereas thermal injection dominates in the high-bias operating range of interest in lasers. Unintentional bandgap grading in the transition between the two sides of the heterojunction further complicates the understanding of the transport processes across heterojunctions.

For our purpose, it suffices to state that the current density J is related to the applied potential across the device V_a by an expression of the form

$$J = J_0\{\exp[A(V_a - V_s)] - 1\}, \qquad (4.2.1)$$

where J_0 is the saturation current density, V_a the applied potential, $V_s = IR_s$ (I = current, R_s = series resistance), and A a function of temperature, junction quality, and operating range. Of course, $A = e/kT$ for an ideal homojunction operating in a regime where the injected carrier density is substantially below the background carrier concentration. In the typical good-quality AlGaAs/GaAs heterojunction, A and J_0 vary with temperature as shown in Fig. 6.

A useful feature of anisotype heterojunctions is the high electron or hole injection efficiency. For example, for high *electron* injection efficiency, we make the bandgap of the *n side* of the heterojunction higher than that of the p side. Conversely, for high *hole* injection, the *p side* of the heterojunction has a higher bandgap energy than the n side.

Consider a n–p heterojunction where the bandgap energy of the n side is ΔE_g greater than that of the p side. In the simplest heterojunction model, the ratio of electron to hole flow across the junction is

$$\frac{J_e}{J_h} = \left(\frac{D_e L_h N_D}{D_h L_e N_A}\right)\left(\frac{m_{e1}^* m_{h1}^*}{m_{e2}^* m_{h2}^*}\right)^{3/2} \exp\left(\frac{\Delta E_g}{kT}\right), \qquad (4.2.2)$$

where D_e and D_h are the diffusivities of electrons and holes, respectively; L_e and L_h the minority carrier diffusion lengths of electrons and holes; N_D

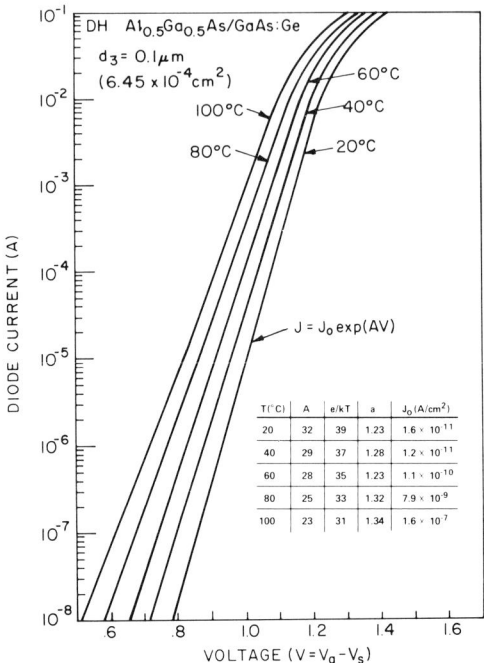

FIG. 6. Current–voltage characteristics in forward bias of an $Al_{0.5}Ga_{0.5}As/GaAs$:Ge double-heterojunction diode having a recombination region width $d = 0.1$ μm. The diode area is 6.45×10^{-4} cm². The insert shows the variation of A and J_0 with temperature and the ratio $a = e/AkT$. (After H. Kressel and M. Ettenberg, unpublished.)

and N_A the donor and acceptor concentrations in the n- and p-type regions, respectively; m_{e1}^*, m_{e2}^*, m_{h1}^*, m_{h2}^* the density of states effective masses for electrons and holes in side 1 and 2. For a typical p–n AlGaAs/GaAs heterojunction, the ratio of electron to hole flow is in excess of 10^3 at room temperature for $\Delta E_g \approx 0.2$ eV. Thus, we can provide very efficient injection into the recombination region of the structure with rather modest bandgap differences.

Irrespective of the detailed current injection process, the applied voltage across a lasing device is approximately

$$V_a \cong IR_s + E_g/e, \quad (4.2.3)$$

where E_g is the bandgap energy of the recombination region. In the simplest model for the semiconductor laser, where the injected carrier population is assumed to remain locked at the threshold value, the *junction* voltage remains constant with drive above threshold because the quasi-Fermi level separation is constant. In practice, the junction voltage itself may increase somewhat with drive, but (4.2.3) still provides a useful approximation. The value of R_s in widely used laser diodes varies from 0.1 to about 2 Ω.

4.3. Carrier Confinement

An isotype heterojunction provides the essential minority carrier confinement within the recombination region. Although a potential barrier several times greater than the thermal carrier energy kT should suffice for effective confinement to the low-bandgap region, a more detailed analysis of the problem[17,18] shows that the loss of injected carriers can sometimes be surprisingly high at elevated temperatures.

Suppose that we have a p–p isotype heterojunction where region 1 contains excess electron–hole pairs and has a bandgap energy ΔE_g lower than region 2. The loss of electrons from region 1 into region 2 by thermal activation over the barrier gives rise to a diffusion current,

$$J_{\text{exc}} \cong \left(\frac{eD_{e2}}{L_{e2}}\right)\left(\frac{N_{v2}N_{c2}}{P_2}\right) \exp\left[\frac{-(\Delta E_g - \delta_{e1})}{kT}\right]. \quad (4.3.1)$$

Region 2 is assumed *not* doped to degeneracy; D_{e2}, L_{e2} are the electron diffusivity and diffusion length in the high bandgap region; P_2 is the hole concentration in that region; N_{v2} and N_{c2} are the density of states for the valence and conduction bands in region 2; δ_{e1} is the energy separation between the bottom of the conduction band and the electron quasi-Fermi level in the lower bandgap region 1.

For $\Delta E_g \geqslant 0.2$ eV, J_{exc} is negligible at room temperature, but the carrier loss can be substantial for steps on the order of 0.1 eV. For small-bandgap steps, it is particularly important to keep the doping level of region 2 high in order to push the Fermi level as close to the valence band as possible and thus maximize the barrier height seen by the electrons in the conduction band of region 1.

The threshold current density includes the current density component needed to inject a carrier pair density N_{th} into the recombination region, the excess current density J_{exc} due to electron loss from the p-type recombination region, and the current density J_{h} due to holes injected from the recombination region into the higher bandgap n-type region,

$$J_{\text{th}} = \frac{eN_{\text{th}}d}{\tau_s} + J_{\text{exc}} + J_{\text{h}}, \quad (4.3.2)$$

[17] D. L. Rode, *J. Appl. Phys.* **45**, 3887 (1974).
[18] A. R. Goodwin, J. R. Peters, M. Pion, G. H. B. Thompson, and J. E. A. Whiteaway, *J. Appl. Phys.* **46**, 3126 (1975).

where τ_s is the spontaneous recombination carrier lifetime in the recombination region of width d; in GaAs devices, $\tau_s = 2$–3 nsec near threshold.

The ratio of electron to hole flow across the n–p heterojunction barrier is given by Eq. (4.2.2), and it is normally very large in typical DH lasers. Likewise, the loss of minority carriers is negligible at room temperature for reasonable barrier heights ($\geqslant 0.2$ eV). However, at elevated temperatures, these effects cannot be neglected unless $\Delta E_g \geqslant 0.4$ eV. [Of course, for an n-type recombination region, J_{exc} is due to holes and J_e replaces J_h in (4.3.2).]

4.4. Radiation Confinement

The basic heterojunction laser is a three-layer dielectric slab with constant refractive index in each layer at the lasing energy (Fig. 3), where the recombination or active region is bracketed by two higher bandgap regions. Maxwell's equations are solved with appropriate boundary conditions to determine the modal properties. This analysis yields the fraction of the radiation confined to the active region Γ and the *transverse* modes capable of propagation in the waveguide. However, which of the modes actually dominates is not determined from such an analysis unless the losses for the various modes are estimated. It is assumed that the mode having the lowest propagation loss reaches threshold first.

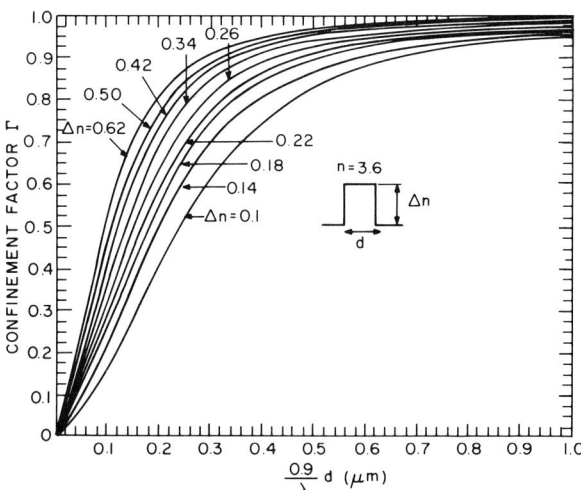

FIG. 7. Fraction of the radiation confined to the recombination region of a double-heterojunction as a function of the width of the region d (normalized for wavelength). The index steps Δn for the various curves are indicated. The index of the recombination region is $n = 3.6$. (After Kressel and Butler.[20])

4.4. RADIATION CONFINEMENT

There is no cut-off condition for a symmetric waveguide, i.e., one with equal index steps at the two walls, but as the waveguide width d is reduced, more of the wave propagates outside the guide. Figure 7 shows the fraction of the radiation confined to the active region of width d for various values of the index step Δn, in a device where the index $n = 3.6$

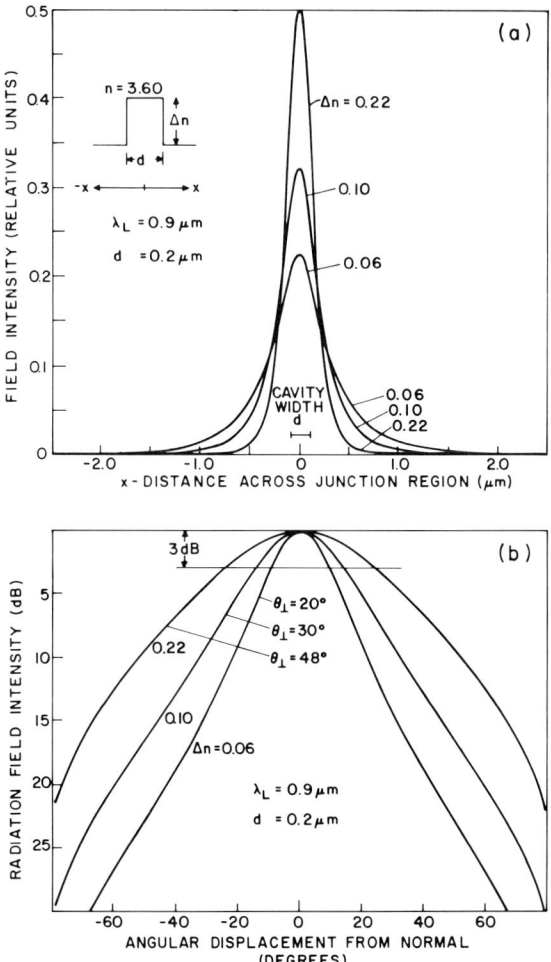

FIG. 8. (a) Radiation distribution in the vicinity of the recombination region of a double-heterojunction laser of width $d = 0.2$ μm for varying index step Δn. (b) Corresponding far-field in the direction perpendicular to the junction plane. The noticeable increase in the beam width with increasing radiation confinement is evident. (After Kressel and Butler.[20])

[19] J. K. Butler and H. Kressel, *RCA Rev.* **38**, 542 (1977).

in the recombination region.[19] As we shall see shortly, Γ is a key quantity in determining the efficiency and threshold of a laser.

The distribution of the radiation in the vicinity of the recombination region of the double-heterojunction laser affects the radiation pattern. Figure 8 shows the effect of changing the index step with the width of the recombination region constant at $d = 0.2$ μm.[20] In Fig. 8a, the areas under the curves are normalized to unity to illustrate the change in the energy distribution as the index step is changing. The corresponding far-field in the direction perpendicular to the junction is shown in Fig. 8b. The beam spread decreases, as expected, with decreasing radiation confinement to the recombination region.

4.5. Gain Coefficient and Threshold Condition

Gallium arsenide is a semiconductor for which extensive theoretical calculations have been made relating the current density to the gain coefficient as a function of bandstructure, doping level, and temperature.[21] These calculations neglect the presence of a strong optical field, which may perturb the inversion level for a given injection rate. However, this does not introduce a serious problem in calculating the *threshold* condition because the field intensity is low.

The calculated relationships are conveniently given in terms of the gain coefficient g as a function of J_{nom}, the unidirectional current density for an active region 1 μm thick. To convert from J_{nom} to the threshold current density J_{th} we need to determine: (1) the gain coefficient value at threshold g_{th}, which must equal the prorated mode loss; (2) the radiation confinement Γ to the recombination region producing the gain; (3) the internal quantum efficiency η_i relevent to stimulated emission; (4) the excess current components [Eq. (4.3.2)], neglected here.

The condition for the laser threshold taking into account the absorption coefficient in the three regions of the laser is easily derived. For simplicity, assume that the absorption coefficient in regions 3 and 1 of the structure of Fig. 3 is equal to α_{out} and that the free carrier absorption coefficient within the recombination region is α_{fc}. Then the gain coefficient at threshold g_{th} is determined from the condition

$$\Gamma(g_{th} - \alpha_{fc}) = \alpha_{out}(1 - \Gamma) + \text{cavity end loss.} \tag{4.5.1}$$

[20] H. Kressel and J. K. Butler, "Semiconductor Lasers and Heterojunction LEDs." Academic Press, New York, 1977.

[21] F. Stern, *IEEE J. Quantum Electron.* **qe-9**, 290 (1973).

4.5. GAIN COEFFICIENT AND THRESHOLD CONDITION

For a Fabry–Perot cavity of length L and facet reflectivity R, the cavity end loss = $(1/L)\ln(1/R)$; $R = 0.32$ in GaAs. Hence,

$$g_{th} = \alpha_{out}\frac{(1-\Gamma)}{\Gamma} + \frac{1}{L\Gamma}\ln\left(\frac{1}{R}\right) + \alpha_{fc}. \tag{4.5.2}$$

A simple expression was derived by Stern[21] relating the gain coefficient to the nominal current density:

$$g = \beta_s (J_{nom} - J_1)^b. \tag{4.5.3}$$

Here b, J_1, and β_s are constants that vary with temperature and doping. Combining (4.5.3) and the definition of J_{nom}, we obtain an expression for the threshold current density taking $b \approx 1$,

$$J_{th} \cong \frac{d}{\eta_i}\left(\frac{g_{th}}{\beta_s} + J_1\right). \tag{4.5.4}$$

For lightly doped GaAs at 300 K, $J_1 = 4100$ A/cm²-μm and $\beta_s = 0.044$ cm-μm/A; at 80 K, $J_1 = 600$ A/cm²-μm and $\beta_s = 0.16$ cm-μm/A.

It is evident that decreasing d reduces J_{th} until the eventual steep Γ decrease; for any heterojunction structure, there is therefore a minimum achievable J_{th}, which is determined by the magnitude of the index steps at the heterojunctions. Figure 9 shows experimental data comparing the variation of the threshold current density at room temperature for symmetric double-heterojunction GaAs/Al$_x$Ga$_{1-x}$As lasers where both d and the index step were varied by changing the Al concentration x.[9] Except where the value of d is reduced below the optimum value, we see that J_{th} varies with d linearly,

$$J_{th}/d \cong 4800 \quad \text{A/cm}^2\text{-}\mu\text{m}. \tag{4.5.5}$$

Other important laser parameters are the differential quantum efficiency above threshold η_{ext}, and the power conversion efficiency η_p:

$$\eta_{ext} = \eta_i'\frac{(1/L)\ln(1/R)}{(1/L)\ln(1/R) + \Gamma\{\alpha_{fc} + [(1-\Gamma)/\Gamma]\alpha_{out}\}}, \tag{4.5.6}$$

$$\eta_p \cong \frac{P_\theta}{I^2R_s + IE_g/e}, \tag{4.5.7}$$

where P_θ is the emitted power at current I. The power conversion efficiency is evidently very low near threshold, but increases rapidly with drive to reach a maximum at a drive current 2 to 3 times the threshold current, if thermal effects or facet damage are not limiting. Power conversion efficiency values up to about 20% have been obtained in the pulsed mode of operation at room temperature, a value in excess of typical solid

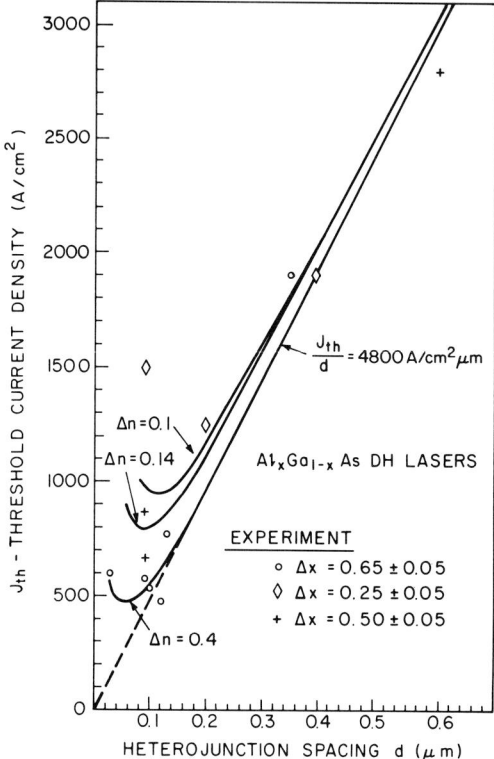

FIG. 9. Threshold current density as a function of double-heterojunction spacing of $Al_xGa_{1-x}As/GaAs$ lasers for varying values of x and corresponding index steps Δn. The lines are calculated. The experimental data are indicated.[9]

state or gas laser efficiencies (with the exception of CO_2 lasers). Differential quantum efficiency values as high as 50% have been attained at room temperature.

4.6. Temperature Dependence of the Threshold Current Density

It is experimentally found that the threshold current density of many laser diodes increases with temperature following an expression of the form

$$J_{th} \propto \exp[(T/T_0)], \qquad (4.6.1)$$

where the value of T_0 and the range of validity of the expression varies. A common observation is that T_0 is in the 75–160 K range, and that (4.6.1) is valid between about 100 and 350 K.

4.6. TEMPERATURE DEPENDENCE OF THE THRESHOLD

Equation (4.6.1) is the combined result of many factors that contribute to increasing J_{th} with temperature. In fact, careful measurements made on some lasers show that (4.6.1) is only a rough approximation. For example, AlGaAs/GaAs DH lasers with lightly doped recombination regions were found to follow a temperature dependence proportional to $T^{1.4}$ between 77 and 300 K.[22]

The factors that contribute to the J_{th} change with temperature consist of the following major elements: (1) the change in carrier and radiation confinement, (2) the change in the average absorption coefficient, and (3) the inherent change in the current density versus gain relationship.

1. Change in Confinement. With relatively small heterojunction barriers, carrier loss with increasing temperature can become substantial and contributes to an increase in J_{th} [see Eq. (4.3.2)]. Radiation confinement changes can also occur if the index steps bounding the mode-guiding region change. For example, in the single-heterojunction laser, the index step at the p–n homojunction interface tends to decrease with increasing temperature, producing a reduction in Γ.

2. Change in Absorption Coefficient. The free-carrier concentration α_{fc} in the recombination region increases with temperature. Furthermore, if Γ shrinks with increasing temperature, the wave spreads into the lossy regions adjoining the recombination region.

3. Current Density versus Gain Relationship. Because the carrier distribution in the conduction and valence bands changes with temperature, an increasing pair density is needed to produce a given gain value at increasing temperatures. The internal quantum efficiency may also decrease with increasing temperature, but in the best lasers this factor is relatively small. The internal quantum efficiency at room temperature of GaAs lasers is generally estimated to be 0.6–0.8, increasing to ~1 at very low temperatures.

It follows from the above summary that the lowest J_{th} change with temperature should occur in devices where neither the radiation nor the carrier confinement are temperature sensitive, and where the defect density in the recombination region is low, resulting in a temperature-insensitive internal quantum efficiency. Experimentally, we indeed find that strongly confined DH lasers have a small temperature dependence of the threshold current density. Figure 10 shows the ratio of J_{th} at 70°C to the value at 22°C for quality AlGaAs/GaAs lasers.[9,18] Note that with increasing Al concentration difference at the heterojunctions (and corresponding increase in the bandgap step) the temperature sensitivity of J_{th} is reduced. Eventually, of course, the barrier becomes so high that the carrier con-

---

[22] H. Kressel and H. F. Lockwood, *Appl. Phys. Lett.* **20**, 175 (1972).

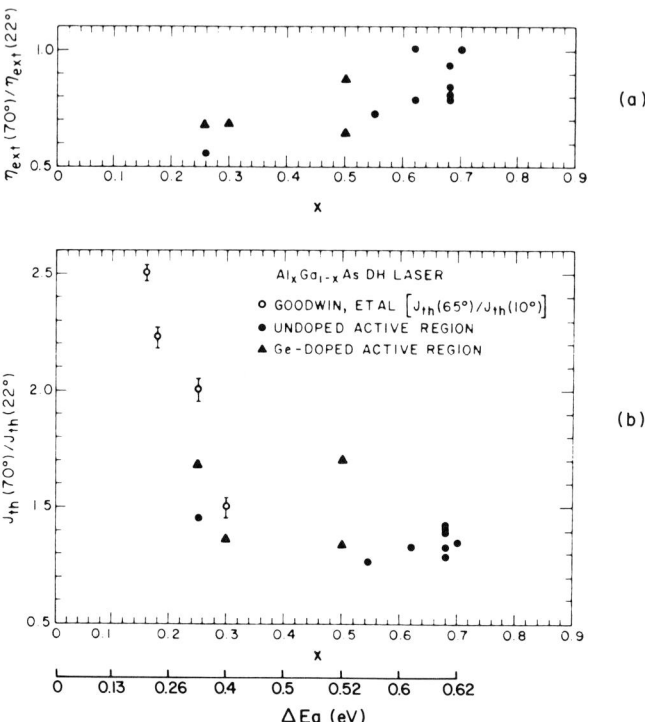

FIG. 10. (a) Ratio of the differential quantum efficiency at 22° and at 70°C for double-heterojunction $Al_xGa_{1-x}As/GaAs$ lasers as a function of x. (b) Ratio of the threshold current density at 22° and at 70°C as a function of x. ○, Data of Goodwin et al.[18] △, ●, Kressel and Ettenberg.[9]

finement loss at high temperature is completely negligible. In the devices shown this occurs for a bandgap step of 0.3–0.4 eV.

4.7. Materials

As discussed in the previous chapters, heterojunction lasers require confining heterojunction barriers with bandgap energy steps of several kT. It is essential, however, that the lattice parameter at the heterojunctions be matched as closely as possible. The tolerable lattice mismatch for reasonable device performance is not generally established because the correlation between lattice misfit, misfit dislocation density, and the resultant density of nonradiative centers differs with the material.

A simplified model provides insight into the impact of nonradiative interfacial recombination on the internal quantum efficiency. For thin double-heterojunction structures, the internal quantum efficiency η for spontaneous recombination is approximately,[23,24]

$$\eta \cong (1 + 2S\tau_r/d)^{-1}, \tag{4.7.1}$$

where τ_r is the radiative carrier lifetime, S the interfacial recombination velocity (assumed equal at the heterojunctions), and the condition $SL/D \ll 1$ is satisfied, where L is the minority carrier diffusion length in the absence of the interfaces and D the minority carrier diffusion constant.

Consider a p-type recombination region; τ_r is typically 3 nsec near threshold, and with $d = 0.2$ μm we need $S \leq 3.3 \times 10^3$ cm/sec to ensure an internal quantum efficiency of at least 50%. The recombination velocity is related to defects at the heterojunction interface. A simple analysis, which yields satisfactory agreement with experiment, starts with an estimate of the density of centers introduced by misfit dislocations.

The surface density of recombination centers, N_{ss}, introduced by misfit dislocations lying in the interfacial plane can be calculated by assuming that each atom terminating an edge dislocation constitutes a recombination center and that the lattice misfit strain is fully relieved by the misfit dislocations. The calculated value of N_{ss} depends on the crystal structure and orientation. For a sphalerite structure on the (100) interfacial plane,[25]

$$N_{ss} = 4 \frac{a_1^2 - a_2^2}{a_1^2 a_2^2} \approx \frac{8 \Delta a_0}{a_0^3} \text{ cm}^{-2}, \tag{4.7.2}$$

and for the (111) plane,

$$N_{ss} = \frac{4}{3^{1/2}} \frac{a_1^2 - a_2^2}{a_1^2 a_2^2} \approx \frac{8}{3} \frac{\Delta a_0}{a_0^3} \text{ cm}^{-2}. \tag{4.7.3}$$

We assume that $\Delta a_0 = a_1 - a_2$, with $a_1 \approx a_2 \approx a_0$. The recombination velocity at the heterojunction interface can be estimated from the above N_{ss} values,[26]

$$S = v_{th} N_{ss} \sigma_t, \tag{4.7.4}$$

---

[23] Interfacial recombination is theoretically treated by A. Many, Y. Goldstein, and N. B. Grover, "Semiconductor Surfaces." North-Holland Publ., Amsterdam, 1965.

[24] Nonradiative recombination at heterojunction interfaces was discussed by R. D. Burnham, P. D. Dapkus, N. Holonyak, Jr., D. L. Keune, and H. R. Zwicker, *Solid-State Electron.* **13**, 199 (1970).

[25] D. B. Holt, *J. Phys. Chem. Solids* **27**, 1053 (1966).

[26] H. Kressel, *J. Electron. Mater.* **4**, 1081 (1975).

where v_{th} is the thermal electron velocity and σ_t the capture cross section for the center. A value of $\sigma_t \cong 10^{-15}$ cm² is believed of the correct order of magnitude for GaAs.

Hence, for the (100) interfacial plane,

$$S \approx 2.5 \times 10^7 (\Delta a_0/a_0), \quad (4.7.5a)$$

and for the (111) plane

$$S \approx 0.7 \times 10^7 (\Delta a_0/a_0). \quad (4.7.5b)$$

These expressions assume a simple model for the abrupt metallurgical interface between the two materials. In practice, matters are considerably more complicated because the interface does not contain a simple array of misfit dislocations, particularly if the misfit becomes substantial. A low interfacial recombination velocity is indeed obtained with $Al_xGa_{1-x}As$/GaAs heterojunctions. The experimentally determined values are $S \cong 4 \times 10^3$ cm/sec for $x = 0.25$ and 8×10^3 cm/sec for $x = 0.5$.[27] Since no misfit dislocations are noted in this alloy system where the lattice parameter is matched at the growth temperature, the origin of the interfacial recombination centers must be other than dislocations, but no theoretical model is as yet available. Experimental studies of (100) InGaP/GaAs interfaces show that S versus $\Delta a_0/a_0$ follows (4.7.5a). Note that for $\Delta a_0/a_0 \approx 1\%$, $S \geq 3 \times 10^5$ cm/sec, approaching in effect the value $S \approx 10^6$ cm/sec found at a free GaAs surface.[28]

Heterojunction lasers require a direct bandgap material in the recombination region but not in the bounding layers. Figure 11 shows the lattice constant, bandgap energy, and corresponding diode emission wavelength of several III–V materials that can be doped both n- and p-type. Of course, AlGaAs alloys are particularly useful because the small (about 0.1%) lattice constant change between AlAs and GaAs makes it possible to grow very nearly lattice-matched structures of AlGaAs on GaAs substrates.

Other heterojunction structures can be deposited on lattice-mismatched binary substrates such as InP, GaAs, GaP, InAs, or InSb. However, to minimize the density of misfit dislocations in the epitaxial structures containing the active region of the device, it is essential to grade the composition between the substrate and the active region. Step grading in vapor-phase epitaxy has been found to be most effective in minimizing dislocation propagation if the lattice constant of the successive step-grading layers is of increasing size, thus placing the layers in compres-

---

[27] M. Ettenberg and H. Kressel, *J. Appl. Phys.* **47**, 1538 (1976).
[28] M. Ettenberg and G. H. Olsen, *J. Appl. Phys.* **48**, 4275 (1977).

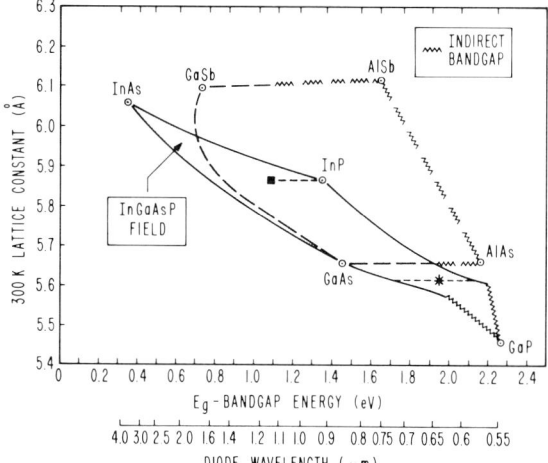

FIG. 11. Lattice constant for various III–V compounds and alloys related to the bandgap energy and the corresponding diode emission wavelength. The dotted line illustrates heterojunction lattice matching possibilities using an InP substrate and an InGaAsP recombination region or a GaAsP substrate.

sion.[29] In this case, the misfit dislocations tend to bend over into the interface and terminate at the crystal edges.

It is also noteworthy that mismatched layers can be grown without misfit dislocations if the layers are kept sufficiently thin, because the strain energy in the layer is less than the energy needed to nucleate the strain-relieving misfit dislocations. The maximum thickness h_c of a dislocation-free layer in a III–V compound structure grown on the usual growth planes is approximately[30]

$$h_c \cong a_0^2/2(2)^{1/2} \Delta a_0. \qquad (4.7.6)$$

Thus, very thin layers (usually well under 1 μm) can be produced without misfit dislocations even if the lattice constant differs between the layers. For example, if the lattice misfit is $\Delta a_0/a_0 = 5 \times 10^{-3}$ and $a_0 = 5.6$ Å, the calculated $h_c \cong 0.4$ μm. This effect is of great benefit in the fabrication of double-heterojunction lasers where the thin recombination region is sandwiched between slightly mismatched bounding layers.

Turning to Fig. 11, note that the quaternary alloys of InGaAsP cover a wide direct-bandgap range with constant lattice constant. For example,

[29] G. H. Olsen, M. S. Abrahams, C. J. Buiocchi, and T. J. Zamerowski, *J. Appl. Phys.* **46**, 1643 (1975).
[30] J. W. Matthews, in "Epitaxial Growth" (J. W. Matthews, ed.), p. 562. Academic Press, New York, 1975.

this alloy can be lattice-matched in InP substrates and thus produce heterojunction devices emitting in the 1 μm region. At the other limit of the same alloy, we note that *visible* heterojunction devices can be produced by lattice-matching to GaAsP grown on GaAs.

Although lattice-matching can be realized at one temperature, differences in the thermal coefficient of expansion of the materials comprising the structure lead to a mismatch at other temperatures. It appears preferable to match the lattice constant at the *growth temperature* rather than at room temperature. The final structures will be somewhat strained because misfit dislocations will usually not be formed while the structure is being cooled at room temperature.

4.8. Heterojunction Lasers of Various Materials

Semiconductor injection lasers incorporating heterojunctions can be conveniently grouped into three spectral regions: near-infrared, visible, and infrared.

4.8.1. Near-Infrared Emission Lasers

These devices cover the spectral region 0.8–0.9 μm and are of the most general current interest, particularly as light sources for fiber optical communications. Structures of AlGaAs/GaAs or $Al_xGa_{1-x}As/Al_yGa_{1-y}As$ are widely used and these are presently highly engineered to produce devices capable of operating lifetimes of several years.

Lasers emitting in the 1–1.2 μm region are also of interest in optical communications using glass fibers because of the low attenuation in that spectral region in state-of-the-art fibers. Lasers of lattice-matched InGaAsP/InP, InGaAs/InGaP, and AlGaAsSb/GaAsSb heterostructures all provide emission in the 1 μm spectral region, although the difficulty of achieving good performance varies. The InGaAsP/InP heterojunction structure has the advantage of being constructed on a lattice-matching substrate, thus eliminating the need for composition grading to reduce the dislocation density in the device.

Figure 12a shows a cross section of an InGaAs/InGaP double-heterojunction laser designed for cw operation at room temperature.[31] Its characteristics are shown in Fig. 12b. The threshold current density, as measured in broad-area form, is ~1000 A/cm².

---

[31] C. J. Nuese, G. H. Olsen, M. Ettenberg, J. J. Gannon, and T. J. Zamerowski, *Appl. Phys. Lett.* **29**, 807 (1976).

4.8. HETEROJUNCTION LASERS OF VARIOUS MATERIALS

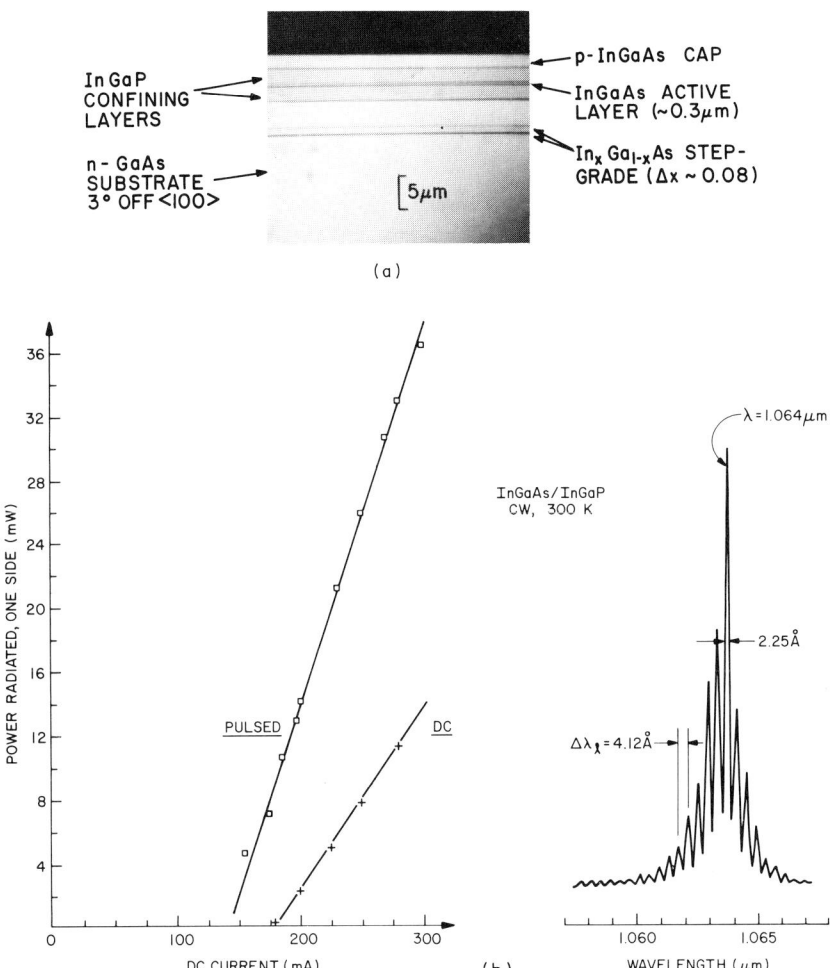

FIG. 12. Lattice-matched double-heterojunction laser of InGaAs/InGaP. (a) Cross section of the structure. (b) Power output as a function of pulsed current and in cw operation at room temperature. The laser emission is at 1.064 μm. (After Nuese et al.[31])

4.8.2. Visible Emission Lasers

At room temperature, the laser diode emission extends into the red portion of the spectrum (about 0.7 μm for cw operation and about 0.65 μm for pulsed operation). Aluminum gallium arsenide structures can operate in the pulsed mode to about 0.69 μm at room temperature and to 0.72 μm in the cw mode of operation.[32] Figure 13 shows the variation of the

---

[32] H. Kressel and F. Z. Hawrylo, *Appl. Phys. Lett.* **28**, 598 (1976); I. Ladany and H. Kressel, *Int. Electron Devices Meet., 1976* p. 129 (1976).

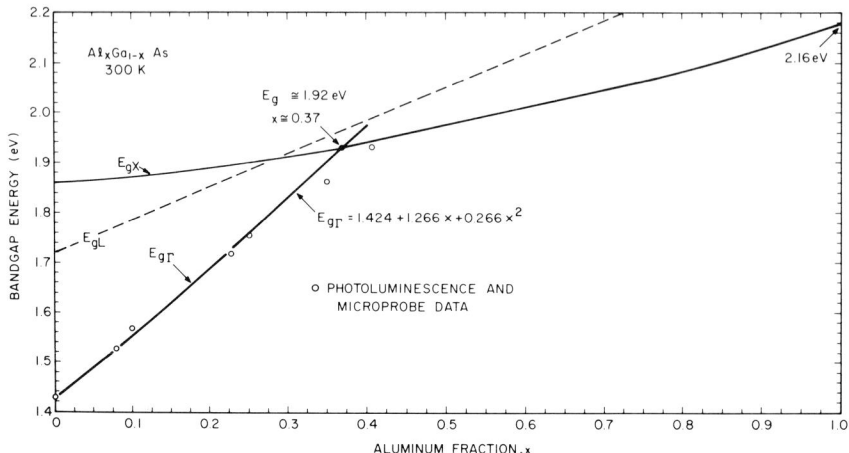

FIG. 13. Bandgap energy of $Al_xGa_{1-x}As$ as a function of x at room temperature. The variation of the L conduction band minima with x is estimated assuming $E_{gL} = 1.72$ eV in GaAs, and 2.4 eV in AlAs.[34]

bandgap energy of $Al_xGa_{1-x}As$ with composition[33,34]; the limit to laser operation is set by the equalization of the direct and indirect bandgaps at about 1.92 eV. The internal quantum efficiency, however, is reduced substantially by the transfer of carriers from the direct to the indirect conduction band minima, resulting in a very steep increase in the threshold current density with increasing bandgap when $E_g \geq 1.75$ eV. This corresponds to a lasing photon energy $h\nu = E_g - 0.03 \approx 1.72$ eV, or $\lambda \cong 7200$ Å.

The use of materials with a more favorable bandstructure for visible emission permits an extension of the emission further into the red. These include InGaAsP/InGaP and GaAsP/InGaP structures, which can be lattice-matched at the heterojunctions. Figure 14a shows a double-heterojunction GaAsP/InGaP laser emitting at about 0.7 μm at 10°C; these have operated cw as shown in Fig. 14b.[35]

4.8.3. Infrared Emission

Heterojunction lasers using IV–VI compounds can provide emission from 2.5 to about 33 μm as shown in Fig. 15, although a substantial lattice

---

[33] O. Berolo and J. C. Wooley, *Can. J. Phys.* **49**, 1335 (1971).

[34] The position of the GaAs L conduction band minima is from D. E. Aspnes, C. G. Olson, and D. W. Lynch, *Phys. Rev. Lett.* **37**, 766 (1976). The L minima in AlAs are assumed to be at 2.4 eV [D. J. Stukel and R. N. Euwema, *Phys. Rev.* **188**, 1193 (1969)].

[35] H. Kressel, G. H. Olsen, and C. J. Nuese, *Appl. Phys. Lett.* **30**, 249 (1977).

4.8. HETEROJUNCTION LASERS OF VARIOUS MATERIALS

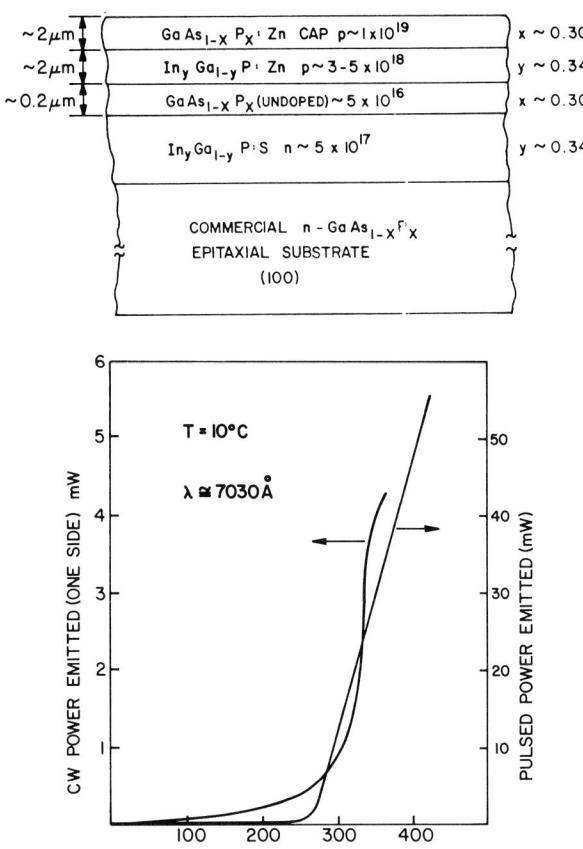

FIG. 14. Lattice-matched double-heterojunction laser of InGaP/GaAsP. (a) Cross section of the structure. (b) Power emitted versus current in pulsed and cw operation at 10°C. The laser emission is at ~0.7 μm. (After Kressel et al.[35])

mismatch is present in the structures. However, it appears that the structures made of these alloys are less susceptible to lattice misfit complications than the III–V compound structures. These heterojunction structures have operated cw at higher temperatures than previously possible with homojunction lasers because of the reduced threshold current density. For example, stripe-contact PbSnTe double-heterojunction lasers have operated cw at heat sink temperatures as high as 114 K. This device can be temperature-tuned between 15.9 and 8.54 μm.[36]

[36] J. N. Walpole, A. R. Calawa, T. C. Harman, and S. H. Groves, *Appl. Phys. Lett.* **28**, 552 (1976).

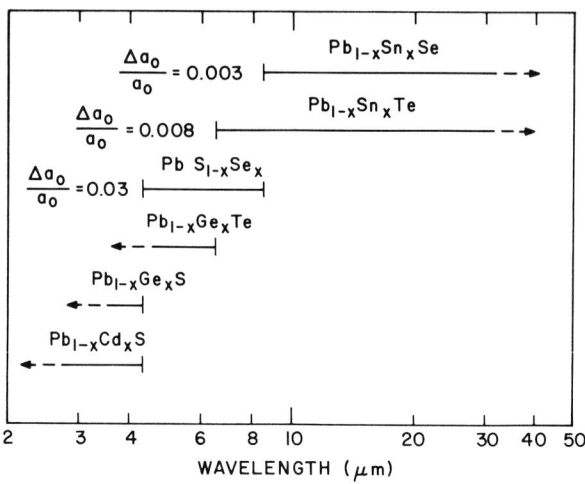

FIG. 15. Heterojunction laser possibilities in the IV–VI compounds. (After A. Grove, unpublished.)

4.9. Performance of Selected Laser Structures

Lasers are designed for either cw operation or pulsed operation by appropriate choice of geometry and internal structure. The useful average power from a specific structure is limited either by thermal effects or degradation, as discussed in Chapter 4.11. In general, $J_{th} < 3000$ A/cm² is needed for room temperature cw operation, although lower values may be required if the electrical or thermal resistance of the device is high or the threshold increase with temperature is steep. The thermal limitations are due to the fact that the junction temperature increases with power dissipation, thus increasing the threshold current density.

The use of relatively narrow stripe-contact structures is advantageous for cw operation because the active region is embedded in the passive (i.e., heat-absorbing) semiconductor, thus minimizing the thermal resistance. In addition, these devices are mounted p-side down on the heat sink as shown in Fig. 16a, thus placing the recombination region within 2–3 μm of the heat sink.[37] The typical AlGaAs double-heterojunction laser shown in Fig. 16 has a thermal resistance of 10–20°C/W. With a pulsed threshold current of 150 mA and resistance $R_s = 0.5\ \Omega$, the power dissipated at threshold in cw operation is $P_{dis} \cong I_{th}^2 R_s + E_g I_{th}/e \cong$

---

[37] I. Ladany and H. Kressel, *Appl. Phys. Lett.* **25**, 708 (1974); H. Kressel and I. Ladany, *RCA Rev.* **36**, 230 (1975).

4.9. PERFORMANCE OF SELECTED LASER STRUCTURES

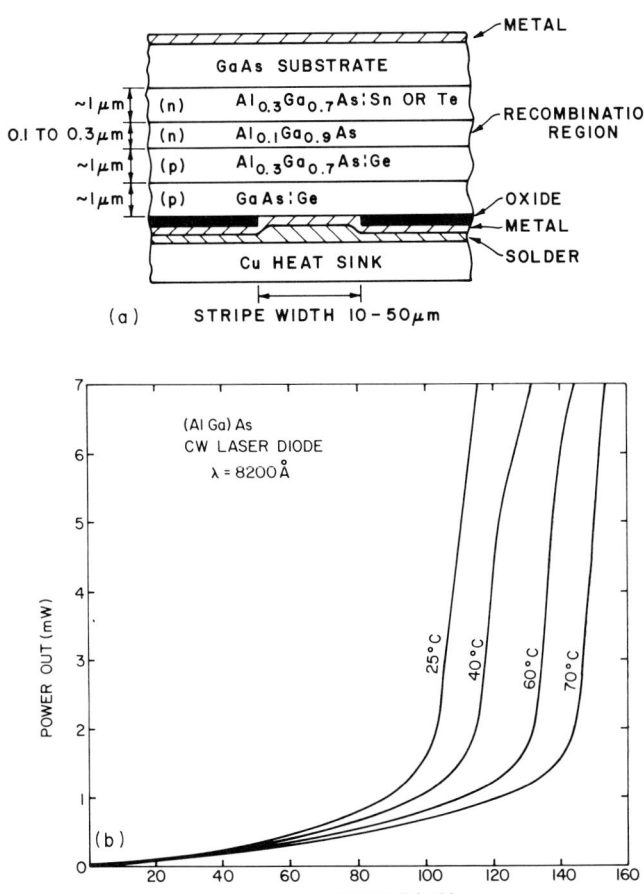

FIG. 16. Typical AlGaAs double-heterojunction laser designed for room temperature cw operation in the 0.82–0.85 μm spectral range. (a) Cross section; (b) temperature dependence of the cw power output as a function of heat sink temperature. (After H. Kressel and I. Ladany, unpublished.)

0.4 W. Assuming a thermal resistance of 20°C/W, the junction temperature is only 8° above the heat sink temperature, and there is little difference in the pulsed and cw threshold currents of such a device. Figure 16b shows the cw power output as a function of temperature of a good quality laser.

A major distinction between the methods of forming stripe-contact lasers in Fig. 4 is in the degree of lateral current and radiation confinement. Ideally, the threshold current should scale with the contact area.

For example, a laser with a stripe width of 10 μm and a length of 300 μm fabricated from material with a threshold current density (measured in broad-area form) of 2000 A/cm² should have a threshold current of 60 mA. In fact, the values obtained with oxide-isolated planar stripe devices are higher by factors of 2 to 4, depending on the spreading resistance of the material near the surface. As the stripe width widens to about 50 μm, however, the effective diode area becomes comparable to the stripe width because the fraction of the "lost" radiation and current becomes negligible.

The stripe formation method that offers the possibility of maximum lateral confinement is the buried-heterojunction one because of the confining barriers introduced by two heterojunctions on either side of the active area. Lasers of this type have provided the lowest threshold current— about 10 mA for stripe widths of only a few micrometers.[38] The fabrication of this device, however, is more difficult than for the planar devices because of the additional processing steps needed.

Laser diodes designed for cw operation at room temperature typically have stripe widths of 10–20 μm and provide useful emission levels of 10–20 mW with currents of 0.1–0.5 A. Diodes designed for high pulsed-power emission, on the other hand, are generally 100–1000 μm wide. Some of these operate at peak power levels of watts at duty cycles $\leqslant 1\%$. Their performance is generally limited by facet damage rather than by junction heating. Because the facet damage limit is related to the power per unit area at the emitting facet, such devices use heterojunction structures with wider mode-guiding regions. The single-heterojunction, large-optical-cavity (LOC), and four-heterojunction configurations with mode-guiding regions of about 2 μm are used for such high-power devices. Of course, the threshold current density is higher than for cw devices; J_{th} values of 10,000 A/cm² are typical at room temperature. However, because the diodes are operated at current densities as high as 50,000 A/cm², the threshold current is less important than the differential quantum efficiency to obtain high peak power with minimum current. A typical figure of merit for reliable operation of high-power lasers is 400 W/cm of emitting facet width (pulse widths of 100–200 nsec). Figure 17 shows the output from a typical commercial single-heterojunction laser, designed to emit about 10 W at room temperature. The displacement of the curves with increasing temperature results from increasing threshold current density and decreasing differential quantum efficiency.

There is also a pulsed-power emission range where the laser design is intermediate between the cw and the high-power device design. For ex-

---

[38] T. Tsukada, *J. Appl. Phys.* **45**, 4899 (1974).

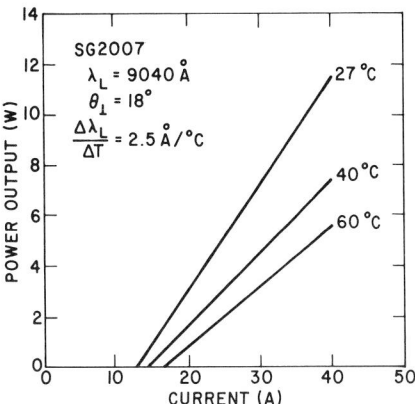

FIG. 17. Power output as a function of diode current (pulsed) for a single-heterojunction laser diode. Data taken with different heat sink temperatures. (Courtesy RCA Solid-State Div. Lancaster, Pa.)

ample, lasers can be designed for 10% duty cycle at peak power values of 100 mW by using low-threshold double-heterojunction material with stripe widths of 50 μm. In this regard, a short pulse length minimizes the possibility of facet damage (Chapter 4.11).

4.10. Radiation Patterns

The radiation pattern depends on the excited-cavity modes, which are grouped into two independent sets of TE (transverse electric) and TM (transverse magnetic) modes. The modes of each set are characterized by three mode numbers, which define the number of field antinodes along the three major axes of the cavity. *Lateral modes* are related to the dielectric profile in the *plane* of the junction; *transverse modes* are related to the dielectric profile in the direction *perpendicular* to the junction plane; and *longitudinal modes* are related to the profile along the junction plane in the direction perpendicular to the Fabry–Perot facets. Many longitudinal modes are typically excited in Fabry–Perot structures. Some stripe-contact lasers do operate, however, mostly in one longitudinal mode, as shown in Fig. 18; also, the distributed-feedback laser was specifically designed to restrict the operation of the device to a single longitudinal mode.

The semiconductor laser's beam width is much broader than the emission from other types of lasers, because of its much smaller emitting area. Figure 19 illustrates the typical radiation pattern from a laser diode with

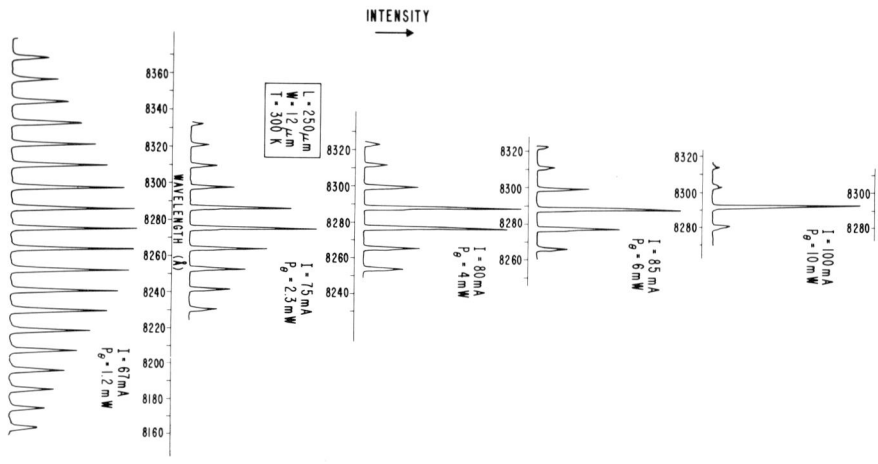

FIG. 18. Spectra from AlGaAs cw laser operating at room temperature showing a relatively pure emission at high power levels. (From H. Kressel and M. Ettenberg, unpublished.)

the beam perpendicular to the junction being much broader than in the junction plane. The full angular widths at the half-intensity points are denoted $\theta_\perp$ and $\theta_\parallel$, respectively. Note that Fig. 19 shows the emission from a laser operating in the *fundamental* transverse and lateral mode, and hence only a single lobe is seen. Higher order mode operation results in multilobed emission, which is sometimes undesirable.

Obtaining fundamental *transverse* mode operation is simple with heterojunction lasers because the structural requirements coincide with those needed for low threshold current densities, i.e., a thin active region. The dielectric profile of the laser in the direction perpendicular to the junction plane (the transverse direction) is controlled by the refractive index steps in heterojunction lasers. For $Al_xGa_{1-x}As/Al_yGa_{1-y}As$ heterojunctions, the dielectric step is approximately $\Delta n \cong 0.62(x - y)$ at the lasing photon energy of the lower bandgap recombination region. From a simple analysis of the critical angle for total internal reflection in a double-heterojunction structure, we find that the *maximum* mode number $m = M$ capable of propagating in the structure is [20]

$$M = \text{Integer}\left[1 + \left(\frac{2nd}{\lambda}\right)\left(\frac{2\Delta n}{n}\right)^{1/2}\right]. \quad (4.10.1)$$

To operate in the fundamental mode, the index step condition is

$$\frac{\Delta n}{n} \leq \frac{1}{8}\left(\frac{\lambda}{nd}\right)^2. \quad (4.10.2)$$

4.10. RADIATION PATTERNS

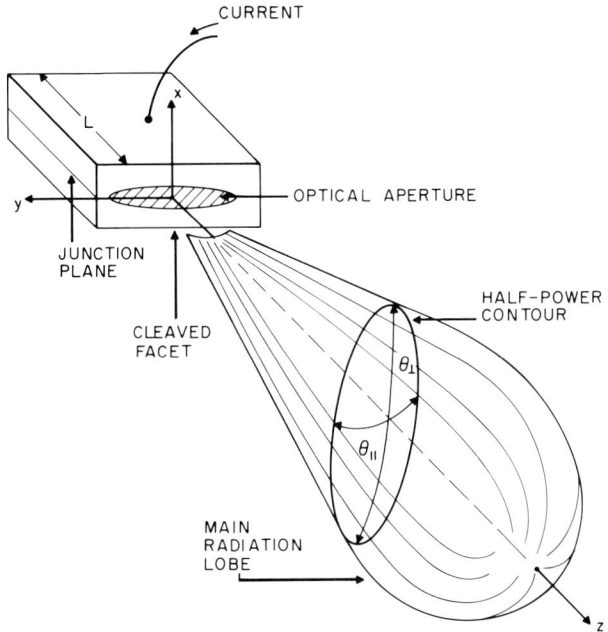

FIG. 19. Far-field from stripe-contact laser diode operating in the fundamental lateral and transverse mode. (After Kressel and Butler.[20])

For GaAs, $n = 3.6$, $\lambda \approx 0.9$ μm, hence

$$\frac{\Delta n}{n} = \frac{7.8 \times 10^{-3}}{d^2}, \qquad (4.10.3)$$

where d is in units of micrometers. For example, with $d = 1$ μm we require $\Delta n < 0.028$; with $Al_xGa_{1-x}As$ bounding layers, $x \leq 0.005$, or $\Delta E_g \approx 0.05$ eV. As the cavity width is decreased to smaller values, however, larger bandgap steps can be used for fundamental mode operation.

A crude approximation of the fundamental mode beam width perpendicular to the junction plane is obtained from simple diffraction theory assuming a uniformly illuminated slit of width d,

$$\theta_\perp \approx 1.2\lambda/d. \qquad (4.10.4)$$

In fact, however, the radiation from the laser deviates from this expression because, as shown in Fig. 8a, the intensity peaks in the center of the active region with "tails" extending into the passive regions to an extent dependent on the value of d and the index step Δn.

Excellent agreement between theory and experiment is obtained for the

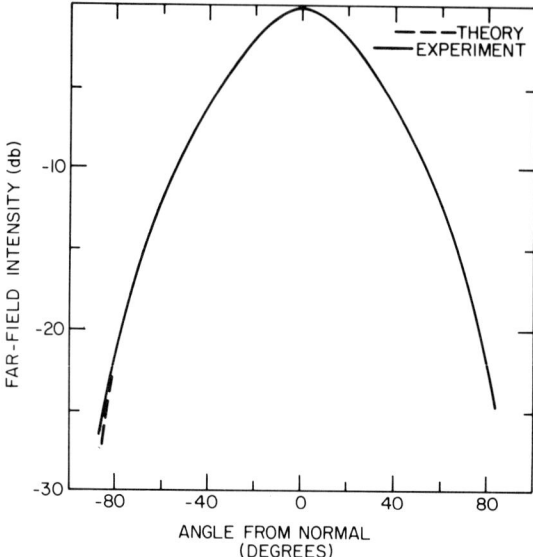

FIG. 20. Comparison of calculated and observed far-field in the direction perpendicular to the junction plane from a double-heterojunction laser. (After Butler and Zoroofchi.[39])

radiation pattern perpendicular to the junction plane of heterojunction structures. Figure 20 shows, for example, the fit of the data for a DH device operating in the fundamental transverse mode.[39] Typical beam widths at half-intensity are $\theta_\perp \approx 40°$ in the most commonly used AlGaAs/GaAs DH lasers. Good agreement is similarly obtained for devices operating in higher order transverse modes.[40]

The *lateral* modes depend on the dielectric profile in the plane of the junction and hence on the technique used for junction area definition. With two strong dielectric steps (independent of current) perpendicular to the junction plane, the device can be modeled in terms of "box modes" and expressions (4.10.1) and (4.10.2) are used, with the substitution of the diode width W for the guide thickness d, to determine the requirements for fundamental mode operation. Such a model is only appropriate for sawed-side or etched-side lasers or buried-heterojunction devices. However, in the planar stripe device the shallow dielectric profile is mostly self-induced by the current distribution, making analysis more difficult.

Fundamental lateral mode operation is frequently obtained in narrow (5–20 μm) planar stripe lasers, but higher order modes generally reach threshold with increasing current producing a complication in the far-field

---

[39] J. K. Butler and J. Zoroofchi, *IEEE J. Quantum Electron.* **qe-10**, 809 (1974).
[40] J. K. Butler and H. Kressel, *J. Appl. Phys.* **43**, 3403 (1972).

4.10. RADIATION PATTERNS

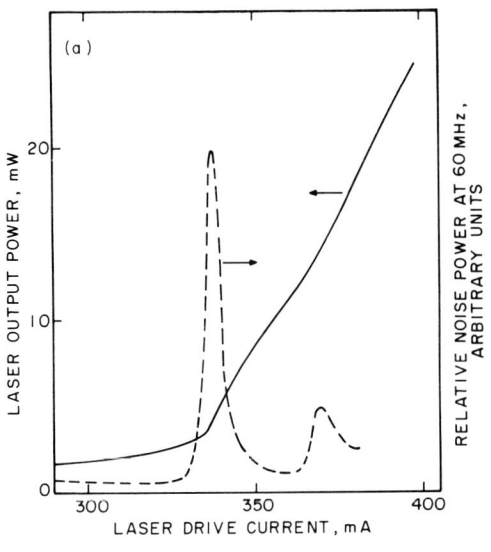

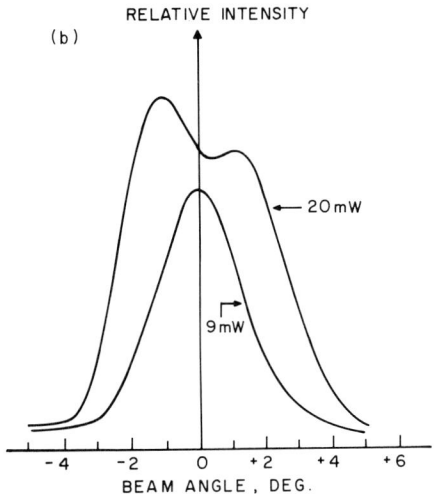

FIG. 21. (a) Optical power output versus current for a cw AlGaAs laser exhibiting a kink, along with the corresponding curve for low frequency (60 MHz) relative noise power. The two noise peaks correspond to threshold for the first (fundamental) and second lateral modes as seen in the lateral far field in (b). (After J. P. Wittke, unpublished.)

pattern. Nevertheless, the value of beam width in the plane of the junction rarely exceeds about 10° at the maximum current level of operation of such lasers. These lateral mode increases are frequently associated with slope changes ("kinks") in the power emission versus current curves as shown in Fig. 21. The laser at threshold operates in the fundamental lateral mode (single lobe). Threshold for the higher order mode is reached with increasing current as evidenced by the increasing complexity of far-field in the plane of the junction.

There are many possible reasons for lateral mode proliferation with increasing current. This phenomenon in sawed-side broad-area devices is believed related to an increase in gain above threshold. In this model the lasing threshold for the various lateral modes with increasing loss was shown to be reached as the current was increased, producing a broadening of the lateral beam width.[41]

In stripe-contact lasers, it is probable that the gain coefficient increases with drive in the "wing" regions of the stripe, thus producing an increasing ability to effectively couple to higher order modes. Hence, modes with higher losses can reach threshold as the current is increased. The restriction of the laser to a single lateral mode requires, therefore, that there be as large a difference as possible in the losses of the fundamental and the higher order lateral modes. Restricting the stripe width to very small values is one method of achieving this objective, although at the expense of the useful power from the device.

4.11. Degradation

Laser diodes operate at high current densities and at high optical fields that can reach the MW/cm^2 level. Consequently, these devices are prone to several degradation processes, which fortunately are becoming better understood and controlled. Basically, the degradation phenomena can be divided into *internal defect formation* processes, whereby nonradiative centers are introduced into the recombination region, and *facet damage* related to the intense optical fields.

A vast amount of literature exists concerning the internal defect formation process, and we restrict the present discussion to a summary of the major effects.[20] It is well established that the process of electron–hole recombination is responsible for the introduction and migration of nonradiative centers in GaAs and related devices. The nature of the defects formed and their rate of introduction varies vastly with the material, the stoichiometry of the crystal, the strain to which the diode is subjected, and the presence of defect sources in the vicinity of the recombination region. Available evidence suggests that the energy released in nonradiative electron–hole recombination can reduce the displacement energy for defects such as vacancies and interstitial atoms. In ideal crystals, this effect would not produce an increasing density of nonradiative centers, be-

[41] H. S. Sommers, Jr. and D. O. North, *Solid-State Electron.* **19**, 675 (1976).

4.11. DEGRADATION

cause the displacement of native atoms is highly unlikely as the required energy is far in excess of the energy released in electron–hole recombination. Therefore, the initial presence of defects in the regions of electron–hole recombination appears to be a requirement for this degradation process. In fact, experimental data do show that GaAs lasers containing dislocations degrade at a higher rate than devices that are dislocation-free. Furthermore, lasers with exposed edges (such as broad-area, sawed-side diodes) that contain edge defects degrade much more rapidly than diodes where the electron–hole recombination process is restricted to internal regions of the crystal.[37]

Although *all* the complicated and interrelated factors controlling the internal degradation process are not identified, the important fact is that lasers exhibiting negligible degradation can be produced. Many thousands of hours of stable cw operation have been obtained from AlGaAs laser diodes emitting in the 0.8–0.86 μm spectral region.[42]

Other experiments conducted by operating the lasers at elevated temperature in order to accelerate the degradation process suggest that extrapolated lifetimes exceed 100,000 hours, although it may be necessary to increase the diode current to keep the output constant.[43] This is not necessarily a major drawback, since an optical feedback loop that senses changes in the diode output and adjusts the current accordingly can be incorporated in practical systems.

Turning our attention to facet damage, complete or partial laser failure may occur as a result of mechanical damage of the facet in the region of intense optical emission. The nature of the damage suggests local dissociation of the material, but the extent of the damaged region and its penetration into the device is highly variable. The damage is unrelated to the operating current density, but it does depend, for a given device, on the pulse length and the ambient conditions. Furthermore, if mechanical flaws exist initially on the laser facet, then the optically induced damage tends to nucleate at such sites producing premature failure. For example, in sawed-side lasers, damage is commonly initiated at the crystal edge.

In general, the optical power density for failure is not easily established if the power distribution in the plane of the junction is not uniform, which is the case, for example, in planar stripe-contact devices. However, the pulse length has a clear relationship to the failure level. Stripe-contact laser studies suggest that the laser damage limit for 100-nsec-long pulses is

---

[42] I. Ladany, M. Ettenberg, H. F. Lockwood, and H. Kressel, *Appl. Phys. Lett.* **30**, 87 (1977).

[43] R. L. Hartman and R. W. Dixon, *Appl. Phys. Lett.* **26**, 239 (1975); H. Kressel, M. Ettenberg, and I. Ladany, *ibid.* **32**, 305 (1978).

TABLE I. Catastrophic Damage Limit of
Uncoated AlGaAs cw Laser Diodes[a,b]

Pulse length (nsec)	Catastrophic damage limit (mW)[c]
100 (1 kHz)	240
400 (0.25 kHz)	90
cw	30

[a] From H. Kressel and I. Ladany, *RCA Rev.* **36**, 230 (1975).
[b] 13 μm stripe width.
[c] Average of four diodes. All diodes selected from the same wafer.

$4-8 \times 10^6$ W/cm^2, but the critical power level is reduced with increasing pulse length Δt as $(\Delta t)^{-1/2}$. Illustrative data for oxide-isolated (AlGa)As DH lasers are presented in Table I. Operation with pulse lengths beyond about 2 μsec are equivalent to cw operation.

In addition to the above failure mode, denoted "catastrophic," which involves a rather rapid onset of facet damage, a slower facet "erosion" process can occur over long operating times. Although facet erosion is much less severe than catastrophic damage, it does reduce the facet reflectivity and hence the laser efficiency. The use of half-wave-thick dielectric coatings of Al_2O_3 on the facet essentially eliminates this damage mode.[42] Thus, by operating lasers below the limit where catastrophic degradation occurs and by coating the facets it is possible to eliminate facet damage as a failure mode.

4.12. Modulation Characteristics

A major advantage of laser diodes compared to other laser types is the ability to conveniently modulate their output at rates exceeding 1 GHz. This is possible because the lifetime for *stimulated* carrier recombination is reduced compared to that for spontaneous emission (~2–5 nsec). Stimulated carrier lifetimes of the order of 10^{-11} sec have indeed been deduced in the lasing regions.[44] Thus, lasers can be modulated at GHz rates, but it is essential that they remain biased to threshold in order to avoid a turn-on delay time related to the spontaneous carrier lifetime. There are, however, many complications in the modulation process due to intrinsic laser resonance effects and phenomena resulting from nonideal laser behavior.

[44] N. G. Basov *et al.*, *Sov. Phys.—Solid State (Engl. Transl.)* **8**, 2254 (1967).

4.12. MODULATION CHARACTERISTICS

Note that the ultimate laser modulation is limited by the photon lifetime in the cavity, which in a Fabry–Perot cavity is given by

$$\frac{1}{\tau_{ph}} = \frac{c}{n}\left[\tilde{\alpha} + \frac{1}{L}\ln\frac{1}{R}\right], \qquad (4.12.1)$$

where c is the velocity of light, n the refractive index in the cavity at λ_L, R the facet reflectivity, and $\tilde{\alpha}$ the averaged internal absorption coefficient. With the quantity in the bracket being typically ~ 50 cm^{-1}, $\tau_{ph} \cong 2 \times 10^{-12}$ sec. Thus, the photon lifetime does not represent a practical limit to the modulation capability of the device.

The analysis of the transient laser properties for semiconductors follows the rate equation approach for the carrier and photon population.[20,45] Such an analysis, based on the assumptions that the cavity is fully and uniformly inverted and that only a single mode is excited, predicts that the injected carrier density in the lasing region is fixed at the threshold value. This means that the gain remains fixed at its threshold value, because the gain coefficient is a function of the carrier density.

Fluctuations in the carrier and photon population about the steady-state value will be dampened in a time dependent on the spontaneous carrier lifetime, the photon lifetime, and the injection level. These interactions and perturbations give rise to fluctuations in the laser output, which we now discuss.

Suppose that we turn the laser diode on with step current I by opening the switch in Fig. 22a. After an initial transient oscillatory effect, the photon density in the cavity will reach the value $\overline{N}_{ph}$ and the carrier pair density will reach $\overline{N}_{th}$. Figure 22c shows the dampened oscillations that are observed after the laser is turned on. Oscillation frequency and the time constant can be simply calculated with simplifying assumptions from the rate equations, which include the assumption that only small deviations occur from the equilibrium photon and carrier densities ΔN_{ph} and ΔN_e, respectively. We then find

$$\Delta N_e = (\Delta N_e)_0 \exp[-(a - i\omega_c)t], \qquad \Delta N_{ph} = (\Delta N_{ph})_0 \exp[-(a - i\omega_c)t]. \qquad (4.12.2)$$

The values of a and ω_c are

$$a \cong \frac{1}{2\tau_s}\left(\frac{J}{J_{th}} + 1\right), \qquad (4.12.3)$$

$$\omega_c \cong 2\pi f_c \cong \left[\frac{1}{\tau_s \tau_{ph}}\left(\frac{J}{J_{th}} - 1\right)\right]^{1/2}. \qquad (4.12.4)$$

---

[45] See, for example, W. V. Smith and P. P. Sorokin, "The Laser." McGraw-Hill, New York, 1966.

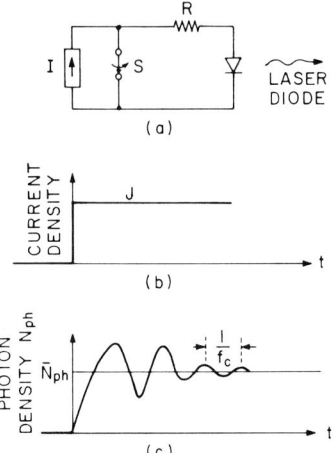

FIG. 22. Transient effects in laser diodes as the switch in (a) is opened resulting in a current density J through the diode. (b) Dampened oscillations in the emitted radiation intensity. The steady-state photon density is $\bar{N}_{\text{ph}}$. (After Kressel and Butler.[20])

From (4.12.3) we see that the oscillations disappear in a time of the order of the spontaneous carrier lifetime, or a few nanoseconds. From (4.12.4) we see that the resonance frequency f_c increases with decreasing spontaneous carrier lifetime and photon lifetime, and with the ratio J/J_{th}. Therefore, the laser should be biased above threshold to minimize both the turn-on delay and oscillations. For example, with $J = 2J_{\text{th}}$, $\tau_s = 2 \times 10^{-9}$ sec, and $\tau_{\text{ph}} = 10^{-12}$ sec, $f_c \cong 4$ GHz. The existence of this dampened oscillation also means that if attempts are made to supply a modulating current density $J = J_0 e^{i\omega t}$ to the diode (with appropriate dc bias) the modulation efficiency will decrease rather steeply with frequency when $\omega > \omega_c$.

The resonant effect in modulation has been seen by Ikegami and Suematsu[46] in homojunction GaAs lasers, and moderate agreement with theory has been obtained using values of $\tau_s = 2$ nsec and $\tau_{\text{ph}} \cong 10^{-12}$ sec.

In addition to the dampened oscillations discussed above, *self-sustaining oscillations* are also observed in lasers. These can have several origins, which include nonuniform inversion of the cavity and shot noise.

Lasher[47] proposed self-sustaining pulse generation from a laser encompassing an emitting region in tandem with an absorbing region within the same Fabry–Perot cavity. The oscillations occur because of saturable

[46] T. Ikegami and Y. Suematsu, *IEEE J. Quantum Electron.* **qe-4**, 148 (1968).
[47] G. J. Lasher, *Solid-State Electron.* **7**, 707 (1964).

absorption, i.e., its absorption coefficient is a function of the optical field magnitude. The basic concept of oscillations due to inhomogeneous population inversion is a general one. It is possible that optical anomalies in devices containing defects within the active region are related to this process, and self-sustaining oscillations increasing in frequency with current in some room temperature DH AlGaAs laser diodes have been attributed to this effect.[48]

Self-sustaining oscillations due to shot noise in the cavity are another source of noise in lasers. Theoretical analysis of this effect predicts that the noise spectrum can peak at frequencies ranging from the MHz region well into the GHz region, dependent on J/J_{th} and the temperature.[49] The magnitude of the noise intensity decreases with increasing J/J_{th}, i.e., the maximum is reached near threshold with a rapid decrease above threshold as J is increased. This effect is illustrated in Fig. 21, where the noise (measured here at 60 MHz) exhibits a maximum in the vicinity of J_{th}, decreasing in intensity as the current is increased until the second lateral mode reaches threshold.

Note that many transient effects that are not easily related to theory are sometimes seen in laser diodes.[20] It is probable that the complicated behavior of various cw lasers is related to their structural perfection, uniformity and mode content stability. The modulation of diodes in a current range that traverses a kink could well result in the emission of random pulses and hence become bothersome in optical communications. However, since devices free of kinks (over a useful power emission level) are fabricated, this is not an inherent limitation.

It is experimentally found that the dc bias applied to a laser that one wishes to modulate at high frequencies is relevant to the quality of the optical pulses obtained. Figure 23 shows the effect of a small increase in the bias current on the "cleanliness" of the light output pulses from a laser diode. In Fig. 23a the laser is biased very close to threshold and significant pulse distortion is seen, whereas in Fig. 23b a small increase in the bias current produces optical pulse shapes that replicate the current pulses.[50]

4.13. Distributed-Feedback Lasers

The distributed-feedback laser provides a means of restricting the laser operation to a single longitudinal mode, although this structure does not affect the transverse or lateral modes.

[48] H. Yanai, M. Yano, and T. Kimiya, *IEEE J. Quantum Electron.* **qe-11**, 519 (1975).
[49] H. Haug, *Phys. Rev.* **184**, 338 (1969).
[50] S. Maslowski, AEG Telefunken (unpublished).

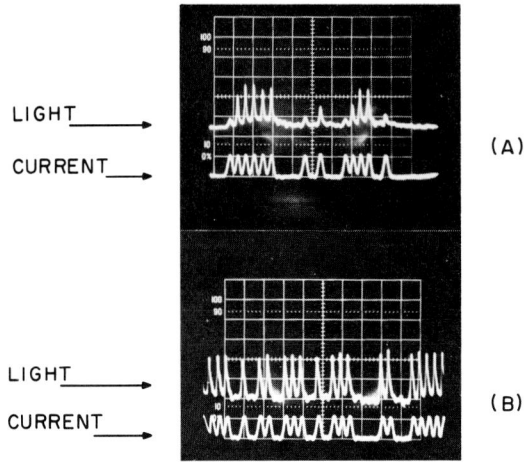

FIG. 23. Comparison of current pulses with optical pulses emitted by AlGaAs DH lasers at room temperature (250 Mbit/sec). (A) The DC bias is too close to the threshold current, resulting in pulse distortion; (B) proper DC bias improves the modulation characteristics. (Courtesy S. Maslowski, AEG-Telefunken.)

The distributed-feedback laser (DFB) incorporates periodic variations of the refractive index along the direction of wave propagation. The energy of wave propagating in one direction is continuously fed back in the opposite direction by Bragg scattering, producing a resonance condition.[51] The greatest interest has focused on injection DFB lasers using AlGaAs/GaAs heterojunctions, where variations in the Al content along the laser axis produce the required dielectric perturbations. The theory of DFB lasers has been reviewed by Yariv.[52]

Figure 24 shows a schematic of a DFB AlGaAs/GaAs laser in which the corrugations are incorporated within the optical cavity defined by two outer heterojunctions, the GaAs recombination region being separate from the corrugated region.[53] The periodicity of the corrugations is about 0.4 μm. Devices of this type have operated at room temperature with threshold current densities of about 5000 A/cm^2, and have been shown capable of cw operation. Operation in a single longitudinal mode has

---

[51] H. Kogelnik and C. V. Shank, *Appl. Phys. Lett.* **18**, 152 (1971).
[52] A. Yariv, "Quantum Electronics," 2nd ed., p. 165. Wiley, New York, 1975.
[53] M. Nakamura, K. Aiki, J. Umeda, and A. Yariv, *Appl. Phys. Lett.* **27**, 403 (1975).

4.13. DISTRIBUTED-FEEDBACK LASERS

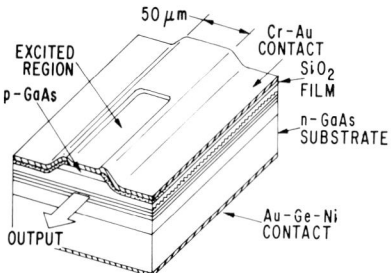

FIG. 24. Distributed-feedback AlGaAs/GaAs laser diode in stripe-contact configuration. The corrugations producing the feedback are shown. (After Nakamura et al.[53])

been obtained, consistent with theory, although the power level that can be reached in a single longitudinal mode is uncertain.

Other distributed-feedback structures have been described in the literature where the corrugations are not internal to the structure. The theory of these devices is reviewed by Wang.[54]

[54] S. Wang, *IEEE J. Quantum Electron.* **qe-13,** 176 (1977).

5. DYE LASERS*

5.1. Introduction

Organic dye lasers have become one of the most versatile and widely used laser devices. This laser is the spectroscopist's dream in that its output can be precisely controlled in every applicable parameter space. Both the output wavelength and bandwidth can be accurately determined. In addition, the time duration of the output is variable from cw to less than a picosecond. The dye laser is capable of achieving these characteristics simultaneously with efficiencies that are competitive with other laser systems operating in the visible region of the spectrum. A dye laser can be used as the primary laser in specific applications or it can be used as a transducer to convert coherent radiation produced by one laser device into laser radiation of a more useful nature. These unique characteristics have placed the dye laser in the forefront of the many fields of laser applications.

The wide variety of uses to which the dye laser can be applied have made it an understandably popular laser device. This popularity has motivated the publication of many excellent bibliographic review articles[1,2] on the subject. No attempt will be made here to duplicate these very complete and useful treatises.

This discussion will be limited to descriptions of models and techniques that are applicable to the different classes of dye laser devices. In developing the models many approximations will be made that reflect the present state of knowledge about the characteristics of the laser-active dye molecules. These approximations will be made to simplify the quantitative descriptions of the dye lasers in specific regions of operation. These approximate models are intended to yield sufficient accuracy to solve most design problems or, short of that, to give a qualitative description of the effects of the more important parameters that must be given consideration.

The description of the dye laser system is introduced by a review of the

---

[1] F. P. Schäfer, ed., "Dye Lasers." Springer-Verlag, Berlin and New York, 1973.
[2] B. B. Snavely, *Proc. IEEE* **57**, 1374 (1969).

* Part 5 is by Otis Granville Peterson.

basic dye molecule and dye solvent properties in Chapter 5.2. This includes the basic rate equations describing the laser properties of the dyes set forth in Section 5.2.2 and application of these equations to laser threshold calculations in Section 5.2.3. The limitations introduced by the thermal properties of the solvent hosts for the dye molecules are outlined in Section 5.2.5.

Dye laser devices are arbitrarily divided into two major classes according to whether or not the time duration of the excitation is greater or less than the lifetime of the dye molecule triplet states. This demarcation point is approximately 100 nsec. Chapter 5.4 describes the short-pulse dye lasers designed to operate with excitation pulses less than 100 nsec. Section 5.4.1 outlines the rate equation description for these lasers. Section 5.4.2 discusses amplified spontaneous emission, which has a significant effect on all the high-gain devices that are characteristic of this time regime. Specific descriptions are included in Sections 5.4.3–5.4.6 of oscillators, amplifiers, and regenerative oscillators designed to perform efficiently with short-pulse excitation.

The last chapter of this part is devoted to the description of devices operating in the steady-state or quasi-steady-state regime. This is initiated by some general discussions concerning the rate equations and distribution of the power delivered to the dye system. The succeeding sections are further divided into one section concerned with flashlamp-excited lasers and another concerned with cw lasers. Descriptions of some of the more popular flashlamp systems are prefaced with some general considerations about the most appropriate plasma temperature for dye laser excitation. The four flashlamp systems selected for discussion are sealed linear flashlamps, ablating lamps, vortex stabilized lamps, and coaxial discharge lamps. Consideration is also given to the reflector geometries required to couple the excitation from the three types of linear lamps into the dye medium. The important characteristics of the dye cell that must be an integral part of the reflector assembly are also outlined. This latter discussion includes a description in Section 5.5.10 of some of the characteristics of free jet dye cells.

Chapter 5.6 is concerned exclusively with the cw dye laser. Section 5.6.1 gives an analytical description of the laser characteristics and is followed by several sections describing in detail the resonator geometries peculiar to this specific device. Among the subjects analyzed are the astigmatic resonators and the compensation of the astigmatism and coma in such devices.

Finally, Chapter 5.7 presents a sample solution of dyes to cover the useful wavelength region and discusses some of their characteristics and applications.

5.2. Basic Dye Molecule and Dye Solvent Properties

5.2.1. Energy States of Dye Molecules

For most purposes, dye lasers can be described by five electronic energy levels and the transitions between these levels. Each of the electronic levels is very wide in energy, being broadened by a continuum of vibrational, rotational, and solvent states. These states are divided into two manifolds of states: first, a singlet manifold that includes the ground state and, of particular interest, the excitation–absorption and the lasing transitions; and second, a triplet manifold of states connected to the singlet manifold by means of spin-forbidden transitions. Transitions within the triplet manifold constitute losses to either the excitation or to the laser emission.

The transitions between these five levels can be divided into those involving stimulated absorption or emission of radiation and those for which radiationless transitions occur either within or between the two manifolds of states.

These levels and transitions are illustrated in Fig. 1. The singlet states are labeled S_0, S_1, and S_2 for the ground, first excited, and second excited states. The lowest and first excited triplet states are correspondingly labeled. The three states that have an appreciable population are labeled n_0 for the ground state of the molecule, n_S for the first excited singlet state, and n_T for the lowest triplet state. These populations are defined as number densities with units of cm^{-3}. The stimulated transitions are illustrated with solid lines, absorptions going up and emissions going down, and are labeled with the cross sections σ_{ij} characteristic of the transition, where the cross sections are in units of cm^2. The cross sections have subscripts to identify the initial state i from which the transition starts and to identify the wavelength j at which the cross section is to be evaluated. Two wavelengths are important to the analysis: the excitation wavelength, labeled X, and the emission wavelength, labeled M. It must be noted that of the cross sections associated with the excited singlet state S_1, the one at the excitation wavelength σ_{SX} is for stimulated absorption, and the other at the emission wavelength σ_{SM} is for stimulated emission.

Many of the states are connected by means of radiationless transitions that occur within a few picoseconds or less. These have been dia-

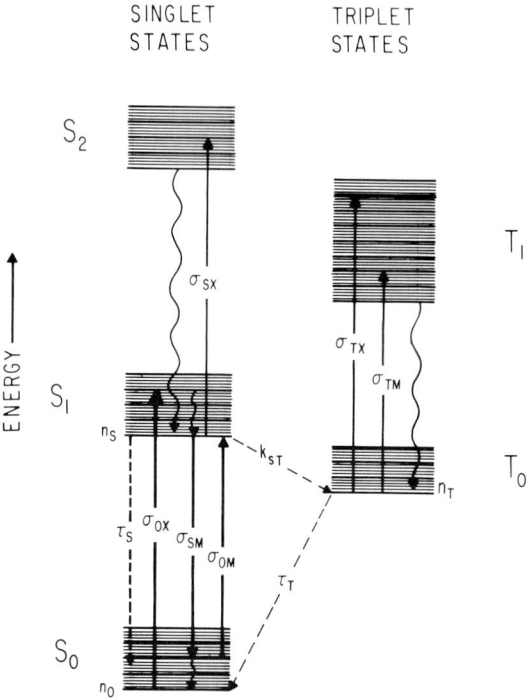

FIG. 1. Dye molecule energy level diagram. The stimulated transitions are labeled with the corresponding cross sections σ_{ij}, the spontaneous transitions between levels are labeled with characteristic lifetimes τ_i or transition probabilities k_{ij}, and radiationless transitions are labeled with wavy lines.

grammed by wavy lines. There are three spontaneous transitions of particular significance to the laser operation, which have been identified with dotted lines. These are the spontaneous transitions labeled τ_s, which are predominantly fluorescence, and the spin-forbidden transitions between the singlet and triplet manifolds labeled k_{sT} and τ_T. The singlet to triplet state intersystem crossing probability k_{sT} is measured in sec^{-1} and the lifetimes τ_i are measured in seconds.

The stimulated optical transitions between the illustrated levels in each manifold are of great importance for the laser operation. The stimulated emission transition is clearly the driving term for laser action. The stimulated absorptions include the transition that populates the excited state and other absorptions that produce losses of either the exciting photons or the emitted photons.

Two of the levels illustrated have lifetimes of particular significance to the operation of the laser. One of these is the first excited singlet state

5.2. BASIC DYE MOLECULE AND DYE SOLVENT PROPERTIES

whose lifetime is of the order of 1 to 5 nsec times the quantum yield for fluorescence for the dye: $\tau_s = (QY)\tau_r$, where τ_r is the radiative lifetime and τ_s the observable lifetime of the state. The quantum yield QY will always be less than one due to radiationless transitions, not labeled in the figure, which compete with the spontaneous fluorescence for depleting the first excited singlet state population n_S.

The second important lifetime is that of the lowest triplet state. This lifetime is usually determined by quenching effects caused by collisions of the molecules with other species, for example, oxygen, dissolved in the dye solution. This lifetime has been measured to be approximately 100 nsec for rhodamine 6G,[3] the best characterized dye at this time. The lifetimes of all other states are expected to be in the picosecond range, a fact that has been verified in a small number of cases.[4,5] This situation is also true for the relaxation processes within each electronic level. For modeling purposes, these lifetimes can be assumed to be zero.

It is important to note that the dye laser is a true four-level system. The excitation is well separated in energy from the emission by essentially infinitely fast radiationless transitions. Therefore, the device can reach laser threshold at very small population inversions, i.e., for values of n_S/n of only 0.01 or less.[6]

There are two forms of self-absorption that significantly affect the dye laser performances. The ground singlet state can absorb the emission I_M, which is illustrated as the transition labeled σ_{0m}. This absorption depends on the thermal population of the high sublevel of the ground electronic state that is shown in Fig. 1 as the initial state for this absorption. This population is extremely small so that the absorption is three to four orders of magnitude less than the peak absorption. The absorption is significant in that it effects the optimum wavelength at which the laser will operate, particularly in flashlamp-excited devices with long lengths of dye solution through which the laser emission must be transmitted. Although this absorption effects the laser gain relation in a significant manner, it does not constitute an energy loss since it repopulates the excited state. In contrast, the power absorbed by the lowest triplet state is totally lost into solution heating. The magnitude of this absorption depends on the population n_T of triplet state molecules. For short-pulse laser devices this absorption can often be ignored since the quantum yield of production of triplet molecules n_T from excited singlet molecules n_S is approxi-

---

[3] J. P. Webb, W. C. McColgin, O. G. Peterson, D. L. Stockman, and J. H. Eberly, *J. Chem. Phys.* **53**, 4227 (1970).
[4] C. Lin and A. Dienes, *Opt. Commun.* **9**, 21 (1973).
[5] A. Penzkofer, W. Falkenstein, and W. Kaiser, *Chem. Phys. Lett.* **44**, 82 (1976).
[6] B. B. Snavely and O. G. Peterson, *IEEE J. Quantum Electron.* **qe-4**, 540 (1968).

mately 0.5%.[7] If the pulse is sufficiently short, the triplet population will not build up to significant levels and the absorption will therefore be negligible. For long-pulse and cw operation, the triplet population will build up to equilibrium values. In this case the absorption is large and becomes the ultimate limitation on the efficiency of the laser.

Figure 2 illustrates the wavelength dependence of these transitions for the popular dye rhodamine 6G.[8] It is clear from the figure that both the ground singlet and triplet states have absorptions that partially overlap the stimulated emission curve. The excited singlet absorption has been shown as data points because of the limited number of measurements that have been made on this absorption spectrum.[9]

The ground singlet absorption has been shown over its entire wavelength extent to 200 nm, the quartz transmission cutoff. Any absorption in this entire region will produce a molecule in the first excited state. The quantum yield for the production of excited states is essentially identical over this region, although the energy efficiency will vary greatly

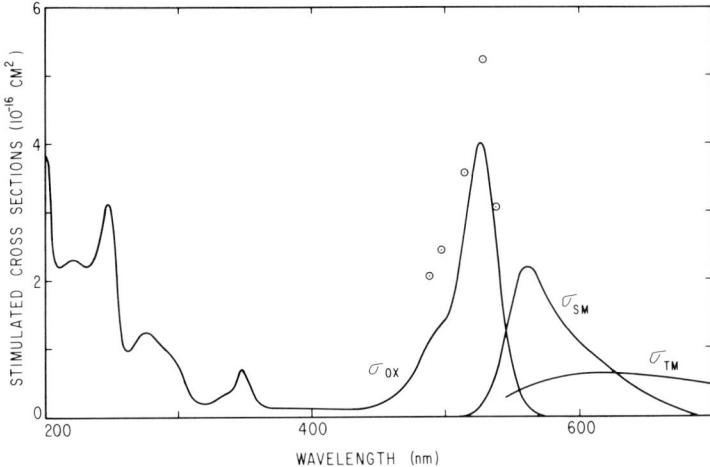

FIG. 2. Wavelength dependence of rhodamine 6G-stimulated cross sections.[8] The cross sections are identified in Fig. 1 and in the text. The data points are the limited measurements available on the excited singlet state absorption.[9]

[7] A. V. Buettner, B. B. Snavely, and O. G. Peterson, in "Molecular Luminescence" (E. C. Lim, ed.), p. 403. Benjamin, New York, 1969.

[8] O. G. Peterson, J. P. Webb, W. C. McColgin, and J. H. Eberly, *J. Appl. Phys.* **42**, 1917 (1971).

[9] O. Teschke, A. Dienes, and J. R. Whinnery, *IEEE J. Quantum Electron.* **qe-12**, 383 (1976).

5.2. BASIC DYE MOLECULE AND DYE SOLVENT PROPERTIES

since the energy discrepancy between the absorbed and emitted photons must be dissipated as heat. For clarification, it should be noted that this excitation transition σ_{0X} was only indicated schematically on the energy level diagram and that no attempt was made to illustrate the energy extent of this absorption.

5.2.2. Rate Equations

A simple schematic of a dye laser is displayed in Fig. 3. The components illustrated include a flashlamp for excitation, a dye cell with window transmissions T_1 and T_2, which contains the active medium of length l, and two mirrors labeled M_0 and M_B to supply the optical feedback required to achieve laser oscillation. As with all lasers, the threshold for oscillation occurs when a given packet of light at one point in the resonator travels the complete round trip, as illustrated, and returns to the same point with the same intensity and phase. It is therefore necessary that the amplification the packet receives in the active region exactly compensates for all the losses the packet incurs during the round trip.

To calculate the conditions required for oscillation, we shall consider an arbitrary infinitesimal volume within the active region. Within this volume the differential rate equations can be written that define the different processes.[8,9] The stimulated processes are all of the same form, being proportional to both the state population n_i and to the stimulating intensity I_j where the proportionality constant is the cross section σ_{ij} for that transition. As the exciting radiation I_X passes through the test volume a distance dy, the intensity will be decreased by absorption from the ground n_0 and first excited singlet state population n_S and the lowest triplet state population n_T:

$$dI_X/dy = -n_0\sigma_{0X}I_X - n_S\sigma_{SX}I_X - n_T\sigma_{TX}I_X. \qquad (5.2.1)$$

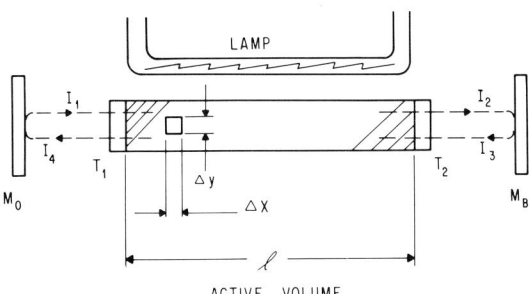

FIG. 3. Elementary laser schematic. The dashed lines represent the circulating laser emission contained within the resonator.

As the laser emission I_M passes through the volume at distance dx it will be amplified by the excited state population n_S and absorbed by the lowest singlet and triplet states n_0 and n_T:

$$dI_M/dx = n_S\sigma_{SM}I_M - n_0\sigma_{0M}I_M - n_T\sigma_{TM}I_M. \qquad (5.2.2)$$

In Eqs. (5.2.1) and (5.2.2) the spatial dimensions are labeled differently to indicate that they are to be defined as parallel to the propagation direction of the respective intensities.

The excited singlet state is pumped by the excitation illumination and the ground-state absorption of the laser intensity. This state is depleted by the laser action in addition to the intrinsic decay to the triplet and ground state. The rate equation defining this population is therefore

$$dn_S/dt = n_0\sigma_{0X}I_X + n_0\sigma_{0M}I_M - n_S\sigma_{SM}I_M - n_S k_{ST} - n_S/\tau_S. \qquad (5.2.3)$$

The triplet level is built up by intersystem crossing from the excited singlet level but has a finite lifetime dependent on collisional quenching:

$$dn_T/dt = n_S k_{ST} - n_T/\tau_T. \qquad (5.2.4)$$

Finally, there must be a conservation of molecules

$$n = n_0 + n_S + n_T, \qquad (5.2.5)$$

where n is defined as the total number density of molecules in a unit volume.

5.2.3. Laser Threshold

Equation (5.2.2) is the intrinsic laser threshold relation, i.e., the condition $dI_M/dx = 0$ must be met for laser action to occur even with no resonator losses. The region where this condition is satisfied is illustrated in Fig. 4. The shaded area represents the graphical difference between the three terms representing the intrinsic characteristics of the laser dyes. For the purposes of illustration, the special case of rhodamine 6G with an excited-state distribution typical for flashlamp operation has been assumed. In this case the triplet population is in a steady-state relationship with respect to the excited singlet population and both populations are equal in magnitude and are approximately 1% of the total dye molecule density.[6] The equality of the excited singlet and triplet population is a consequence of the fact that the product $k_{ST}\tau_T$ is approximately 1 for rhodamine 6G.[3] The ground singlet absorption has been multiplied by 100 to bring it into the correct proportion with respect to the triplet and emission cross-section curves. Both the wavelength range and the intensity maximum for laser action can be easily identified from the graph.

5.2. BASIC DYE MOLECULE AND DYE SOLVENT PROPERTIES

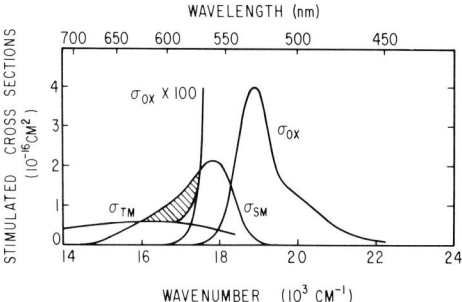

FIG. 4. Gain region for rhodamine 6G. The shaded area is the region where rhodamine 6G exhibits positive gain for an $n_S/n = 0.01$ inversion.

For a real laser, the threshold equation must include the resonator losses. Threshold is defined as that set of conditions at which the laser intensity within the resonator upon making one round trip through the resonator has the same initial I_M^i and final I_M^f values. To determine this condition it is necessary to integrate Eq. (5.2.2) over the length of the amplifying region. It will be assumed that the populations of the various molecular states are spatially uniform throughout the volume to simplify this integration. This assumption implies that the excitation and laser intensities are also uniform. This uniformity can occur only for weak absorption of the excitation and for a minimum of resonator losses for the laser emission. Laser oscillation will occur in the illustrated device when

$$I_M^i = I_M^f = I_M^i e^{2gl} T_1 T_2 R_2 T_2 T_1 R_1, \quad (5.2.6)$$

where g is the gain coefficient given by

$$g = n_S \sigma_{SM} - n_0 \sigma_{0M} - n_T \sigma_{TM}, \quad (5.2.7)$$

where L is the length of the active region, T_i is the transmission of an internal resonator element, e.g., dye cell window, and R_i the reflectiviy of a resonator mirror. Rearranging yields

$$g = n_S \sigma_{SM} - n_0 \sigma_{0M} - n_T \sigma_{TM} = L/2l, \quad (5.2.8)$$

where L is defined as the total round trip, dye-independent resonator losses as follows:

$$L = - \sum_i \ln T_i - \sum_i \ln R_i. \quad (5.2.9)$$

If the transmissivities and reflectivities are close to unity, they can be approximated using the relation

$$\ln(1 - x) \approx -x, \quad (5.2.10)$$

In this limit, the total resonator loss function L is the simple sum of all the individual losses L_i introduced by each element in the resonator

$$L = \sum_i L_i, \qquad (5.2.11)$$

where each loss is added once for each time laser emission encounters the element during a round-trip traversal of the resonator.

Imposing the steady-state relationship between the triplet n_T and excited singlet n_S state populations, which can be obtained by setting Eq. (5.2.4) equal to zero,

$$n_T = k_{ST}\tau_T n_S, \qquad (5.2.12)$$

and the conservation of molecules relation Eq. (5.2.5) yields a relation defining the population inversion n_S/n required to achieve laser threshold:

$$\frac{n_S}{n} = \frac{L/2ln + \sigma_{0M}}{\sigma_{SM} + \sigma_{0M}(1 + k_{ST}\tau_T) - \sigma_{TM}k_{ST}\tau_{TM}}. \qquad (5.2.13)$$

This equation defines the fraction of the molecules that must be excited to the first excited singlet state to reach laser threshold as a function of only two parameters. These parameters are the wavelength, which is implicitly contained within the cross sections σ_{ij}, and the normalized resonator loss $L/2ln$.

A three-dimensional plot (J. P. Webb diagram[8]) of this function is displayed in Fig. 5. It is seen that there is an optimum wavelength (min-

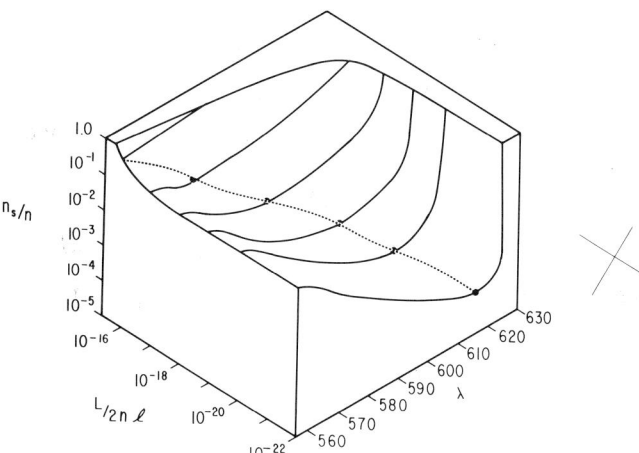

FIG. 5. Laser threshold surface for rhodamine 6G. The surface shows the value of the inversion n_S/n required to achieve laser threshold as a function of wavelength λ and normalized losses $L/2nl$.[8]

5.2. BASIC DYE MOLECULE AND DYE SOLVENT PROPERTIES

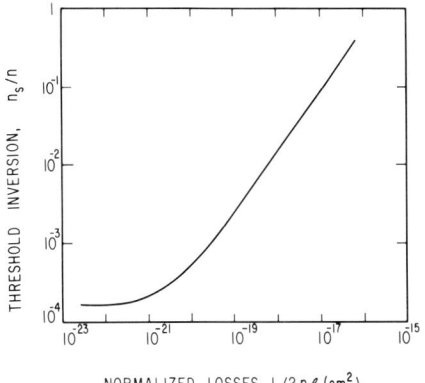

FIG. 6. Minimum inversion required for laser threshold in rhodamine 6G. This is the minimum of the surface illustrated in Fig. 5.[8]

imum n_S/n) for laser action with any specific loss. This optimum is seen to move to shorter wavelengths as the resonator losses increase. It is also clear how much additional excitation power is required to reach laser threshold if one wishes to force or tune the laser to operate away from this minimum. The locus of points defining the minimum of this surface is displayed in Fig. 6 as a guide to the excitation required to obtain lasing for a given resonator loss. The corresponding optimum wavelength for rhodamine 6G is shown in Fig. 7. The graph of wavelength has the interesting feature that the relationship between the wavelength and the loss parameter is approximately logarithmic with a slope identical to a similar plot of σ_{0M} vs. wavelength. The identical slopes lead to the conclusion that the dye laser, if allowed to select its optimum wavelength, will do so such that the ratio between the resonator losses and the losses intrinsic to the dye will remain a constant.

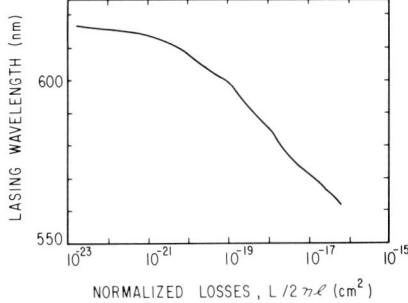

FIG. 7. Lasing wavelength for rhodamine 6G as a function of normalized losses at the minimum inversion.[8]

The graphs in Figs. 6 and 7 exhibit two distinct regions of laser operation. In the first region the threshold population inversion and the emission wavelength are dependent on the normalized loss value. This is the operational regime of greatest interest and importance since essentially all dye lasers operate in this range. In the second region of operation the inversion and wavelength have reached their respective minimum and maximum values and are dependent only on the intrinsic properties of the dye molecules.

In the first region of operation the near linearity of the threshold population inversion allows some useful approximations to be made. While the intrinsic self-absorption of the dye $2\sigma_{0M}nl$ is very important for determining the optimum wavelength for lasing, it is usually small compared to the resonator losses L. To the extent that this term can be neglected, the independently adjustable parameters n_S and l associated with the active media in Eq. (5.2.13) can be gathered together:

$$n_S l = \frac{1}{2} \frac{L}{\sigma_{SM} + (1 + k_{ST}\tau_T)\sigma_{0M} - k_{ST}\tau_T\sigma_{TM}}. \qquad (5.2.14)$$

This equation allows us to make some useful generalizations. It is seen that the important criterion for achieving laser threshold is the excitation of a critical concentration of molecules per cross-sectional area of the laser bore. For rhodamine 6G, which is probably as good a dye as can be synthesized, this threshold value is 85 kW/cm² if the excitation is at 530 nm and the resonator round trip loss is 10%. The implication of this generalization is that the operating characteristics of the laser are independent of the length of the active medium to a first approximation. This independence permits the length to be adjusted to satisfy other criteria, for example, the use of long dye cells and flashlamps to reduce the plasma temperature within the flashlamps and thus increase the lamp lifetimes.

There are several factors that reduce the accuracy of this approximation. Some resonator losses depend on the length, e.g., diffraction losses and scattering losses within the dye solution. Also, the ground-state absorption has a definite effect on the threshold. This effect can be seen from Fig. 6. If the approximation were strictly true, the straight line portion of Fig. 6 would show a linear relationship between the loss function and the critical inversion. The relation between these is found to be $n_S/n \propto (L/2nl)^{0.78}$. For constant resonator parameters the threshold will thus increase as a low power of the dye cell length for the special case of rhodamine 6G, i.e., $n_S l \propto (nl)^{0.22}$. Figure 6 also shows the extent of the region where this approximation is applicable.

In the second region of operation at very low losses it is evident that the population inversion n_S/n reaches a minimum and becomes independent of the resonator losses. In this region the laser threshold inversion is dependent only on the intrinsic properties of the dye molecule and is no longer affected by changes in the dye cell length. This condition defines a brightness limitation for the excitation source. If the arbitrary assumption is made that 10% of the intensity emitted by a black body between 200 and 500 nm can be converted into excited molecules within the laser mode, then the minimum brightness that a dye laser excitation source can have is equivalent to a 5700 K black body. This temperature is identical to the surface temperature of the sun.

5.2.4. Stimulated Emission Cross Section

In the previous discussion, the dye characteristics are described by lifetimes and cross sections. To obtain these values for the excited-state parameters is in most cases straightforward, although difficult. It should be pointed out, however, that the stimulated emission cross-section determination requires three independent measurements.[8] These measurements are the fluorescence lifetime τ_S, the fluorescence quantum yield QY, and the wavelength dependence of the fluorescence $E(\lambda)$. The wavelength function is normalized so that the integral over all wavelengths is equal to one:

$$\int_0^\infty E(\lambda)\, d\lambda = 1. \tag{5.2.15}$$

The probability that a molecule will fluoresce in a given wavelength interval $d\lambda$ is then the product of this wavelength function times the total probability that the molecule will emit. The total probability of emission is the inverse of the radiative lifetime τ_r, where

$$\tau_r = \tau_S/QY. \tag{5.2.16}$$

The spontaneous emission probability P_{sp} at a given wavelength is then

$$P_{sp}(\lambda)\, d\lambda = (QY/\tau_S)E(\lambda)\, d\lambda. \tag{5.2.17}$$

This probability can also be shown to be equal to the product of a probability function $M(\lambda)$ and the density of states in free space $\rho(\lambda)\, d\lambda$ in the wavelength interval into which the photon is emitted[8]:

$$P_{sp}(\lambda)\, d\lambda = M(\lambda)\rho(\lambda)\, d\lambda. \tag{5.2.18}$$

The function $M(\lambda)$ is defined as being dependent only on parameters characteristic of the molecule, such as internal densities of states and the transition matrix element.

The probability of stimulated emission into a given resonator mode dI_M/dt is equal to the same molecular function $M(\lambda)$ times the intensity of photons incident on the molecule $I_M(\lambda)$ times the volume density of excited molecules n_S:

$$dI_M(\lambda)/dt = M(\lambda)I_M(\lambda)n_S. \qquad (5.2.19)$$

Substituting,

$$\frac{dI_M(\lambda)}{dt} = \frac{QY}{\tau_S} \frac{E(\lambda)I_M(\lambda)n_S}{\rho(\lambda)}. \qquad (5.2.20)$$

The density of states in free space is

$$\rho(\lambda) = 8\pi\eta^3/\lambda^4, \qquad (5.2.21)$$

where η is the index of refraction. Since the cross section is defined as a function of length we use

$$\frac{dI_M}{dx} = \frac{1}{v}\frac{dI_M}{dt},$$

where v is the propagation velocity, which equals c/η, where c is the velocity of light. The relation between the experimentally measured values and the desired emission cross section σ_{SM} is[8]

$$\sigma_{SM} = \frac{QY}{\tau_S} \frac{E(\lambda)\lambda^4}{8\pi\eta^2 c}. \qquad (5.2.22)$$

5.2.5. Thermal Limitations

The previous sections have outlined the minimum excitation required to obtain lasing with dyes. At the other extreme, there are limitations imposed by thermal effects.[10-12] The most sensitive effect of the heat input is on the lasing wavelength, which would conflict with the most outstanding characteristic inherent to the dye laser—the selectivity of its output wavelength. A shift in wavelength can be caused by a change in the resonator optical pathlength z due to changes in the active medium index of refraction. The wavelength at which the resonator will operate will therefore shift according to the following relation:

$$d\lambda/\lambda = d\left(\sum_i \eta_i l_i\right)/z = \sum_i l_i d\eta_i/z, \qquad (5.2.23)$$

where λ is the wavelength, l_i the length of a resonator element, z the total resonator optical pathlength, and η_i the index of refraction of the element. Since the only resonator element with a substantial heat input is the dye

---

[10] O. G. Peterson, A. A. Pease, and W. M. Pearson, *IEEE/OSA Conf. Laser Eng. Appl.* p. 4 (1975).
[11] U. Balucani and V. Tognetti, *Opt. Acta* **23**, 923 (1976).
[12] R. Pratesi and L. Ronchi, *Opt. Acta* **23**, 933 (1976).

5.2. BASIC DYE MOLECULE AND DYE SOLVENT PROPERTIES

solution, this relation reduces to a single term characteristic of the dye solution alone,

$$\frac{d\lambda}{\lambda} = \frac{l}{z} d\eta, \quad (5.2.24)$$

where l is defined as the dye cell length.

The change in index $d\eta$ can be calculated in a straightforward manner from the heat per unit volume dH/dV put into the dye solvent during lasing:

$$d\eta = \frac{d\eta/dT}{\rho C_p} \frac{dH}{dV}. \quad (5.2.25)$$

The heat input per unit volume dH/dV is the total excitation absorbed per unit volume minus the laser emission and fluoresence emitted from the volume. The heat capacity C_p, the density ρ, and the change of index with temperature $d\eta/dT$ are properties of the dye solvent. The maximum heat input for a given value of $d\lambda/\lambda$ is therefore

$$\frac{dH}{dV} = \frac{z}{l} \frac{\rho C_p}{d\eta/dT} \frac{d\lambda}{\lambda}. \quad (5.2.26)$$

If the heat load is uniform throughout the active medium, this relation reduces to

$$\frac{\Delta H}{A} = z \frac{\rho C_p}{d\eta/dT} \frac{d\lambda}{\lambda}, \quad (5.2.27)$$

where ΔH is the total heat injected into the solvent within the active medium and A the cross-sectional area of the active region.

It is instructive to note that the high-power limit is similar in functional dependence to the threshold limitation in that both depend on the cross-sectional area of the active mode volume. An immediate conclusion is that neither transverse excitation of the active medium nor longitudinal excitation has any particular advantage over the other with respect to high average power operation. In addition, the active length of the medium can be adjusted to satisfy criteria other than the laser threshold or the thermal limitations.

The thermal properties of several popular solvents are listed in Table I.[13-23] Equation (5.2.27) leads to a definition of a figure of merit M_H for solvents

[13] "Selected Values of Properties of Hydrocarbons and Related Compounds." Thermodyn. Res. Cent., Texas A&M University, College Station, 1977.

[14] J. Timmermans, "Physico-Chemical Constants of Pure Organic Compounds," Vol. 1. Am. Elsevier, New York, 1950.

TABLE I. Solvent Characteristics at 20°C

	$10^5\, dn/dT$	C_p(J/gm)	ρ(gm/cm^3)	ν(cm^2/sec) × 100	M_H(J/cm^3)	M_L(J cm^{-1} sec^{-1})
H$_2$O[a]	8.9[f]	4.1819[h]	0.9982	1.004	47,000	470
D$_2$O[a]	6.6[a]	4.15[i]	1.1053	1.129	69,000	780
Methanol[b]	39	2.50[j]	0.791	0.75[j]	5,100	38
Ethanol[b]	36	2.40[j]	0.789	1.52[j]	5,300	80
Propanol (iso)[b]	34	2.49	0.785	3.1[j]	5,700	180
Ethylene glycol[b]	26	2.34[k]	1.1134	17.9[j]	10,000	1,800
Glycerol[b]	22	1.5[j]	1.2613	1120.0	8,600	96,000
Dimethyl sulfoxide[c]	126[e]	1.97[j]	1.1014	2.0	1,700	34
Dimethyl formamide[d]	43[e]	2.05	0.9493	0.84	4,500	38

[a] Unless otherwise noted, data from Thermodynamics Research Center.[13]
[b] Unless otherwise noted, data from Timmermans.[14]
[c] Data from Crown Zellerbach.[15]
[d] Data from du Pont de Nemours & Co.[16]
[e] Calculated from Gladstone–Dale relation $\eta = 1 + K\rho$, where K is a constant and ρ is the density.
[f] Data from Eisenberg,[17] Steckel and Szapiro,[18] and Hawkes and Astheimer.[19]
[g] Data from Reisler and Eisenberg[20] and Steckel and Szapiro.[18]
[h] Data from Chemical Rubber Co.[21]
[i] Data from Long and Kemp.[22]
[j] Extrapolated values.
[k] Data from Union Carbide Corp.[23]
[l] 25°C value.

5.2. BASIC DYE MOLECULE AND DYE SOLVENT PROPERTIES

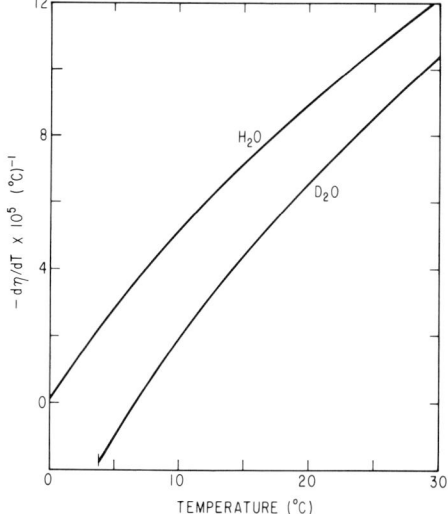

FIG. 8. Variation in the change in refractive index as a function of temperature $d\eta/dT$ plotted as a function of temperature for both water and heavy water.[17-20]

$$M_H = \frac{\rho C_p}{d\eta/dT}, \qquad (5.2.28)$$

which is also included in the table. The solvents with the highest figure of merit M_H are normal and heavy water, which are therefore the solvents of choice for high-power applications where laminar flow conditions are not required. This selection is true even if the dye must be solubilized with a surfactant. The change in refractive index with temperature $d\eta/dT$ is displayed in Fig. 8 for the two types of water.[17-20] It is seen that the thermal characteristics dramatically improve as the solution temperature is de-

---

[15] "Dimethyl Sulfoxide Technical Bulletin." Crown Zellerbach, Camas, Washington, 1966.

[16] "Dimethyl Formamide Technical Bulletin," Ind. Chem. Dep., E. I. duPont de Nemours & Co., Wilmington, Delaware, 1976.

[17] H. Eisenberg, *J. Chem. Phys.* **43**, 3887 (1965).

[18] F. Steckel and S. Szapiro, *Trans. Faraday Soc.* **59**, 331 (1963).

[19] J. B. Hawkes and R. W. Astheimer, *J. Opt. Soc. Am.* **38**, 804 (1948).

[20] E. Reisler and H. Eisenberg, *J. Chem. Phys.* **43**, 3875 (1965).

[21] "Handbook of Chemistry and Physics." 53rd ed. Chem. Rubber Co., Cleveland, Ohio, (1972).

[22] E. A. Long and J. D. Kemp, *J. Am. Chem. Soc.* **58**, 1829 (1936).

[23] "Union Carbide Chemicals and Plastics Booklet," No. F41515. Union Carbide Corp., New York, 1971.

creased, especially for the case of heavy water where the change of index goes to zero at 6.6°C, which is above the freezing point 3.8°C for this material.

It is useful to insert some arbitrary values into the thermal limitation equation to obtain some limit-of-operation estimates of the dye laser. For this purpose 22°C water values will be assumed. Another arbitrary choice must be made as to an acceptable value for the relative wavelength shift $d\lambda/\lambda$. A specific, but widely applicable choice would be the Doppler width of the absorption of a vaporizable species. This value is remarkably universal and is approximately $d\lambda/\lambda = 10^{-6}$.[24] This result leads to the useful approximation that for narrow-line operation only 4 J/cm² can be absorbed by an aqueous solution of laser dye where the area is that of the laser bore cross section.

For incoherently excited lasers, where it can be assumed that all the excitation goes into solution heating, this thermal limit quite severely restricts the average power at which an oscillator can operate. The limit makes lasers using coherent excitation look especially attractive for narrow linewidth applications. For both excitation methods the thermal effects place a great importance on the design of the dye solution flow system to minimize the transit time of the solution through the active volume.

The use of water as a solvent for dyes is significantly restricted by the fact that most dyes have a low solubility in this solution. Even if the dyes will dissolve into water they usually dimerize, which quenches the fluorescence and thereby destroys any chance of laser action. This problem becomes even more severe as the temperature is reduced. The dye molecules must either be specially synthesized with substituents that increase their solubility or must be solubilized with an additional solvent or a surfactant added to the solution. Adding extra substituents to the fluorescent molecules often reduces the fluorescence quantum yield of the molecules, which would reduce the laser efficiency. Any additives put into the dye solution must be carefully selected so as not to introduce any additional losses for the laser emission or the excitation intensity. For flashlamp-excited devices this additive must be transparent throughout the visible and ultraviolet portions of the spectrum. Surfactants have an advantage in this regard in that they can be effective in very low concentrations, 1 or 2%, of the solution. However, the surfactants dissolve the dye by forming small micelles in the water solution. These micelles are of sufficient size to scatter visibly the laser emission. Over a large length this scattering can become a significant loss for the laser emission. The effect of the surfactants is also temperature dependent, with

[24] A. C. G. Mitchell and M. W. Zemansky, "Resonance Radiation and Excited Atoms." Cambridge Univ. Press, London and New York, 1971.

their solubility being reduced as the temperature decreases. One of the most effective surfactants is Ammonyx LO™,[25] which has greater UV transparency and smaller micelle formation than many other such materials. It is clear that for a specific application the optimum choice of solvent system and conditions can be complex and will require considerable experimental effort.

5.3. Laser Devices

Following the discussion given earlier concerning the lifetimes of the states characteristic of the lasing dyes, it is useful to classify dye laser devices according to the time domain in which they operate. The demarcation point is the triplet state lifetime, which is of the order of 100 nsec.[3,26,27] Devices with excitation pulses shorter than this will be defined as short-pulse lasers and are characterized completely by transient nonequilibrium behavior. Devices with excitation pulses longer than the triplet state lifetime, which includes cw, are characterized by steady-state or quasi-steady-state kinetics and will be so labeled. There are design considerations specific to each of these regions of operation, which will be discussed in some detail in the following chapters. In particular, the approximations that are appropriate for generating analytical models of the devices are significantly different. These approximations can greatly simplify the model calculations at these two extremes of excitation pulse widths and are also presented in the following chapters.

The operating regime around 100-nsec pulse widths has been ignored in this discussion because analyses of systems operating in this time period will require the solution of the complete time-dependent equations as set forth in Chapter 5.2. The solutions will be tedious but straightforward using numerical methods. There is a relative scarcity of excitation sources with pulse widths in this time period with the anticipated scarcity of descriptions of dye lasers operating in this regime.

5.4. Short-Pulse Dye Lasers

5.4.1. Rate Equation Description

Laser operation in the time regime under 100 nsec is dominated by the relationship between the excitation pulse width and the round trip time for

---

[25] A. H. Herz, W. C. McColgin, J. S. Hayward, and O. G. Peterson, U.S. Patent 3,818,371 (1974).
[26] O. Teschke, A. Dienes, and G. Holtom, *Opt. Commun.* **13**, 318 (1975).
[27] O. Teschke and A. Dienes, *Opt. Commun.* **9**, 128 (1973).

light contained within the resonator. The limitation on the number of round trips imposed by the short excitation pulses necessitates the production of very high gains in the active media.[28] The high gain requires large densities of excited states. These densities, in turn, make the excited-state absorption very important for efficiency considerations.[9,29–33] For most calculations in this time regime the triplet population n_T can be assumed to be zero or at worst a perturbation on a calculation for zero triplet state population. The excitation used to obtain such short pulses is the monochromatic output from other lasers. It will be assumed, therefore, that the excitation and dye absorption will be appropriately matched in wavelength so that excessive dye concentrations will not be required to absorb the excitation. This assumption permits an additional simplification to be made, i.e., the ground-state absorption, $n_0 \sigma_{0M} I_M$ of the laser emission will be ignored in the forthcoming analyses.

The kinetic equations for the short-pulse approximation can be easily obtained from Eqs. (5.2.1)–(5.2.5):

$$dI_X/dy = -[n\sigma_{0X} + n_S(\sigma_{SX} - \sigma_{0X})]I_X, \quad (5.4.1)$$

$$dI_M/dx = n_S \sigma_{SM} I_M, \quad (5.4.2)$$

$$dn_S/dt = n\sigma_{0X} I_X - n_S(\sigma_{0X} I_X + \sigma_{SM} I_M + 1/\tau_S). \quad (5.4.3)$$

It is useful to convert these equations to dimensionless variables. The intensities can be redefined in terms of saturation fluxes $\phi_j = I_j/I_j^s$, where the saturation flux I_j^s is defined as $I_j^s = 1/\sigma_{ij}\tau_S$ and the cross section σ_{ij} is evaluated at the wavelength of the specific light intensity and the lifetime τ_S is that of the first excited singlet state. The excited-state population n_S can be normalized with respect to the total dye molecule density n to yield the excited state fraction β and the time can be normalized with respect to the fluorescence lifetime τ_S. The distances x and y can be similarly expressed in terms of an absorption length. To simplify the equations it is useful to use the characteristic length $\alpha_M = \sigma_{SM} n$ associated with the laser emission for the normalization. The normalized set of variables then becomes:

---

[28] P. P. Sorokin, J. R. Lankard, E. C. Hammond, and V. L. Moruzzi, *IBM J. Res. Dev.* **11**, 130 (1967).
[29] I. Wieder, *Appl. Phys. Lett.* **21**, 318 (1972).
[30] E. Sahar, D. Treves, and I. Wieder, *Opt. Commun.* **16**, 124 (1976).
[31] R. F. Leheny and J. Shah, *IEEE J. Quantum Electron.* **qe-11**, 70 (1975).
[32] C. D. Decker, *Appl. Phys. Lett.* **27**, 607 (1975).
[33] G. Dolan and C. R. Goldschmidt, *Chem. Phys. Lett.* **39**, 320 (1976).

5.4. SHORT-PULSE DYE LASERS

$$\phi_X = I_X/I_X^s, \quad \Phi_M = I_M/I_M^s,$$
$$\beta = n_S/n, \quad \tau = t/\tau_S,$$
$$z_X = x\alpha_M, \quad z_M = y\alpha_M,$$
$$\alpha_M = \sigma_{SM}n, \quad \alpha_X = \sigma_{0X}n, \quad \alpha_{SX} = \sigma_{SX}n,$$
$$I_X^s = 1/\sigma_{0X}\tau_S, \quad I_M^s = 1/\sigma_{SM}\tau_S, \quad (5.4.4)$$

By using the normalized variables together with the approximations appropriate for the short-pulse regime, the kinetic equations then may be written

$$\frac{d\phi_X}{dz_X} = -\phi_X \left[1 + \left(\frac{\alpha_{SX}}{\alpha_X} - 1\right)\beta\right]\frac{\alpha_X}{\alpha_M}, \quad (5.4.5)$$

$$\frac{d\phi_M}{dz_M} = \phi_M\beta, \quad (5.4.6)$$

$$\frac{d\beta}{d\tau} = \phi_X - (\phi_X + \phi_M + 1)\beta. \quad (5.4.7)$$

These equations should adequately describe most devices operating this short time regime. The equations do not include the resonator losses. Since the equations are usually solved by computer techniques, these losses can be included as simple fractional multipliers operating on ϕ_M.

5.4.2. Amplified Spontaneous Emission

All devices excited by short pulses are of necessity very high gain devices with very high excited-state population densities. This introduces a complication in that amplified spontaneous emission (ASE) can become sufficiently large so as to compete with the desired laser emission.[34] It is obviously desirable to minimize this ASE intensity and very important to keep it significantly less than the saturation value. If the amplified spontaneous emission is permitted to reach higher intensities it will begin bleaching the excited-state population and thus reduce the useful gain of the amplifying medium.

The value of this intensity can be estimated in the low-intensity limit. The transit time of spontaneous emission through the amplifying region can be assumed to be very much shorter than the half-width of the excitation pulse. A steady-state calculation can therefore be employed. The intensity at an arbitrary point is then the integral over the excited volume of the fluorescence emitted by each differential volume into the solid angle subtended by the test volume multiplied by the amplification it

---

[34] U. Ganiel, A. Hardy, G. Neumann, and D. Treves, *IEEE J. Quantum Electron.* **qe-11**, 881 (1975).

receives as it travels the distance between the two points. The fluorescence is emitted over a wide-wavelength range and both the fluorescence probability and the cross section for amplification vary greatly over this range. The fluorescence emitted by a differential volume dV into a specific solid angle $d\Omega = \delta A/4\pi r^2$ and wavelength increment $d\lambda$ is equal to

$$dF = \frac{n_S E(\lambda)\, d\lambda}{\tau_S} \frac{\delta A\, dV}{4\pi r^2}, \qquad (5.4.8)$$

where δA is the detection area and $E(\lambda)$ is defined in Section 5.2.4 as the experimentally measured fluorescence emission function normalized so that $\int E(\lambda)\, d\lambda = 1$. If the desired condition of small ASE is assumed, the excited-state density n_S will not be affected by this flux. In this limit, the values of the flux $F_{ASE}(\lambda)$ at different wavelengths will be independent of each other and can be calculated separately. Also, for the purposes of this calculation the value of n_S will be assumed to be constant and homogenous, which is also dependent on the assumption of small ASE. The point of view will be taken that the important independent variable is the net gain characteristic of the amplifying medium. In a laboratory device this gain will depend not only on the excitation rate but also on the amount of gain saturation produced by the amplified signal.

The maximum value of the ASE will occur at the ends of the amplifying region, which will be taken to be a circular cylinder as illustrated in Fig. 9. The value of the total ASE flux impinging on a differential volume of cross-sectional area δA that is on the axis of the cylindrical volume will be calculated. The portion of this flux dF_{ASE} that starts from fluorescence radiated by an arbitrary differential volume dV is the fluorescence emitted by that volume as evaluated with Eq. (5.4.8), multiplied by the amplification it receives as it traverses the distance r between the emission volume dV and the measurement point dA:

$$dF_{ASE} = \frac{n_S}{\tau_S} E(\lambda)\, d\lambda\, dV\, \frac{\delta A}{4\pi r^2} \exp(n_S \sigma_{SM} r), \qquad (5.4.9)$$

where $dV = r^2 \sin\psi\, dr\, d\theta\, d\psi$ in spherical coordinates.

The value of a saturation flux falling on this same test volume would be

$$F_S = \delta A/\sigma_{SM}\tau_S. \qquad (5.4.10)$$

By normalizing the ASE flux to this value and integrating over the radius r and polar angle θ coordinates the following result is obtained:

$$\phi_{ASE} = \frac{1}{2}\int_0^\infty d\lambda E(\lambda) \left[\int_0^{\psi_A} d\psi\, \sin\psi \left(\exp\frac{G(\lambda)}{\cos\psi} - 1\right) \right.$$
$$\left. + \int_{\psi_A}^{\pi/2} d\psi\, \sin\psi \left(\exp\frac{G(\lambda)}{2R_A \sin\psi} - 1\right) \right], \qquad (5.4.11)$$

5.4. SHORT-PULSE DYE LASERS

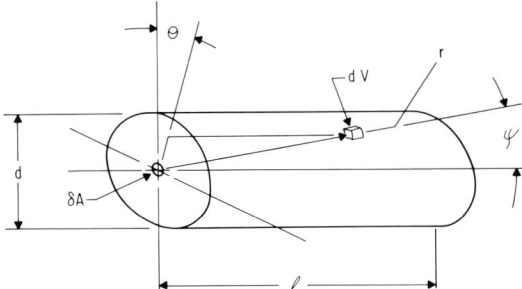

FIG. 9. Representative geometry for amplified spontaneous emission calculation.

where ϕ_{ASE} is the amplified spontaneous emission at one end of the gain region measured in saturation units. In this equation $G(\lambda)$ is the logarithmic gain $G(\lambda) = n_S \sigma_{SM} l$, l is the length of the amplifying region, R_A is the aspect ratio of the amplifying region $R_A = l/d$, ψ is the azimuthal angle, and $\psi_A = \tan^{-1}(1/2R_A)$.

The integral over the azimuthal angle is a function of two parameters, the aspect ratio and logarithmic gain value. This can be evaluated numerically in straightforward manner. The integral over wavelength is difficult since both $E(\lambda)$ and $G(\lambda)$ are wavelength dependent. This function will be sharply peaked near the point of maximum σ_{SM}. Only a small fraction of the fluorescence in the region near this peak will contribute to the final value of the ASE. For the purposes of this low-intensity estimate, the effective fraction of the fluorescence can be graphically evaluated. This fraction was taken to be that which is contained between the two points where the ASE had dropped to half of its peak value. This fraction was found to be approximately $\frac{1}{4}$ of the total emitted over all wavelengths. The results of these estimates are displayed in Fig. 10, where the value of the maximum ASE has been plotted as a function of the logarithmic gain for selected values of the amplifying region aspect ratio.

Caution must be exercised in interpreting these graphs. While the high-aspect-ratio devices will support higher gains, it must be noted that the volume of the amplifying region is decreasing as the square of the aspect ratio. The sources for the ASE are therefore also decreasing as the square of the aspect ratio. Also, this calculation only identifies the limits where an amplifying region will start to deplete its own population of excited states. ASE also introduces noise into signals being amplified, which can reduce the effectiveness of an amplifier at much lower levels of ASE.

It is seen that the results presented in Fig. 10 are not as conservative as would be predicted using the rule of thumb that the gain must be kept to

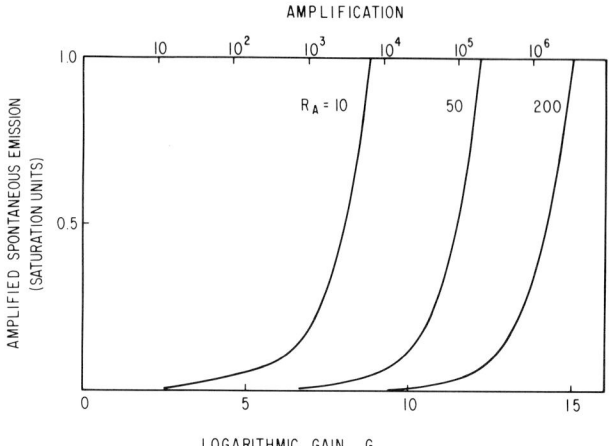

FIG. 10. Amplified spontaneous emission as a function of the gain of the amplifying medium. The curves are for different active volumes characterized by aspect ratios R_A defined as the ratio of the volume length to diameter.

values less than $(l/R)^2$, where R is the radius of the excited volume.[35] The $(l/R)^2$ relation is more universal in its applicability but more approximate in derivation. Its simplicity in form and conservative predictions make it a most useful guide for the design of laser amplifiers.

If the excited volume is contained in a resonator so that a fraction of the ASE can be reflected back for further amplification, the amount of ASE will be greatly increased. An estimate of the ASE produced when there is optical feedback is very geometry dependent and must be estimated for each application. This condition is also true for any estimates of ASE generated noise on amplified signals. Calculations of these effects are similar to the calculation presented here with some minor simplifications. It should be necessary to calculate only the amount of ASE that passes through a particular aperture in the optical system. Equation (5.4.11) can then be reduced to

$$\phi_{\text{ASE}} = \frac{1}{2} \int_0^{\psi'} \{\exp[G(\lambda)/\cos \psi] - 1\} \sin \psi \, d\psi, \qquad (5.4.12)$$

where ψ' is the half-angle subtended by the aperture.

A complete calculation of the ASE emitted by an amplifying medium into all solid angles in the regime where ϕ_{ASE} is greater than one is indeed a formidable task. In this case the excited-state population density in any arbitrary volume will be dependent on the value of total ϕ_{ASE} intensity in-

---

[35] M. H. Gassman and H. Weber, *Opto-Electron.* **3**, 177 (1971).

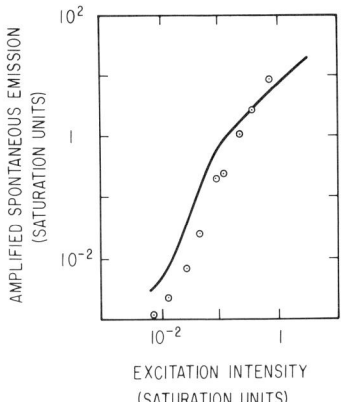

FIG. 11. Amplified spontaneous emission as a function of excitation intensity.[34]

cident on that volume. This couples together all of the variables in the previous calculation. The emission intensities at all wavelengths and emission angles are now interdependent. This problem has been solved for the special case of a very large aspect ratio geometry so that the ASE is emitted predominantly in the axial direction.[34] At the higher intensities it was found that the emission was further narrowed in wavelength so that 70% of the ASE was contained in a 5-nm bandwidth. The results of the calculation are displayed in Fig. 11, where the ASE has been normalized to a saturation flux and is plotted as a function of the excitation intensity incident on a 1.0-cm-long dye volume. The excitation has also been normalized to a saturation flux. It was found that the ASE increases linearly with respect to the excitation at low intensities, where the output intensity is predominantly fluorescence. At higher excitations the effects of the amplification become evident as the ASE increases exponentially. The exponential increase continues until the ASE reaches a value of approximately one in saturation units and then the output again becomes linear as the gain is saturated by the large ASE intensities. Also included are experimentally measured values of ASE obtained by the same authors using a nitrogen-laser-excited system.

In the laboratory, devices based on this amplified fluorescence make very effective light sources.[34] They are highly directional but provide illumination that is very broad in wavelength by laser standards. This emission is used in short pulse oscillators to provide the initial intensity, which is filtered and subsequently amplified to produce narrow-band, wavelength-tunable laser output.[36] This will be further discussed in Section 5.4.3.

[36] T. W. Hänsch, *Appl. Opt.* **11**, 895 (1972).

The importance of effects of ASE on the operation dye lasers is one of the continuing fields of investigation; one can look forward to more definitive experimental and theoretical investigations of this effect.[36a,b]

5.4.3. Short-Pulse Oscillator

The excitation sources for short-pulse dye lasers have often been Q-switched lasers or other short-pulse discharge lasers such as the nitrogen laser.[36-46] The output pulses from these devices are usually in the range from 10 to 30 nsec. These short-pulse durations limit the number of round trips that the laser emission can make in an oscillator of any significant size. The limit on the number of round trips can put a rather severe limit on the efficiency of such a device. The oscillator would have to be a very high gain device requiring high excited-state inversions, which would exaggerate the losses due to excited-state absorption of the excitation.

A wavelength-tunable oscillator must be sufficiently long to contain the required tuning elements and therefore is particularly subject to these efficiency restrictions. To obtain wavelength-selective output, it is therefore useful to separate the wavelength selection and power production into an oscillator plus amplifier chain. The inefficiency of the initial oscillator can be ignored if a greater fraction of the excitation power can be utilized in a high-efficiency amplifier chain.[41]

A tunable short-pulse oscillator is of necessity a very high gain laser to allow stimulated emission to build up in the small number of round trips permitted by the pulse length.[37] The high gain requires that the tunable elements have high discrimination to limit the output wavelength. One of the first such designs is illustrated in Fig. 12.[36] The wavelength discrimi-

---

[36a] G. Dujardin and P. Flamant, *Opt. Commun.* **24**, 243 (1978).
[36b] W. Heudorfer, G. Marowsky, and F. K. Tittel, *Z. Naturforsch., Teil* **33A**, 1062 (1978).
[37] R. Wallenstein, *Opt. Acta* **23**, 887 (1976).
[38] I. Itzkan and F. W. Cunningham, *IEEE J. Quantum Electron.* **qe-8**, 101 (1972).
[39] J. L. Carlsten and T. J. McIlrath, *Opt. Commun.* **8**, 52 (1973).
[40] U. Ganiel and G. Neumann, *Opt. Commun.* **12**, 5 (1974).
[41] R. Wallenstein and T. W. Hänsch, *Opt. Commun.* **14**, 353 (1975).
[42] E. D. Stokes, F. B. Dunning, R. F. Stebbings, G. K. Walters, and R. D. Rundel, *Opt. Commun.* **5**, 267 (1972).
[43] S. A. Myers, *Opt. Commun.* **4**, 187 (1971).
[44] D. C. Hanna, P. A. Kärkkäinen, and R. Wyatt, *Opt. & Quantum Electron.* **7**, 115 (1975).
[45] G. K. Klauminzer, *IEEE J. Quantum Electron.* **qe-13**, 92D (1977).
[46] M. A. Novikov and A. D. Tertyshnik, *Sov. J. Quantum Electron.* (*Engl. Transl.*) **5**, 848 (1975).

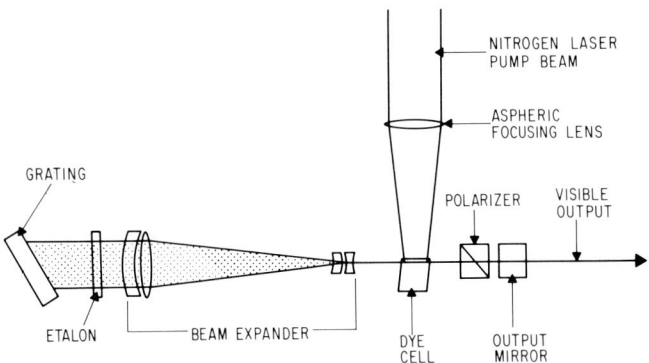

FIG. 12. Short-pulse, wavelength-tunable oscillator. This oscillator uses a high-magnification telescope to expand the intraresonator beam to increase the resolution of the wavelength tuning elements (illustration courtesy of Molectron Corp., Sunnyvale, CA).

nation is achieved by expanding the beam by a factor of 20–50 with a low-loss telescope before it reaches the dispersive elements. The beam expander serves several purposes. By increasing the number of illuminated lines on the Littrow mounted grating, the resolution of this coarse tuning element is greatly enhanced. It also reduces the walkoff losses of the line-narrowing etalon. These effects are considered in more detail in Section 5.5.13 which discusses wavelength-tuning elements. The telescope also reduces the intensity of the laser light falling on the grating, which would otherwise be destroyed by the original beam. The telescope then collects the wavelength-selected emission and focuses it back through the active volume for amplification to useful intensities. The gain of the device is fixed at such a high level that the stability of the resonator is not critical. Efficiency considerations dictate, however, that the telescope focus be carefully adjusted to return a maximum amount of light through the gain region.

A distinct disadvantage of this design is the length to which the resonator must be extended to accommodate the telescope. This length, which is approximately 40 cm, limits the number of round trips to four for laser emission produced by a 10-nsec-long excitation pulse. This small number of round trips does not permit the intraresonator intensity to build up to the saturation levels required to couple effectively the energy out of the active medium. It is therefore useful to consider alternative schemes for expanding the beam.

Since the resolution of the grating depends only on the number of grating lines illuminated, it is necessary to expand the beam in only one dimension. This can be done by directing the beam through an optical sur-

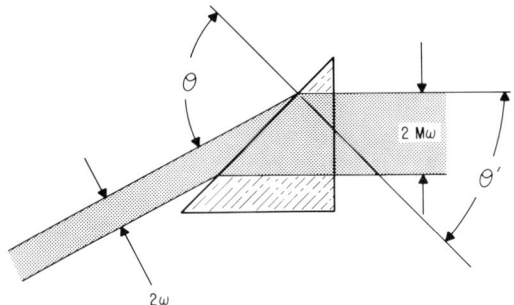

Fig. 13. Beam expansion using high-incidence-angle prism.

face at a high angle of incidence as shown in Fig. 13. A single such surface, if aligned to produce an equivalent beam expansion, would have a very high reflectivity. This high-incidence angle introduces equivalently high loss into the resonator. The gain in efficiency obtained by shortening the resonator in this manner nearly compensates for the loss introduced by the surface. The resulting oscillator is a much more easily aligned and maintained device, and is therefore an attractive alternative to the telescope design.[42–44]

The reflection loss can be greatly reduced by using several elements at smaller incidence angles, which together yield the desired magnification.[45,46] Such a geometry is illustrated in Fig. 14. In this design four prisms that were arranged in achromatic pairs were employed. This pairing of the prisms permits alignment of the laser at wavelengths other than the operating wavelength. Since a grating is to be used as the pri-

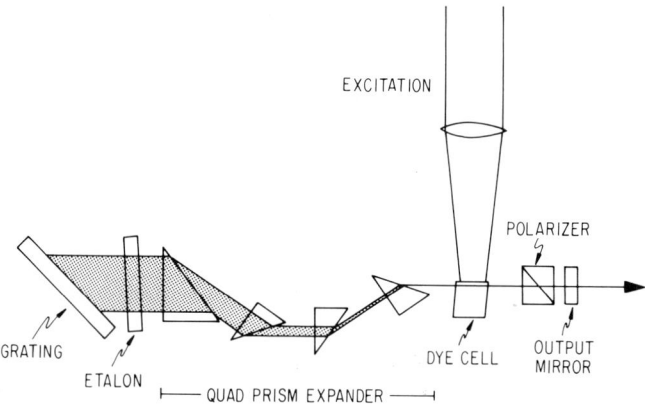

Fig. 14. Short-pulse, wavelength-tunable oscillator using high-incidence-angle refraction to expand the intraresonator beam.[45]

mary dispersive element, the additional dispersion that could be introduced by the prisms is not significant. The prisms are cut with one surface perpendicular to the beam propagation direction to maximize the expansion. This surface should be antireflection coated.

The intraresonator beam will be expanded as shown in Fig. 13 by a factor M, which is given by[44]

$$M = [1 - (\sin^2 \theta)/\eta^2]^{1/2}/\cos \theta, \qquad (5.4.13)$$

where θ is the angle of incidence of the beam. The reflection loss for light with E-vector in the plane of Fig. 13 (p-polarization) can be calculated from[44]

$$R = \tan^2(\theta - \theta')/\tan^2(\theta + \theta'), \qquad (5.4.14)$$

where θ' is the refracted angle as illustrated in Fig. 13. These relations for M and R have been evaluated for an index of refraction of $\eta = 1.54$, and are displayed in Fig. 15.[45] The graph shows the magnification and the transmission as a function of incident angle for one-, two-, and four-prism designs. In each case the two-prism value was the square of the single-prism number and similarly for the four-prism quantity. It is clear that increasing the number of elements allows one to obtain arbitrarily high magnifications simultaneously with low resonator losses. The major advantage of this arrangement, however, is the possibility that the desired spectral resolution can be achieved with a much shorter resonator than is needed to contain a telescope: therefore, the number of elements must be limited. A four-prism device with a 40× magnification appears to be an effective compromise for maximizing the oscillator efficiency.

In Fig. 15 the transmission illustrated is for p-polarized light. The transmission for s-polarized light is much smaller, which makes the tuned laser output highly polarized. It is particularly useful, therefore, to introduce a polarizer between the dye cell and the output mirror to prevent s-polarized fluorescence from being reflected back through the gain medium, where it would compete with the wavelength-selected radiation for the available gain.

The original design included a transversely excited active volume with the dimensions of this volume fixed by adjusting the focusing of the excitation and adjusting the dye concentration so that the excitation would be absorbed in a depth approximately equal to the width of the focal spot. This adjustment would yield an approximately cylindrical excited volume. The excitation was focused into a long, narrow volume by the use of spherical or cylindrical lenses as required by the excitation beam geometry. The length of the focal spot reduced the intensity of the excitation incident on the dye, thus decreasing excited single-state absorption, and thereby improving the laser efficiency.

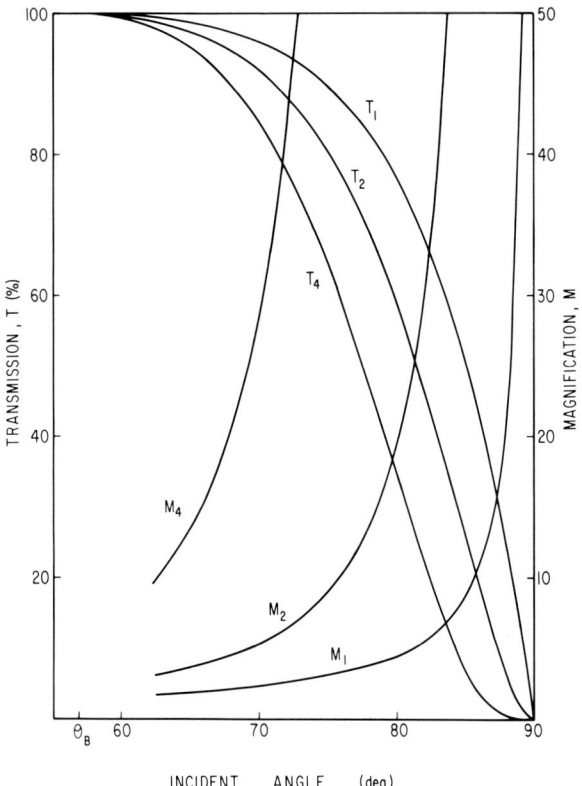

FIG. 15. Beam magnification M and transmission T as a function of incidence angle. The subscripts indicate the number of prisms in the beam expander assembly.[45]

The size of the excitation volume was chosen[36] such that the absorbed power per cross-sectional area of the laser mode was 500 MW/cm². If all of this excitation were efficiently converted into excited molecules, the logarithmic gain of the active volume would be 400. This gain is an order of magnitude higher than that calculated for the onset of amplified spontaneous emission. Therefore, the device is essentially a single-pass amplifier for filtered amplified spontaneous emission.

The dye cell containing the active region was specially designed with nonparallel windows, as shown in Fig. 12, to prevent parasitic laser action due to Fresnel reflection from these windows. The output mirror has extremely high transmission, 50–95%. This high transmission restricts the number of round trips that stimulated emission can make through the resonator to just slightly over one. Additional round trips would not be useful because of the short excitation pulse duration and are not necessary because of the extremely high gain of the active volume. The small amount of feedback from the output mirror, however, does have an effect, as is evidenced by the mode structure such devices possess.[47]

5.4.4. Short-Pulse Amplifiers

Amplifiers for the output of the oscillator must be designed such that they can be locked in wavelength to that of the oscillator without permitting the buildup of parasitic wavelengths.[37–39,41] The amplifiers can be divided into two classes, single-pass and regenerative oscillators.[48–52] The single-pass amplifiers are the simplest and most straightforward to use of the two devices. They can give an accurate reproduction of the input signal but can be inefficient if used for an amplification of too great a magnitude. Regenerative oscillators are devices that include a finite amount of feedback. This feedback allows the intraresonator intensity to build up to saturation levels during the excitation period, which permits efficient extraction of the energy stored in the active medium. However, this amplifier will exhibit a very distinct mode structure, which must be carefully matched to that of the oscillator. These two types of devices will be considered in more detail in the following sections.

5.4.5. Single-Pass, Short-Pulse Amplifier

To obtain significant output and efficiency from a single-pass amplifier, the gain length must be sufficiently long for the laser emission to be amplified up to saturation intensities. It therefore can be assumed that the amplifier will be transversely excited with an aspect ratio of at least one order of magnitude, with a geometry similar to that illustrated schematically in Fig. 16. Note that the windows have been purposely aligned antiparallel to each other to prevent parasitic laser action.

This system can be easily analyzed using Eqs. (5.4.6) and (5.4.7). Since the transit times of the excitation and laser pulses through the amplifier are very short compared to other times characteristic of the system, i.e., fluorescence lifetime and excitation pulse width, the excited-state population can be assumed to be in equilibrium with the light intensities. Within this approximation, we can set $d\beta/d\tau = 0$, eliminate β between the two

---

[47] A. A. Pease and W. M. Pearson, *Appl. Opt.* **16**, 57 (1977).
[48] J. E. Bjorkholm and H. G. Danielmeyer, *Appl. Phys. Lett.* **15**, 171 (1969).
[49] L. E. Erickson and A. Szabo, *Appl. Phys. Lett.* **18**, 433 (1971).
[50] C. J. Buczek, R. J. Freiberg, and M. L. Skolnick, *Proc. IEEE* **61**, 1411 (1973).
[51] U. Ganiel, A. Hardy, and D. Treves, *IEEE J. Quantum Electron.* **qe-12**, 704 (1976).
[52] P. Juramy, P. Flamant, and Y. H. Meyer, *IEEE. J. Quantum Electron.* **qe-13**, 855 (1977).

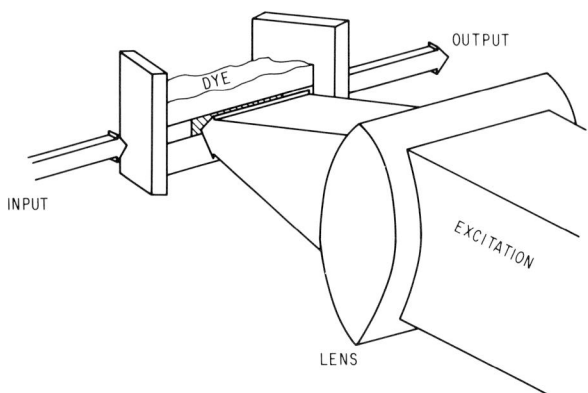

FIG. 16. Single-pass, short-pulse amplifier. The windows are aligned nonparallel to prevent parasitic oscillation from occurring.

equations, and solve analytically for the value of the laser intensity as a function of the amplification length. The result is

$$\phi_M(z_M) = \phi_M(0) \exp\left[\frac{z_M \phi_X + \phi_M(0) - \phi_M(z_M)}{\phi_X + 1}\right]. \quad (5.4.15)$$

It has been assumed that the excitation is absorbed uniformly throughout the active medium. Implicit in the assumption of uniform excitation absorption is the approximation that the absorption cross section for excitation is the same in both the ground and excited singlet states. Geometrical effects have been ignored in this treatment, specifically diffraction and the effect of imperfect matching of the excitation volume to the volume irradiated by the input beam that is to be amplified. These latter effects depend strongly on the characteristics of the excitation source, i.e., beam dimensions, transverse mode content, etc., which must be experimentally determined. To the extent that diffraction can be ignored, this equation should be applicable point by point across the face of the amplifier perpendicular to the propagation direction of the amplified beam. To make the model conform more closely to experimental conditions, the calculations were performed for sine function excitation and input pulses where the initial pulsewidths were identical and synchronized. The performance of the amplifier will be enhanced if the excitation rate for the excited state is much greater than the spontaneous decay rate for this state. This condition occurs if $\phi_X \gg 1$ as seen by setting $d\beta/d\tau = 0$ in Eq. (5.4.7). There is an obvious limit to the effect of increasing ϕ_X where the dye molecules have been totally inverted, $\beta = 1$. This limit implies that all highly excited short-pulse amplifiers will have a small signal logarithmic gain of $G = n\sigma_{SM}l$, where n is the total molecular density in the dye solution.

5.4. SHORT-PULSE DYE LASERS

Because of these considerations, an arbitrary value of $\phi_X = 100$ was chosen and Eq. (5.4.15) evaluated iteratively. The results are displayed in Fig. 17A for three values of the input intensity $\phi_M(0) = 0.001, 0.1$, and 10.

It is seen that the graph of output energy vs. normalized length increases exponentially for short amplification lengths until it is approximately equal to the excitation pulse energy measured in saturation units. The gain then saturates and the output eventually increases at a rate linear with normalized distance.

Since the amplification does not start to saturate until the amplified beam approaches the intensity of the excitation beam, it is possible to push the previously discussed limitations imposed by ASE. The ASE will not significantly deplete the excited state population until it also reaches an intensity comparable to the excitation intensity. In addition, the small signal amplification and the sources for ASE obviously reach a maximum at total inversion of the molecules. Therefore, the signal-to-noise ratio should be significantly enhanced by using an amplifier with $\phi_X \gg 1$ to obtain a specified gain, provided that the input signal can be focused to sufficient intensity to take advantage of this amplifier.

The amplifier will be efficient only if it is operating in the saturation region. To illustrate this contention, the point-by-point efficiency E of the device is displayed in Fig. 17B, again for three values of the input intensity $\phi_M(0) = 0.001, 0.1$, and 10. The efficiency was calculated from the relation

$$E = \frac{\Delta I_M}{\Delta I_X}\bigg|_{\substack{\text{unit}\\\text{volume}}} = \frac{\phi_M}{\phi_M + \phi_X + 1}, \quad (5.4.16)$$

again assuming that the excited state absorption of ϕ_X is approximately equal to the ground state absorption. In the event that the excited state absorption is negligible, Eq. (5.4.16) reduces to $E = \phi_M/(\phi_M + 1)$. Equation (5.4.16) can be easily derived from Eqs. (5.4.4)–(5.4.7) and has been evaluated at each value of z_M and averaged over the pulse width. It is clear from the equation that an efficient amplifier will have $\phi_X \gg 1$. Also, for a single device to be both amplifying and efficient, it must be sufficiently long that the major portion of the device will be operating in the efficient saturated regime.

The maximum length that can be used will be limited by the buildup of ASE. If one imposes the limitations discussed in the section on ASE, it is possible to consider the amplifier in a little more detail. The ASE limits the amplifications to the range between 10^4 and 10^5 under the assumption previously discussed. By referring to Fig. 17A, it can be seen that such an amplification can be obtained if the input signal is 0.001 of the excita-

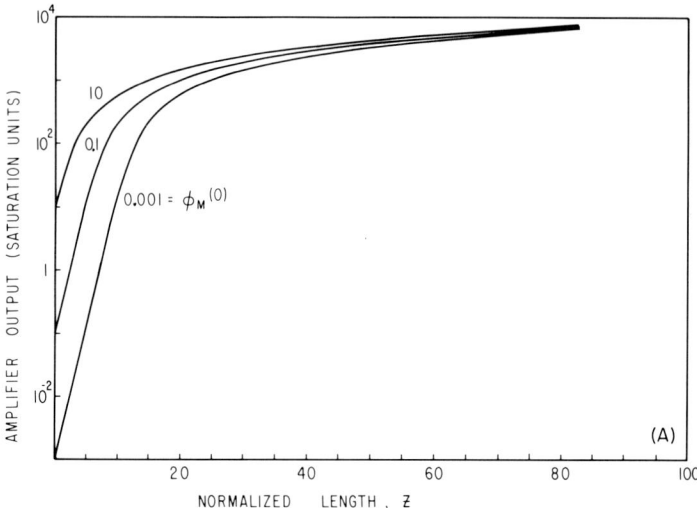

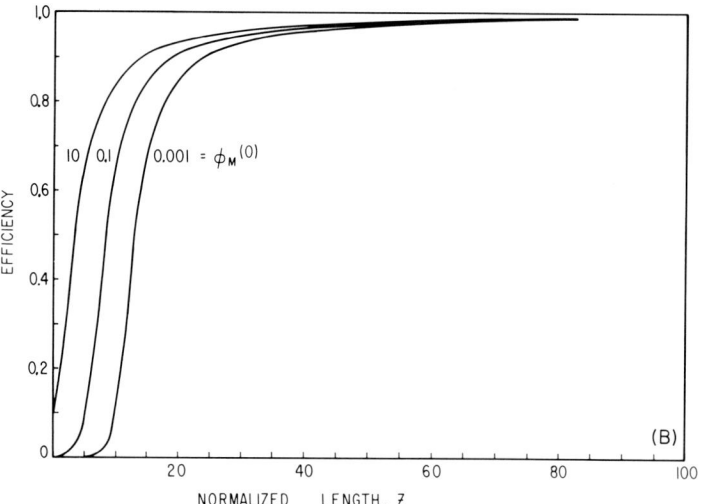

FIG. 17. Output (A) and point-by-point efficiency (B) of short-pulse, single-pass amplifier as a function of the normalized length $z = l\sigma_{SM}n$. The curves were calculated for an excitation intensity of $\phi_X = 100$ measured in saturation units and three values of the input intensity $\phi_M(0)$ as listed in the figure.

tion intensity, i.e., $\phi_M = 0.001\phi_X$ and the amplifier length is approximately $l = 50/n\sigma_{SM}$. By averaging the point-by-point efficiency over this length, a total efficiency in photon units of 77% is obtained for the amplifier. The realization of this efficiency in practice depends not only on fixing the excitation and input intensities at the appropriate values but also on cleverly matching the excitation volume to the mode volume of input beam. It is also clear that the same amplification can be obtained for several values of the input intensity but that the greatest efficiency is obtained for the highest input intensity and resultant longest amplification length. It is therefore advantageous to design the amplifier for the largest aspect ratio l/d consistent with the diffraction characteristics of the amplified beam.

It is useful at this point to note that the transverse dimensions of the amplifier will probably be fixed so that $d = 1/n\sigma_{0X}$ for effective absorption of the excitation beam. The aspect ratio of the amplifier can be expressed in terms of the amplification length parameter

$$l/d = z_M \sigma_{0X}/\sigma_{SM}. \tag{5.4.17}$$

It is clear from this equation that employing an excitation source with an emission wavelength such that the absorption cross section for the excitation σ_{0X} is significantly smaller than the cross section for stimulated emission σ_{SM}, will yield a longer gain length z_M for a specific physical aspect ratio l/d. Substantial quantum efficiencies can be obtained in this manner from devices with smaller aspect ratios than would have been anticipated. This fact has been used to advantage in nitrogen laser-pumped amplifiers, where the ratio of cross sections σ_{SM}/σ_{0X} can be as large as 10.

It is clear that the linear amplifier can be quite efficient. Its greatest advantage lies in the simplicity in design and use and in the fact that it requires no tuning for operation. Its use is restricted by the fact that it must be excited transversely for high amplifications. The beam to be amplified will in all probability be circular in cross section. The excited volume, in contrast, will appear more nearly rectangular as viewed down the amplifier axis. Also, to obtain uniform amplification, the dye concentration must be adjusted so that the excitation intensity has only decreased to approximately $1/e$ at the edge of the amplifying region. To the extent that these simple approximations are true, the overlap of the laser beam with the excited volume will be reduced from unity by the product of these two factors. The resultant efficiency E without clever manipulation of the geometry can not exceed $E = \frac{1}{4}\pi(1 - e^{-1}) \approx 50\%$.

For small amplifications the laser amplifier can be excited longitudinally. This type of excitation reduces the geometrical restrictions on the efficiency. The efficiency relation (5.4.16) is valid in this case also. However, since both the excitation and emission beams will have the same cross-sectional area, the emission beam power must be comparable to the excitation beam power to get the device up to 50% efficiency as long as excited state absorption is significant. If the two beam powers are equal, the maximum amplification the device can have is two.

In any amplifier chain it is very important to minimize the amount of amplified spontaneous emission produced in one stage from reaching the succeeding amplification stage. This can be accomplished by filtering the emission beam in both wavelength and spatial dimensions. A simple filter consisting of a direct vision prism and a pinhole spatial filter has been shown to be very effective in reducing the amount of ASE pro-

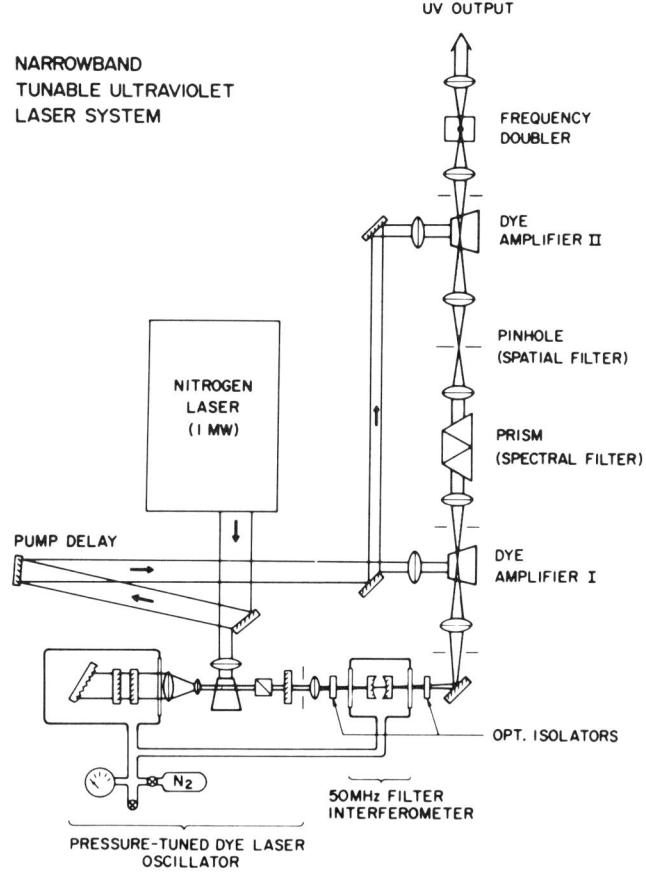

FIG. 18. Components of a dye laser oscillator–amplifier system.[41]

pagated down the amplifier chain.[41] The experimental arrangement is shown in Fig. 18.

5.4.6. Regenerative Oscillators

Regenerative oscillators are simple oscillators with the means provided for injecting an external signal to lock the wavelength of oscillation.[48-53] Such oscillators will accurately amplify an injected signal for a specific length of time before switching in wavelength to that wavelength at which the amplification is a maximum.[52] They can make very efficient short-pulse amplifiers because they can be constructed with very small dimensions that allow the intraresonator intensity to build up quickly to saturation levels. This rapid increase in intensity is possible since the round trip transit time for this intensity can be very small compared to the excitation pulse width. In contrast to the linear amplifier, regenerative oscillators can be efficiently excited either longitudinally or transversely. The restrictions on the maximum excitation intensity that forced linear amplifiers to have large aspect ratios have been relieved by the short buildup time of the emission intensity to saturation levels. The emission intensity can then effectively extract the energy deposited into the dye by the excitation.

Two examples of this device have been illustrated in Fig. 19. The simple two-mirror resonator (Fig. 19A) is the easiest to construct and tune. However, the intraresonator emission intensity in a single-mode amplifier will be in the form of a standing wave because of the interference between the oppositely directed intensities. The standing wave can only stimulate emission from a fraction of the molecules in the excited region, namely, those molecules located at the antinodes of the standing wave pattern. The efficiency of the amplifier will be limited to this fraction, which will be approximately one-half but will be dependent upon the intraresonator intensity. The remaining fraction of excited molecules can amplify spontaneous emission at wavelengths other than that of the injected signal. The appearance of any such subsidiary wavelengths in the output would greatly reduce the utility of the amplifier. The likelihood of such lines appearing can be minimized if the amplifier medium fills a large fraction of the intraresonator space. The dye solution then becomes a distributed feedback tuning element that inhibits the buildup of any parasitic oscillations.

The device shown in Fig. 19A has both the input injection signal and the output going through the same front mirror M_0. These two beams must

---

[53] A. A. Pease, O. G. Peterson, M. L. Spaeth, and W. M. Pearson, *IEEE J. Quantum Electron.* **qe-13,** 28D (1977).

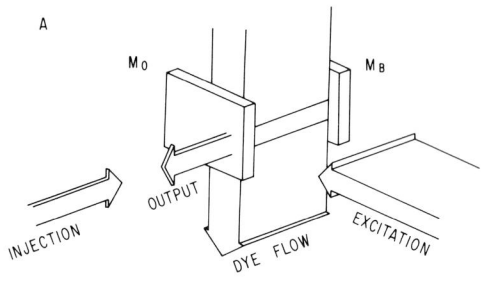

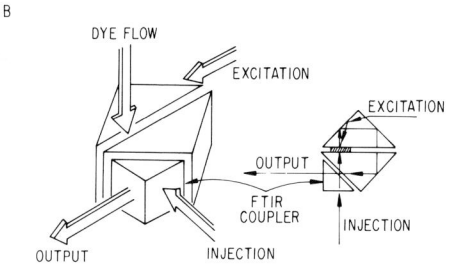

FIG. 19. Regenerative oscillators for amplifying short-pulse signals. (A) Simple two-mirror oscillator with the input and output passing through the same partially transmitting mirror M_0. (B) Ring geometry oscillator constructed of three prisms, where one of these prisms is used to provide the input and output coupling by frustrating the total internal reflection used to reflect the laser intensity around the ring.[53]

be separated so that the output can be utilized and feedback from the amplifier into the oscillator can be prevented. Any such feedback would change the emission wavelength of the master oscillator. The simplest isolator for separating the two beams is a Glan–Thompson prism with a quarter-wave plate between it and the amplifier. The laser intensity within the resonator therefore will be circularly polarized, which will cause no difficulties in this simple geometry.

The second regenerative oscillator illustrated in Fig. 19B is a ring oscillator.[53] In this device the injected signal not only determines the wavelength of the emission but also the direction of travel around the oscillator. Since the radiation is all in the form of a traveling wave, all the excited molecules are accessible to the laser emission for amplification and the probability of parasitic oscillation is negligible.

The device shown consists of two small 90° prisms aligned so that the total internal reflection at the prism surfaces could be used as the resonator mirrors. The excitation is focused into the dye cell, which is sandwiched between the two prisms. The excitation enters the laser through one prism surface and is reflected from an adjacent surface so that it will enter the dye cell at a small angle with respect to the emission. The dye solution is flowed through the active region in a direction perpendicular to the plane of the resonator. The injection of the initial beam into and subsequent extraction out of the resonator were achieved by frustrating the total internal reflection on one prism surface using an evanescent wave coupler.

This oscillator, however, is more difficult to fabricate and requires careful alignment. Any misalignment of the prisms, either translational or rotational, will make the resonator unstable. However, the rotation-induced instability may prove useful for ensuring single-mode output by making the device a variable-magnification unstable resonator. The size of the focal spot of the excitation effectively determines the mode size for the laser. The fact that the only aperture in the laser is the boundary of this focal spot is a great advantage for the device. This condition eliminates entirely the diffraction losses caused by physical apertures that exist in most lasers.

The adjustment of the evanescent coupler is extremely critical but easily effected.[54-56] The spacing between the surfaces of the coupler was set by evaporating on one surface a metal ring a fraction of one wavelength thick $(0.3-1.0\ \lambda)$ depending on the transmission desired. Final adjustments in the transmission are accomplished by compressing the unit, which deforms the output coupler prism and metal ring sufficiently to produce changes of greater than 10% in the transmission, depending on the ring diameter and prism elasticity. Experimentally a chromium ring of thickness equal to the light wavelength gave easily adjustable transmissions starting at $T = 10\%$. The ideal coupler will have a transmission that can be calculated from the following relations:

$$\frac{R_p}{T_p} = \frac{(\eta^2 - 1)^4}{4\eta^2(\eta^2 - 2)} \left\{ \sinh \left[2\pi \frac{w}{\lambda} \left(\frac{\eta^2 - 2}{2} \right)^{1/2} \right] \right\}^2, \quad (5.4.18)$$

$$\frac{R_s}{T_s} = \frac{(\eta^2 - 1)^2}{\eta^2(\eta^2 - 2)} \left\{ \sinh \left[2\pi \frac{w}{\lambda} \left(\frac{\eta^2 - 2}{2} \right)^{1/2} \right] \right\}^2, \quad (5.4.19)$$

$$R_i + T_i = 1, \quad (5.4.20)$$

where R_p and T_p are the reflectivity and transmissivity for light polarized parallel to the plane of incidence, and R_s and T_s are the corresponding val-

---

[54] E. L. Steele, W. C. Davis, and R. L. Treuthart, *Appl. Opt.* **5**, 5 (1966).
[55] G. Marowsky, *Z. Naturforsch.*, Teil A **29**, 536 (1974).
[56] R. Polloni, *IEEE J. Quantum Electron.* **qe-8**, 428 (1972).

ues for light polarized perpendicular to the plane of incidence; w is the spacing between the prisms; λ is the wavelength of the light; and η is the index of refraction of the prisms, which are assumed to be equal. The relations were calculated for a 45° angle of incidence. It is important to notice that the terms $(\eta^2 - 2)$ are very sensitive to the exact value of η. This makes it essential to determine very accurately the value of η in order to estimate the chromium ring thickness needed to produce a specified reflectivity.

This oscillator is a totally transient device. Its characteristics can be determined by solving Eqs. (5.4.5)–(5.4.7) and including resonator losses. The solutions can be most easily obtained by computer techniques. The equations can be arranged so that the calculations require a minimum set of parameters. These parameters are the excitation and emission intensities in saturation units ϕ_X and ϕ_M, the resonator losses including the output mirror transmission, the optical density of the dye solution for the excitation radiation, the ratio of the absorption and emission cross sections σ_{0X}/σ_{SM}, and the relative ratios of the time constants characteristic of the device. The important time constants that affect the laser operation are the dye molecule fluorescent lifetime, the excitation pulse width, and the round trip time through the resonator. The assumptions appropriate to this time domain and geometry have been made. These include neglecting the ground singlet and triplet state absorptions of the emission, neglecting diffraction effects, and equating the absorption cross sections for the ground σ_{0X} and excited σ_{SX} singlet states at the excitation wavelength.

The transient equations (5.4.5)–(5.4.7) have been solved for a selected set of parameters and the results of two calculations are shown in Fig. 20. The fluorescence lifetime and the excitation pulse width were taken to be equal and the round trip time was fixed at 1/30 of this value. This choice of conditions is equivalent to having a ring resonator of 5-cm round trip optical path length excited by a 5-nsec pulse. The excitation was given an arbitrary half-sine wave shape with a maximum value of $\phi_X = 9$, and a near optimum value of the output mirror reflectivity $R = 0.9$ was selected. The time development of the excitation pulse ϕ_X, the excited state fraction β, and the output pulse ϕ_M are presented for two values of the injection signal $\phi_M(0) = 0.0001$ and 0.01. It is seen that the excited-state population rises quickly at first in proportion to the excitation pulse and then falls dramatically as the laser pulse builds up and depletes the inversion. It is also evident that the intraresonator intensity required 40 round trips through the resonator to build up to its peak value even though the amplifier gain was very high as evidenced by the fact that the excited-state population reached 50% at one point. This emphasizes

5.4. SHORT-PULSE DYE LASERS

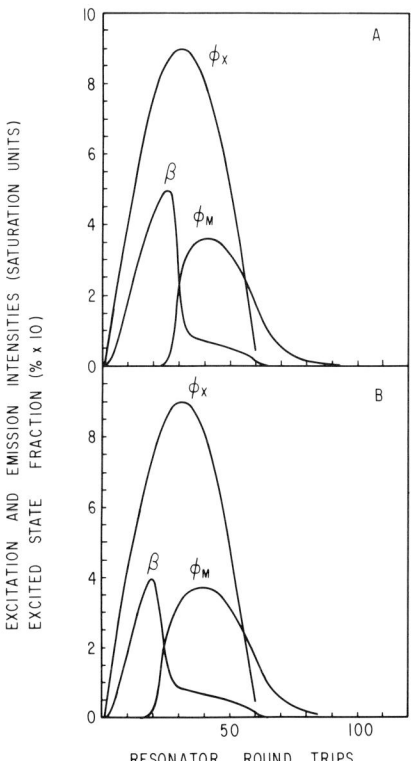

FIG. 20. Time evolution of ring oscillator parameters. These parameters are the excitation ϕ_X, the laser emission ϕ_M, and the inversion $\beta = n_S/n$, which have been plotted as a function of the number of round trips the emission makes through the oscillator. The initial injected signal was $\phi_M(0) = 0.0001$ (A) and 0.01 (B).[53]

in a dramatic way the importance of constructing such devices with small dimensions and short round trip transit times.

Laser threshold is that point where the excited-state population is at a value such that the amplification of the laser is unity. This point is easily identified in the figure to be the point where the laser output has reached its peak. The population inversion is seen to be approximately 7% at this point. It is of interest to note that the excited-state population reaches values 5–10 times the threshold value in the initial part of the pulse.

This device was calculated to be 54% efficient for an input of $\phi_M(0) = 0.0001$ and 64% efficient for $\phi_M(0) = 0.01$. The loss to excited-state absorption was found to be 31 and 22%, respectively, for the two cases. These were calculated in photon units.

The algorithm used to obtain these results divided the intraresonator intensity and the active volume each into a large number of segments, 50 and 20, respectively. The intraresonator intensity ϕ_M is assumed to be constant within each segment and the resonator segment lengths were chosen to be equal to the dye cell dimension. For each of the length segments in the amplifying region, several parameters were stored and updated. These were the excitation ϕ_X and emission ϕ_M intensities, and the population inversion β characteristic of each segment. A selected time segment of ϕ_M in the resonator is amplified step by step by the divisions within the dye cell. The values of ϕ_M in each length segment in the active medium are then updated with these newly calculated values. The values of ϕ_X are then recalculated and the values for the population inversion determined from the new intensities for the next time increment. The final amplified value of ϕ_M is stored as the new resonator segment value. The whole process is repeated for successive time segments of the intraresonator intensity. This separation of the time and distance calculations into serial operation is possible because the active region is very small, approximately 1/50 of the round trip distance, so the transit time for ϕ_M through this region is extremely short compared to all other time constants in the calculation. The resonator losses are included in the model as simple multiplicative factors operating on the intraresonator intensity.

The amount of amplified spontaneous emission produced by this device has not been calculated here but its effects should be quite analogous to those discussed for the linear amplifier. The similarities between the two devices are quite striking but dimensionally transformed. Where the linear amplifier has very high population inversions at the input end of the amplifier, the regenerative oscillator has very high population inversions during the early part of the excitation pulse. The efficiency of both devices depends upon a significant fraction of the amplification occurring in an operational regime where the amplifier has high efficiency. The regenerative oscillator has an advantage in this regard since the dye molecules can store energy on these short time scales, some of which is still available for use when the intraresonator intensity has built up to saturation levels.

The ring geometry regenerative oscillator has been experimentally demonstrated to have a 60% efficiency.[53] It is free from many of the geometric restrictions that plague the linear amplifier. All of the excitation is accessible to the laser emission as contrasted to transversely excited amplifiers. However, since the regenerative amplifier has a resonant structure, it will impose its own mode structure on the output. This means that the primary oscillator and the regenerative amplifier must be mode-matched for single-mode operation. Since the amplifier is a ring, it will not feed back into the oscillator, which simplifies the mode matching of the two devices and also simplifies the separation of the input and output beams. The line width and transverse mode structure will also be characteristic of the amplifier and only marginally affected by the input beam conditions.

As mentioned earlier, this oscillator will remain locked on the wavelength of the injected signal for a specific length of time. The wavelength locking time is determined by the buildup of amplified spontaneous emission. When this emission becomes sufficiently large, it will dominate the laser dynamics. Before this time has elapsed the amplified spontaneous emission will have introduced sufficient noise onto the laser output to degrade the amplifier performance. It is important, therefore, to estimate the magnitude of this intensity by using Eq. (5.4.11). The calculations must include the increase in amplification length and corresponding reduction in the solid angle due to the multiple reflections within the resonator.

5.4.7. Thermal Limitations

The average power emitted by most of the excitation lasers that have been used to drive short-pulse dye lasers has been relatively small. Consequently, there have not been severe problems affecting the laser performance due to thermally induced optical inhomogenities in the dye solutions. As the technology advances, this will not remain a negligible effect. The techniques required to obtain high average power from all types of dye lasers are essentially the same. The discussion of these techniques will be taken up in the sections on steady-state lasers where thermal effects are presently a severe problem.

5.5. Steady-State Laser

5.5.1. General Description

The second class of dye lasers consists of those with excitation times sufficiently long so that the triplet state population can be considered to be in equilibrium with the excited singlet state population. This region of operation includes two distinct and very interesting cases: the flashlamp-excited dye laser and the cw dye laser. These devices are characterized by steady-state or near steady-state operation. It therefore can be assumed that the triplet state population n_T is in equilibrium with the excited singlet state, so that setting Eq. (5.2.4) equal to zero yields

$$n_T = k_{ST}\tau_T n_S \tag{5.5.1}$$

and that the excited singlet population n_S is fixed at the threshold value so that $dn_S/dt = 0$. With these constraints, Eqs. (5.2.1)–(5.2.3) become

$$dI_X/dy = -I_X[\sigma_{0X}n_0 + (\sigma_{SX} + k_{ST}\tau_T\sigma_{TX})n_S], \tag{5.5.2}$$

$$dI_M/dx = I_M[(\sigma_{SM} - k_{ST}\tau_T\sigma_{TM})n_S - \sigma_{0M}n_0], \tag{5.5.3}$$

$$n_S/n_0 = (I_X\sigma_{0X} + I_M\sigma_{0M})/(I_M\sigma_{SM} + k_{ST} + 1/\tau_S). \tag{5.5.4}$$

The special case where n_S and, therefore, I_X are taken to be uniform throughout the active medium has been considered in the introductory section and the laser threshold conditions presented in Figs. 5, 6, and 7.

5.5.2. Steady-State Power Balance

In the steady-state or near steady-state regime, the dye molecule excited states remain fixed at values very near the laser threshold levels. The output of the laser can then be determined by calculating the power balance between the input power, the output power, and the various power losses characteristic of the laser.[10,57] It is instructive to perform this exercise for the familiar case of complete uniformity of all variables within the active media.

Some of the excitation power will be absorbed by the excited singlet-state population. This power becomes significant if a sizable fraction of the molecules has been excited to this state. Since such excitation levels are characteristic only of coherently pumped systems, the excited singlet-state absorption will be neglected for the present but will be considered in more detail in the section on cw devices.

The absorbed power in photon units P_I will be distributed into four parts: output P_O, spontaneous fluorescence P_F, triplet absorption P_T, and internal resonator losses P_L. These powers can be calculated from the following relations:

$$\begin{aligned} P_I &= P_O + P_F + P_T + P_L, \\ P_O &= I_M AT, \qquad P_F = n_S Al/\tau_S, \\ P_T &= I_M \sigma_{TM} n_T Al, \qquad P_L = I_M AL'. \end{aligned} \tag{5.5.5}$$

The powers are measured in photons/sec, I_M is the intraresonator laser intensity also in photon units, A and l are the active volume cross-sectional area and length, n_S and n_T are the population densities of the excited singlet and triplet states, τ_S is the fluorescence lifetime, σ_{TM} is the triplet state absorption cross section for the emission, and L' is the round trip resonator loss exclusive of T, which is the output mirror transmission, $L' = L - T$ where L is defined by Eq. (5.2.9).

[57] R. Polloni, *Appl. Phys.* **7**, 131 (1975).

The triplet population is evaluated from the equilibrium condition, Eq. (5.5.1). The approximation will be made that the excited singlet population is a linear function of the total resonator losses as was done to obtain Equation (5.2.14):

$$n_S = (L' + T)\tau_S \alpha/l. \qquad (5.5.6)$$

This approximation is valid only for small changes in the loss function $L/(2nl)$. Therefore, the proportionality constant α must be evaluated from Fig. 6 in the selected region of operation. A representative value of 5×10^{24} (cm² sec)$^{-1}$ has been selected for this discussion.

To evaluate the fraction of the input power that goes into each branch it is necessary first to find the optimum output coupling for the laser. This coupling is found by maximizing the output power with respect to the output mirror transmission. The solution to this is the familiar equation[58]

$$T + L' = [P_I L'/A\alpha]^{1/2}. \qquad (5.5.7)$$

It is useful to note that the optimum output coupling is dependent on dye-triplet properties only to the extent that it affects the values of the proportionality constant α. A more explicit dependence is not to be expected since an increase in the output coupling raises the excited-state population and therefore the triplet-state population, but at the same time it reduces the laser intensity within the cavity. Therefore, within the linear approximation, the product $I_M n_T$ should remain constant and the loss to triplet absorption will not change.

To generalize the results we shall rewrite the equations in terms of a unitless parameter ϕ_I representative of the input power. The power will be normalized with respect to the input power required to achieve lasing threshold with no output, i.e., 100% reflectivity mirrors, $P_{TH} = A\alpha L'$. Therefore,

$$\phi_I = P_I/P_{TH} = P_I/A\alpha L'. \qquad (5.5.8)$$

The optimum mirror transmission in terms of this parameter is

$$T = L'(\phi_I^{1/2} - 1), \qquad (5.5.9)$$

and the intraresonator laser intensity becomes

$$I_M = \frac{\alpha(\phi_I^{1/2} - 1)}{\xi}, \qquad (5.5.10)$$

where $\xi = 1 + \alpha k_{ST}\tau_T\sigma_{TM}\tau_S$. By evaluating this equation for rhodamine

---

[58] A. E. Siegman, "An Introduction to Lasers and Masers." McGraw-Hill, New York, 1971.

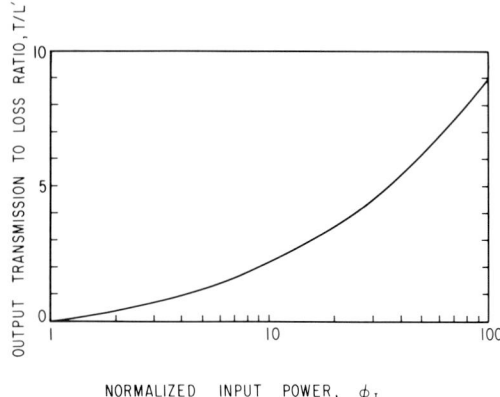

FIG. 21. Ratio of optimum output coupling to resonator losses as a function of normalized excitation power.

6G it is found that $\xi = 2.0$. By substituting these relations into the original equation the following relations are obtained:

$$P_O/P_I = (\phi_I^{1/2} - 1)^2/\xi\phi_I,$$
$$P_F/P_I = \phi_I^{-1/2},$$
$$P_T/P_I = (1 - \phi_I^{-1/2})(1 - \xi^{-1}), \quad (5.5.11)$$
$$P_L/P_I = (\phi_I^{1/2} - 1)/\xi\phi_I,$$

In Fig. 21 the ratio of the optimum output mirror transmission to the resonator losses T/L' has been plotted as a function of the normalized input power. The intensity internal to the resonator has the same functional dependence and therefore can be obtained from the same graph by using the appropriate proportionality constant $I_M = (T/L')(\alpha/\xi)$.

The fraction of the input photons that goes into each of the various channels is displayed in Fig. 22 as a function of the dimensionless parameter ϕ_I representing input power. The figure is drawn to show how the input power is divided into the four parts and how these fractions change relative to each other as the input power is varied.

There are several useful conclusions that can be drawn from these rather simplistic considerations. First, if one is not limited in available excitation intensity, then internal resonator losses can be reduced to insignificance. In this case, the losses due to internal tuning elements will not be significant. The efficiency will continue to improve as the excitation is increased up to a maximum of 30–35%. For limited intensity excitation, i.e., incoherent, a reasonable design goal is seen to be the point where the input power is approximately ten times the zero output threshold, i.e., $\phi_I = 10$. At this point, the internal resonator flux is approxi-

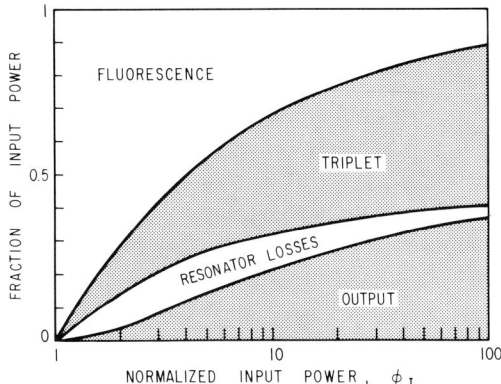

FIG. 22. Distribution of input photons into output and various loss mechanisms as a function of normalized excitation power for steady-state laser.

mately twice the saturation value and the output is approximately 20% of the input photons. The output transmission should be twice the round trip cavity loss. Above this point the laser efficiency improvement diminishes.

It is obvious from the figure that minimizing the triplet population of the lasing dye is of the utmost importance. The triplet loss is greater than the laser output and at higher powers is the dominant loss. The extension of these concepts to high-intensity excitation is not strictly valid since other loss mechanisms that have been ignored here will effect the device efficiency, specifically, excited-state absorption. These efficiency considerations have been treated in more detail[57] and have been applied to an experimental cw laser to test the validity of the concepts.[59]

5.5.3. Flashlamp-Excited Laser

Flashlamp-excited dye lasers range in complexity from the simplest and easiest laser for a novice to assemble to high-repetition-rate, high-average-power lasers designed for isotope separation. Flash-pumped dye lasers have defied many attempts at modeling their characteristics because of the many uncertainties in their operation. The major obstacles to definitive characterization of their output are the difficulties in predicting the effects of thermal heating and of predicting the number of excited dye molecules that will be produced from the energy stored in the flashlamp discharge current.

Dye lasers require very intense excitation. The dye molecules have

[59] G. Marowsky and R. Polloni, *Appl. Phys.* **8**, 29 (1975).

very short, less than 5 nsec, fluorescent lifetimes. The molecules will not store energy for longer than this lifetime so that the flashlamps must excite the molecules on a steady-state basis. The power required to maintain a specified excited-state population density n_S is immense. It is given by $P_I = (h\nu n_S/\tau_S)V$, where $h\nu$ is the excitation photon energy and V the volume of the excited region. In Section 5.2.3 this power was shown to be 85 kW for a 1-cm² cross section active volume at laser threshold. The fluorescence transitions in dye molecules are strong dipole transitions with high oscillator strengths. In spite of this, the density of excited molecules n_S required to achieve laser threshold is very high because this transition probability has been spread over a large wavelength region. This condition is responsible for the dye laser's unique attribute of wavelength tunability but it makes the cross section for stimulated emission at any given wavelength surprisingly low. The result is that the flashlamps must be driven very hard to obtain sufficient illumination from them to excite organic dyes to laser action. It is necessary to drive them with as short an electrical pulse as can be obtained, which requires specially designed low-inductance discharge circuits. Typical circuits discharge between 10 and 100 J through the lamps in 1–2 μsec. These severe requirements imposed by dye lasers were beyond the capabilities of many of the flashlamps available at the inception of dye lasers. This laser has motivated many novel improvements in flashlamps, some of which will be discussed below.

5.5.4. Flashlamp Plasma Temperature

All the lamps must be operated at very high discharge plasma temperatures to obtain efficient dye laser operation. The temperature at which the lamp must be operated to excite a dye to laser action is dependent on a large number of variables. These include the lamp geometry, the fill gas and pressure, the electrical discharge characteristics, the reflector cavity efficiency, and the discharge plasma opacity. The plasma temperature is sufficiently high that a significant fraction of the emission will be at very short wavelengths, which will be absorbed by the lamp envelope and dye solvent. These photons will not reach the dye molecules so that the system efficiency will be proportionately reduced. It is therefore useful to keep this temperature at a minimum that will still permit relatively efficient laser operation.

An accurate estimate of the required plasma temperature is essentially impossible to calculate. However, it is possible to determine the manner in which this temperature scales with one important laser parameter, the aspect ratio of the flashlamp and dye cell volumes. This determination leads to some very useful and interesting conclusions.

An approximate value for the plasma temperature can be obtained by first calculating the number of photons that must be incident on the surface of a cylindrical dye cell to obtain a specified excitation level within the dye. This number will be proportional to the threshold power parameter, which was defined earlier and is equal to 85 kW/cm² for rhodamine 6G. It is inversely proportional to the aspect ratio of the cell R_A, which is defined as the ratio of length to diameter. The number of photons absorbed by the dye solution per unit area of surface must be equal to that number emitted by the plasma times some efficiency factor. It has been assumed that the lamp and cell have identical dimensions and are imaged onto each other by the reflector cavity. The photon emission of the black body between specific frequency limits can be calculated from

$$W = \frac{2\pi}{c^2} \int_{\nu_1}^{\nu_2} \frac{\nu^2 d\nu}{e^{h\nu/kT} - 1}, \quad (5.5.12)$$

where ν_1 and ν_2 are chosen to be the limits of the dye absorption spectrum, c is the velocity of light, and the other parameters have their usual definitions.

The relationship between the aspect ratio R_A and the plasma temperature then becomes

$$R_A = \frac{P_{TH}N}{Q_c 4} \left[\frac{2\pi}{c^2} \int_{\nu_1}^{\nu_2} \frac{\nu^2 d\nu}{e^{h\nu/kT} - 1} \right]^{-1}, \quad (5.5.13)$$

where P_{TH} is the threshold power density parameter for the selected dye and resonator loss as discussed in Section 5.2.3, N an arbitrary multiplier of the threshold parameter to select the operating point of the laser, and Q_c is the total efficiency for producing excited molecules within the lasing mode volume from the photons emitted by the black body. This equation has been evaluated for an arbitrary excitation level of 10^{25} photons/cm² sec = $P_{TH}N/Q_c$. The parameter $P_{TH}N/Q_c$ is the number of photons absorbed within the dye cell divided by the cross-sectional area of the dye cell bore. This arbitrary value was selected by converting the threshold parameter for rhodamine 6G, 85 kW/cm², to photon units, by choosing $N = 5$ and using an estimate for Q_c of 10%. This efficiency is obtained from estimates of the plasma emissivity, the reflector cavity efficiency, the dye absorption, and the overlap between the laser mode volume and the uniformly excited volume. The black-body radiation function was integrated between 20,000 and 50,000 cm⁻¹. These limits were chosen to correspond to the center of the visible region 500 nm, and the quartz cutoff of 200 nm.

The results of this calculation are presented in Fig. 23. The values for the temperature are in themselves not very significant because of the sim-

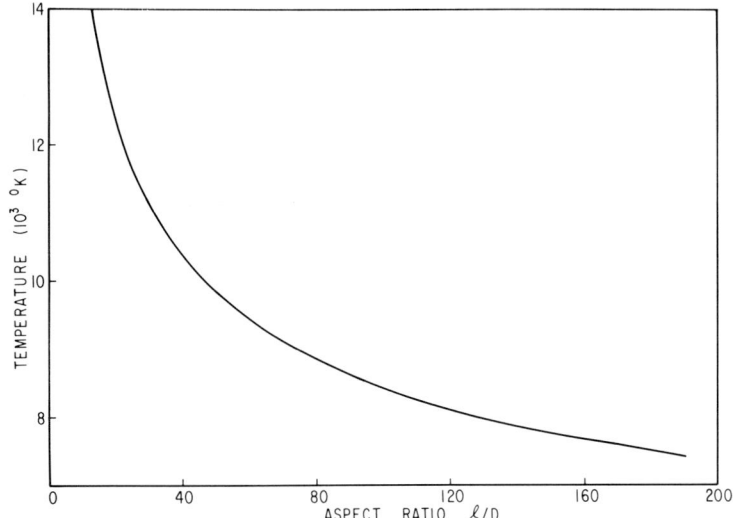

FIG. 23. Flashlamp plasma temperature required to excite idealized dye laser as a function of dye cell and flashlamp aspect ratio.

plicity of the assumptions that went into the calculations. What is important is the dramatic decrease in the required temperature as the dye cell and flashlamp aspect ratio are increased. The temperature decrease should produce an equally dramatic increase in laser efficiency as a greater fraction of the photons emitted by the black body fall in the wavelength region where the dye can absorb and utilize them. It was shown earlier that the laser threshold was only slightly increased by lengthening the active region. Therefore, it is clear that large aspect ratios should yield more efficient flashlamp-excited laser devices.

The importance of these temperature considerations is further illustrated in Fig. 24. Here is plotted the fraction of the photons emitted by a black body of a specific temperature within the selected wavelength range between 200 and 500 nm. It is clear that the laser efficiency will depend critically on the flashlamp plasma temperature, which in turn depends on the judicious choice of lamp and dye cell aspect ratios. Also plotted in Fig. 24 is the efficiency curve for the black-body emission, where the optical system transmission is taken to be between 300 and 500 nm. The 300-nm wavelength is the cutoff for transmission through Pyrex®.[60] By removing the lamp emission between the wavelengths of 200 and 300 nm, the efficiency of the device is significantly reduced but the useful life of the organic dye solution used as the active medium is greatly increased.

[60] "American Institute of Physics Handbook." McGraw-Hill, New York, 1972.

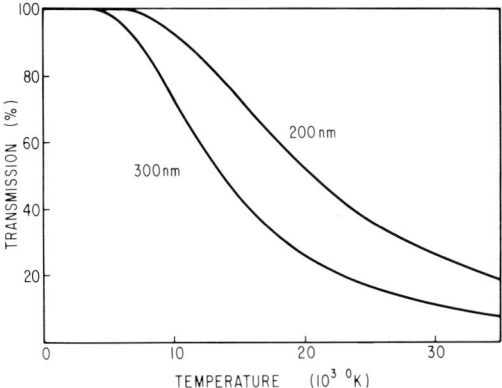

FIG. 24. Fraction of photons radiated by a black body that is emitted between 500 nm and the short-wavelength cutoff for quartz, 200 nm, and for Pyrex, 300 nm, as a function of the black body temperature.

Organic dyes absorb continuously from their fundamental absorption band up to the point where they decompose. The absorption cross section, however, varies greatly over the entire range as is easily seen in Fig. 2. The emission spectra of flashlamps also are very wavelength dependent and are effected strongly by the line spectra of the fill gas, the power loading, and the opacity of the plasma. Any attempt to calculate from these spectra an optimum dye concentration to absorb effectively the flashlamp emission and at the same time to produce a uniform concentration of excited states is a formidable, if not impossible, task.[61] Fortunately, since the active medium is a liquid, the experimental determination of the optimum concentration is extremely simple. Changes of a factor of two in a near optimum dye concentration have rarely been found to alter the laser output by more than 10%. Attempts to estimate the overlap integrals of these spectra at the optimum concentration have indicated that approximately half of the excitation is absorbed.[61]

The experimentally determined optimum dye concentrations have usually been in the range from 10^{-4} to 10^{-3} molar for dye cells with diameters of several millimeters. Higher concentrations are required for some classes of dyes, e.g., coumarins. At the optimum concentration the excitation in the region of the fundamental absorption band is absorbed in the outermost few tenths of a millimeter at the edge of the dye chamber. The fundamental band therefore cannot contribute significantly to the volume excitation of the laser-active medium. This circumstance emphasizes the importance of the breadth of the absorption even if the absorption cross

[61] S. Blit and U. Ganiel, *Opt. & Quantum Electron.* **7**, 87 (1975).

section over much of this range is quite low. All dyes have these broad absorptions so that the optimum dye concentrations are remarkably similar even though the detailed spectra may vary greatly between different molecules. The universality of these absorptions explains the disappointing results of adding extra chromophores to dyes or mixing dyes together for flashlamp excitation applications. It is also clear that care must be exercised in selecting solvents and additives, such as surfactants and triplet quenchers, for the dyes so that these do not absorb significant portions of the excitation.

5.5.5. Linear Flashlamps

Commercially available, xenon-filled flashlamps are the excitation source of choice for most flashlamp-excited dye laser applications. They can produce the plasma temperatures required to obtain laser action from efficient dyes and they can be used in high average power applications. Their convenience, reproducibility, and reliability far exceed that of the other special-purpose lamps often used for dye lasers.

Linear flashlamps have been quite well characterized for flash durations between 30 μsec and 10 msec. In this time regime the current–voltage relationship is[62–65]

$$V = K_0|i|^{1/2}, \qquad (5.5.14)$$

where K_0 is a constant characteristic of the flashlamp with the units of $\Omega A^{1/2}$. The constant K_0 scales with the lamp geometry in the following manner:

$$K_0 = kl/d, \qquad (5.5.15)$$

where l is the electrode spacing, d the bore diameter of the lamp, and k a constant dependent on the pressure and type of fill gas used in the tube. Typical values for k are in the range from 1.1 to 1.3. These relationships have been found to be valid for current densities up to 10^5 A/cm² in the time domain discussed. For short discharge pulses in the range between 1 and 5 μsec, the use of the I–V relation in the differential equation describing the discharge circuit will predict discharge times that are short by approximately a factor of two.

A schematic of a simplified discharge circuit is illustrated in Fig. 25.

---

[62] J. P. Markiewicz and J. L. Emmett, *IEEE J. Quantum Electron.* **qe-2**, 707 (1966).
[63] I. S. Marshak, *Sov. Phys.—Usp. (Engl. Transl.)* **5**, 478 (1962).
[64] J. H. Goncz, *J. Appl. Phys.* **36**, 742 (1965).
[65] R. H. Dishington, W. R. Hook, and R. P. Hilberg, *Appl. Opt.* **13**, 2300 (1974).

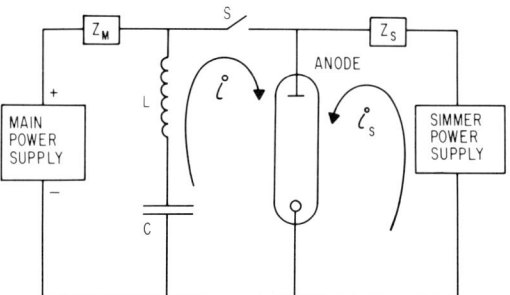

FIG. 25. Discharge circuit schematic for flashlamp excitation. The current loop labeled i is the discharge circuit that determined the width of the excitation current pulse. The schematic shows both the discharge circuit and the circuit that supplies the continuous simmer current to maintain the ionization within the lamp.

The differential equation describing the discharge characteristics is

$$V_0 = L\frac{di}{dt} \pm K_0|i|^{1/2} + \frac{1}{C}\int_0^t i\, dt', \qquad (5.5.16)$$

where V_0 is the initial voltage on the storage capacitor, L the total discharge loop inductance, and C the value of the energy storage capacitor.

If this equation is normalized with the relations

$$\begin{aligned}\tau &= t/(LC)^{1/2}, & I &= iZ_0/V_0, \\ Z_0 &= (L/C)^{1/2}, & \alpha_d &= K_0/(V_0 Z_0)^{1/2},\end{aligned} \qquad (5.5.17)$$

where Z_0 is the discharge circuit impedance and α_d is the damping factor,

$$\frac{dI}{d\tau} \pm \alpha_d|I|^{1/2} + \int_0^\tau I\, d\tau' = 1. \qquad (5.5.18)$$

It is seen that the damping factor α_d is voltage dependent. Critical damping of the circuit can be obtained only at one operating voltage. Computer solutions of this equation have shown that the critically damped discharge condition occurs for $\alpha_d = 0.75$. At this point the maximum peak power is dissipated, which is the most desirable operating point.

The average number of discharge pulses N_p that a flashlamp can withstand before it breaks is a very strong function of the energy in these pulses. The relation is[66]

$$N_p = (E_x/E_p)^{8.6} \qquad (5.5.19)$$

[66] J. H. Goncz, *ISA Trans.* **5**, 28 (1966).

where N is the number of pulses the lamp can endure, E_p the energy dissipated in the lamp during each pulse, and E_x the explosion energy characteristic of the lamp. The explosion energy is defined as the energy that destroys one-half of the lamps tested. This parameter has been found to scale as

$$E_X = E_0 l d T^{1/2}, \qquad (5.5.20)$$

where l and d were previously defined, E_0 is an empirical constant having a value approximately 2.2×10^4 J/cm²(sec)$^{1/2}$, and T is the pulse width measured between the 10% current points.[66] For a critically damped circuit $T = 3(LC)^{1/2}$.

Equation (5.5.19) is an empirically determined relation that was based on measurements with 100-μsec discharge pulses. This relation has been found to be conservative when it has been applied to systems with approximately 1.0-μsec discharge pulses. The lamp lifetime can be even further extended by maintaining a continuous ionization within the lamp by means of a simmer current.[67]

The maximum average power that can be dissipated by these lamps is limited by the thermal conductivity of quartz. The accepted practice within the industry is to limit the usual 2-mm wall lamps to a dissipation of 200 W/cm² of wall area.[67] It is, of course, necessary to provide effective water cooling to the lamps to maintain high average power loading. For high average power applications, electrode evaporation is another limitation on the flashlamp lifetime. The electrodes must be effectively cooled to minimize the surface evaporation, which will eventually plate the inside bore of the lamp. The tungsten deposition is a relatively localized problem, obviously being most severe near the electrodes. The reduced transmission of the lamp envelope produces one more motivation for making the laser reasonably long.

The gas in these lamps must be ionized so that the lamp will be conducting for the discharge pulse. The triggering of the lamp can be accomplished in several ways. With a fortuitious combination of discharge circuit parameters, the main discharge may have a high enough voltage to break down the gas in the lamp by itself. Such a circuit must have a switch, such as a spark gap or thyratron, and must be designed for operation at many tens of kilovolts. The breakdown voltage for the flashlamp depends on the proximity of a ground plane to enhance the electric field within the discharge volume. Voltages in excess of 10 kV are often sufficient to ionize the gas.[67]

The lamp may also be ionized by an auxiliary high-voltage trigger pulse,

---

[67] J. Moffett, ILC Corporation, Sunnyvale, California (private communication).

which is usually applied to a fine wire wrapped around the outside of the lamp. This scheme is particularly advantageous for low-duty cycle and for relatively simple laser designs because the lamp can now also serve as the high-voltage switch. In this case, the discharge circuit may consist only of a high-voltage, low-inductance capacitor wired directly to the flashlamp. This trigger method is difficult to apply to laser designs where the lamps and current leads are built into low inductance or coaxial geometries. The high-voltage, high-frequency trigger pulse is easily shorted to ground capacitively and is therefore difficult to transmit into any such closed chamber. This circuit must be designed for voltages significantly less than that required to ionize the fill gas. As previously mentioned, xenon-filled lamps can not be expected to hold off more than 10 kV.

For repetitively pulsed devices, the preferred method of ionizing the lamp is to maintain the ionization continuously by the use of a simmer current supplied by a power supply in parallel with the discharge circuit.[68] There are several advantages to this method, including the fact that more power can be applied to the gas so that a larger fraction of the atoms will be ionized before the main pulse. It has been found experimentally that this method of ionizing the gas increases the flash risetime, improves the reproducibility of the flashlamp output, and increases the lamp lifetime. Currents in the range from 50 mA to 1.0 A have been found useful, with the higher currents yielding slightly improved lamp discharge characteristics. The circuit to supply this current contains several special features. An example schematic was shown in Fig. 25. The high-voltage supply must deliver voltages many times higher, e.g., factors of 4 to 20, than the minimum required to maintain the desired current under steady-state conditions. Without this extra voltage the lamp will go out following the main discharge pulse. It is also necessary to protect the simmer circuit from the main discharge. This protection can be accomplished by the inclusion of a high-voltage diode in the circuit.

Finally, one of the distinct advantages of these commercially available flashlamps is the quantities of data available from the manufacturers describing the characteristics of the lamps.

5.5.6. Ablating Flashlamps

This lamp is a special case of linear flashlamp.[69,70] It is designed for extreme brightness at the expense of lamp lifetime. The lamps are extremely simple as is illustrated in Fig. 26. The lamp consists of a

---

[68] J. Jethwa and F. P. Schäfer, *Appl. Phys.* **4**, 299 (1974).
[69] C. M. Ferrar, *Rev. Sci. Instrum.* **40**, 1436 (1969).
[70] R. Goldstein and F. N. Mastrup, *IEEE J. Quantum Electron.* **qe-3**, 521 (1967).

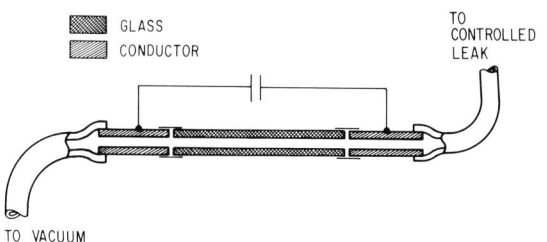

FIG. 26. Ablating lamp schematic.

thick-walled quartz tube, e.g., 3-mm i.d. and 3-mm wall, connected to hollow metal electrodes and subsequently to vacuum connections. The quartz tube and the electrodes are connected together with some simple vacuum seals, which need be no more sophisticated than the shrinkable tubing used in electronics applications. The type of fill gas used in the lamp is not important since, as the name suggests, the major constituent in the discharge plasma is vaporized quartz. Air therefore has been the obvious choice for fill gas. The gas fill pressure is then controlled by balancing a vacuum pump or aspirator on one line against a controlled leak on the other line.

These lamps are used in an attempt to obtain the highest temperatures and therefore the highest brightness available from a wall-stabilized arc. Hence, the very thick walls on the quartz envelope and hollow electrodes allow the discharge-generated shock wave to be vented into the vacuum system, where it is safely dissipated. The simplicity of the design permits the quartz envelope to be easily and quickly replaced, as is often necessary with this device. The continuous ablation of the inner bore of the quartz keeps it free from electrode deposits, which reduces the importance of the choice of materials for the electrodes.

Since the lamp fill gas pressure is a variable controlled by the operator, this lamp can also serve as a self-triggered, high-voltage switch. In this mode of operation, the electrodes are connected directly to the high-voltage capacitor. The capacitor is charged up to the desired voltage while the air pressure in the lamp is maintained at a pressure well above the breakdown point for that voltage. When it is desired to fire the lamp the air is pumped out until the pressure drops below the breakdown point and the discharge occurs.

The most outstanding advantage of this lamp is its extreme brightness. This makes the ablating lamp an ideal choice for testing novel dyes and dye–solvent systems or obtaining laser emission on the extremes of the wavelength range covered by dye lasers. Additional advantages of this design are clearly the extreme simplicity and economy of the device.

It offers distinct advantages to the economy minded not only because of this simplicity but because it can eliminate the need for high-voltage triggers and switches.

5.5.7. Vortex Stabilized Lamps

In an attempt to circumvent the limitations imposed by wall stabilization of the flashlamp arc, several successful devices have been constructed that employ a vortex generated in the lamp fill gas to achieve this stabilization.[71-75] These devices are able to utilize extremely high pulse energies and are capable of very high repetition rates. The pulse energies are typically in the range from 50 to 300 J and repetition rates up to 350 Hz have been achieved. These are limited by the driving circuitry at the present state of development. With this device, it is possible to deliver more excitation to the dye laser solution than the solution can absorb without thermally induced optical inhomogenities reducing the efficiency. In the devices that have been studied, the laser pulse has often been terminated prematurely compared to the excitation pulse and the average output power has saturated at pulse repetition rates of 250 Hz, well below the maximum for the lamp.

The operation of the device is based on increasing the E/P (electrical field/gas pressure) at the center of a large cylindrical lamp assembly by reducing the pressure at this point by means of a high-velocity gas vortex. A pressure drop of approximately 35% has been achieved using a gas injection velocity of approximately 30 m/sec. This requires a gas flow rate of 5 liters/sec STP for a device 4 cm in diameter built to support a 6-cm arc. A representative design is illustrated in Fig. 27. The gas vortex is established by injecting argon gas tangentially along the chamber perimeter at one end of the chamber. The injection ports are well separated from the discharge region so that the gas flow will be uniform when it reaches the discharge region. The gas is extracted on the chamber axis through the hollow electrodes. Argon has been used almost exclusively in these lamps for reasons of economy. Heavier gases that would improve the performance significantly would require a closed-loop recycling and cooling system. All components exposed to the irradiation must be water cooled with the electrodes and quartz envelope requiring the most cooling. The device has been operated with the axis of the arc mounted

[71] M. E. Mack, *Appl. Opt.* **13**, 46 (1974).
[72] M. E. Mack, *Appl. Phys. Lett.* **19**, 108 (1971).
[73] W. W. Morey and W. H. Glenn, *IEEE J. Quantum Electron.* **qe-12**, 311 (1976).
[74] W. W. Morey and W. H. Glenn, *Opt. Acta* **23**, 873 (1976).
[75] W. W. Morey, W. H. Glenn, and C. M. Ferrar, ARPA Report No. R75-921617-13 (1975).

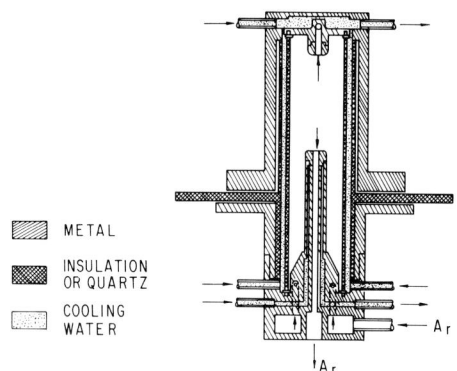

FIG. 27. Vortex stabilized lamp schematic. The vortex is generated by the injection of the Ar gas through tangential ports at the base of the lamp and the extraction of this gas through the hollow-lamp electrodes.[74]

vertically to prevent the buoyant hot gases from distorting the pressure gradient symmetry.

In contrast to other lamp designs, the discharge circuit is purposely fabricated with a selected amount of inductance to limit the current rise time. This added inductance was found to be necessary to limit the shock wave produced by the rapidly expanding gas and thereby prevent the quartz envelope from being destroyed. A total circuit inductance of 150 nh was found to be adequate for the purpose. Even with this inductance the rise time of the light output was in the range between 1 and 2 μsec. The shock wave is still large enough to cause some operator discomfort at high repetition rates. The lamp has been operated in both exfocal elliptical reflectors and the more familiar cylindrical reflectors with elliptical cross sections (see Fig. 28). The original systems used the exfocal geometry to ensure complete symmetry of all conducting materials around the arc. It was anticipated that any asymmetry would pull the arc away from the lamp axis. It was later observed that the arc path was dependent more on the position of the ionization path created by the lamp trigger than the electric field symmetry. For high-repetition-rate lamps that would use a simmer current to establish the ionization channel, it is necessary to inspect the stability of this position and make appropriate corrections.

The arc will grow from the size of the original ionization region as a function of time and input power.[71,75] The ultimate diameter is in the range from 0.3 to 1.0 cm for input energies between 50 and 300 J. The surface temperature of a 6-cm-long arc was measured to be 24,500 K. When the arc was extended to 10 cm, a power balance was determined for

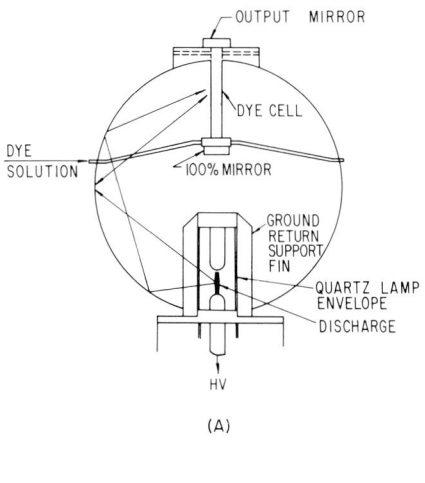

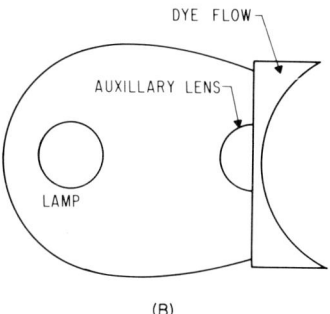

FIG. 28. Elliptical reflectors for focusing vortex-stabilized lamp emission into a dye cell. (A) Exfocal geometry; (B) cylindrical geometry.[75]

the lamp assembly. It was found that approximately 17% of the input power was converted into radiation transmitted by the quartz and 60% was lost into the hot exhaust gas. The remaining power went into miscellaneous heating of the quartz envelope, electrodes, spark gap, etc.[74,75]

This lamp is a very attractive device for producing very high average power although the dye lasers that have been excited by the device have only exhibited efficiencies in the order of 0.5%. The most apparent limitations on the lamp include the excessively high plasma temperatures and the rapid erosion of the lamp electrodes, both effects that would benefit if the technology could be extended to longer lamps. Improvements can also be expected in the dye laser systems. To take advantage of the high input powers, dye laser oscillators must be large-aperture, high-transverse-mode-number devices. To minimize the premature termination of the output, they must have large volumes of dye solution to absorb the pulse energy and to take advantage of the potential high pulse repetition rate, they must be designed to have dye volumetric flow rates bordering on the astronomical.

5.5.8. Linear Flashlamp Reflector Cavities

Linear flashlamps have been used extensively to excite solid-state lasers. Very thorough analyses have been made on various geometries for reflector cavities that couple the flashlamp emission into cylindrical laser media. The reflectors used for dye lasers are for the most part logical extensions of geometries used for pumping other lasers. Therefore, mention will only be made of generalities plus features that are unique to dye lasers.

Reflector cavities can be divided into two general classes. One class consists of imaging geometries of which the most common are composed of elliptical-cross-section cylinders. The other class is characterized by nonimaged, often diffuse reflection of the excitation. This latter class can be considered to be a "white" cell where the excitation is randomly reflected until it is absorbed by the laser medium, the flashlamp, or some other absorbing element. It is important to recall that thermodynamics limits a passive element, such as a reflector cavity, from producing an image of greater brightness than the source. This means that multilamp imaging cavities cannot deliver more power than a single-lamp cavity to a laser-active volume with dimensions identical with the lamp plasma.[76] Such multielement cavities produce much more uniform excitation of the active volume and can be used to pump effectively laser media of dimensions greater than the lamp plasma. Also, elliptical reflectors can produce only theoretically perfect images for line sources.[76,77] The image of finite objects is generally imperfect so that it is useful to design the laser such that the active medium is somewhat larger than the flashlamp bore.

In nonfocused geometries, it is very important to minimize the number of absorbing elements other than the lamp plasma and the dye volume and to select the cavity components for minimum absorption. Aluminum is often used for the reflector surfaces because its reflectivity is 90% or greater from 200 nm into the infrared. Diffuse reflectors are often used in these devices because their reflectivity is very close to 100%.[78] These may be sintered materials, such as high-purity alumina, or powdered

---

[76] C. Bowness, *Appl. Opt.* **4**, 103 (1965).
[77] D. Röss, "Lasers, Light Amplifiers, and Oscillators." Academic Press, New York, 1969.
[78] F. Grum and G. W. Luckey, *Appl. Opt.* **7**, 2289 (1968).

materials, such as barium sulfate, contained behind quartz liners. To minimize the reflector absorption, nonfocused cavities are usually close-coupled, which makes compactness another of their advantages. The efficiency of the nonfocused cavities can be estimated by comparing the area-absorption product for the absorbing dye to the total sum of area-absorption products for all elements in the cavity. These elements include the flashlamp plasma, which will re-emit a fraction of the absorbed radiation but at low efficiency.

It is clear that all reflective cavities must include the means for flowing the dye solution through the active region. For low-duty-cycle operation it has been customary to use a cylindrical dye cell inside the reflective cavity and to flow the dye solution longitudinally through this cell. This construction represents the near-optimum geometry since focusing cavities produce nearly cylindrical images.

For high average power application, it becomes necessary to flow the dye in a transverse direction to minimize the transit time through the active region. A representative geometry is illustrated in Fig. 29. Transverse-flow systems have the disadvantage that the incoming and outflowing dye solution shadows the cylindrically symmetric mode volume from a significant fraction of the excitation. This shielding can be partially compensated by altering the resonator so that the lasing mode is asymmetric with the wide dimension along the liquid flow direction.[79,80] The width of the excited region will be determined by the focusing system. The excitation distribution produced by an elliptical reflector focusing a cylindrical lamp is shown in Fig. 30.[77] It is seen that in this particular case where the axis ratio for the prolate focusing ellipse was 1.1

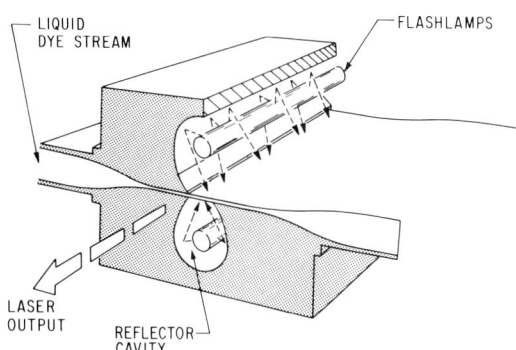

FIG. 29. Flashlamp-excited laser with transverse dye flow.

[79] P. Burlamacchi, R. Pratesi, and R. Salimbeni, *Appl. Opt.* **14**, 1311 (1975).
[80] P. Burlamacchi, R. Pratesi, and R. Salimbeni, *Opt. Commun.* **11**, 109 (1974).

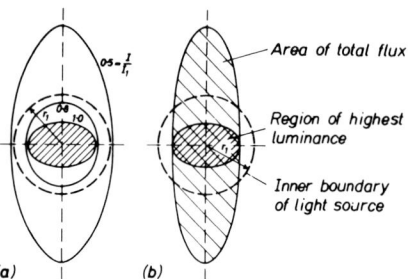

FIG. 30. (a) Calculated curves of equal luminescence for a cylindrical reflector of elliptical cross section. (b) Areas of maximum luminescence and of total flux: schematic for axes ratio of ellipse of 1.1:1.[77]

that the one-half power point for the image is approximately twice the size of the flashlamp arc. The dye channel thickness can now be adjusted for maximum output by balancing laser threshold considerations against thermal loss considerations.

5.5.9. Dye Cell for Flashlamp Excitation

The fact that the active medium in dye lasers is liquid requires special consideration. The optical homogeneity of the liquid medium is very susceptible to thermal and shock-wave-induced disturbances. It is necessary that the solution be completely uniform in temperature before the initiation of the excitation pulse. This uniformity can be achieved only if the solution is in thermal equilibrium with its surroundings.

To minimize the thermally induced inhomogeneities, it is useful to flow the liquid medium rapidly through the active region. It is helpful to have the solution flow rate well into the turbulent region. This turbulence will increase the solution mixing and thereby improve the temperature uniformity. The pressure variations caused by turbulent flow of the dye solution through the cell introduce index of refraction inhomogeneities that are insignificant compared to those introduced by temperature variations. The index of refraction of liquids is dominated by their density. The difference in sensitivity of the index of refraction to changes in pressure and temperature therefore is a consequence of the small coefficient of compressibility of most liquids as contrasted with relatively large coefficients of thermal expansion.

In any dye flow system, it is useful to make the active region the major restriction in the flow circuit. This restriction is to ensure that the pressure is high at this point to minimize the possibility of dissolved air coming out of solution and producing bubbles. In this regard, it is also necessary to design the flow channels with large radius bends and no sharp corners to prevent any localized pressure drops which could induce bubble formation.

In most laser systems the dominant design consideration has been the achievement of laser threshold and the minimizing of the power required to obtain this goal. To minimize the threshold it is necessary to focus carefully the excitation into the smallest dye cell consistent with the limitations of the focusing cavity. With some flashlamp systems, particularly high-intensity ones, the excitation levels are sufficiently high that the focusing of this power into a minimum-diameter cell causes severe thermal problems. If the excitation level is too high, the laser emission will terminate prematurely. This can be determined by simultaneously recording the laser and flashlamp waveforms. The ideal operating condition would be achieved if the laser pulse both started and stopped at points where the excitation intensity had the same value. In practice, the laser emission invariably terminates early while the excitation is still greater than that which was required to initiate laser action. This is indicative of some thermally induced optical disturbance in the dye volume. In a situation such as this, the dye cell volume should be increased in an attempt to increase the output by providing additional dye solution to absorb the heat. This enlarged volume will increase the intensity required to achieve laser threshold. There will be an optimum size for the dye cell that maximizes the utilization of the excitation. It should be possible to make a first estimate of the optimum cell size from the before-mentioned laser and flashlamp waveforms. It was shown in Section 5.2.5 that both the laser threshold and the thermal limitation were a function of the excitation absorbed by the active volume per cross-sectional area of the dye cell. If the cross-sectional area of the dye cell is doubled, the laser threshold will double but the solution should be able to absorb twice the amount of excitation before the thermal cutoff point. It should be possible to perform some simple graphical integrations on the waveforms and to obtain a quite resonable estimate for the optimum cell volume. Complete elimination of the optical disturbance is usually not possible so the maximum laser output will probably be obtained using a dye cell with dimensions intermediate between the minimum size calculated from focusing considerations and the size as estimated from this threshold approximation.

One of the more severe limitations on all high average power dye lasers is heat stored in the dye cell walls. The dye cell is heated by the dye solution, which has been excited to laser action. Since the materials of choice for the cells have often been amorphous, e.g., fused quartz, the thermal conductivity has been very low. Thus, the heat can be removed only by subsequent volumes of dye solution flowing through the cell.

This situation makes the experimentally determined upper limit on the average power of a laser several times less than anticipated from a simple calculation of the frequency at which the dye volume is exchanged.

For broad-band operation, this difficulty has often been ignored. The small region near the liquid–solid interface would exhibit large refractive index inhomogenities. If the dye cell is large compared to the resonator mode dimensions, the loss of the interface region may not be significant. This circumstance has been found to be true in systems with dye cells approximately 1 cm in diameter.

The dye cell can be fabricated of high thermal conductivity crystalline material, e.g., sapphire, to remove the heat. Such fabrication is difficult and expensive. However, it is then possible to flow cooling water on the outside of the cell to remove the excess stored heat. To be successful, this scheme requires that the dye and cooling water have essentially identical temperatures, which can be assured by initially pumping both fluids through a heat exchanger with a contact area significantly greater than the contact area within the laser.

5.5.10. Free Jet Dye Cell

An alternative solution to the thermal storage difficulty is to remove the offending solid completely and to have the excitation radiation enter the dye solution through a liquid–air interface. Such a jet of dye solution has its own set of unique problems.[81] The surface of a large-size liquid jet probably will not have an optical-quality surface, so that the laser emission must be injected and extracted from the active medium by some unusual means. To prevent losses of the laser emission, the dye jet has been made to flow between two glass windows placed at each edge of the stream. A perspective of an experimental device is shown in Fig. 31.

The nozzle channel must be polished to optical perfection with particular emphasis on the edge where the liquid breaks away from the nozzle. Any imperfections in the surface and edge will create surface waves on the jet. A major source of surface waves in this design is the corner where the fluid breaks away from the nozzle but still wets the window. To prevent this wave from limiting the clear aperture through the jet, the nozzle must be flared out at the ends where this wave originates. This is illustrated in Fig. 31.

An additional characteristic of jets that is illustrated in the diagram is that the thickness of the jet will diminish as the fluid first leaves the nozzle. This narrowing is a consequence of the fact that within the

---

[81] R. J. Foley, T. C. Kuklo, and O. G. Peterson, *IEEE /OSA Conf. Laser Eng. & Appl.* p. 5 (1975).

5.5. STEADY-STATE LASER

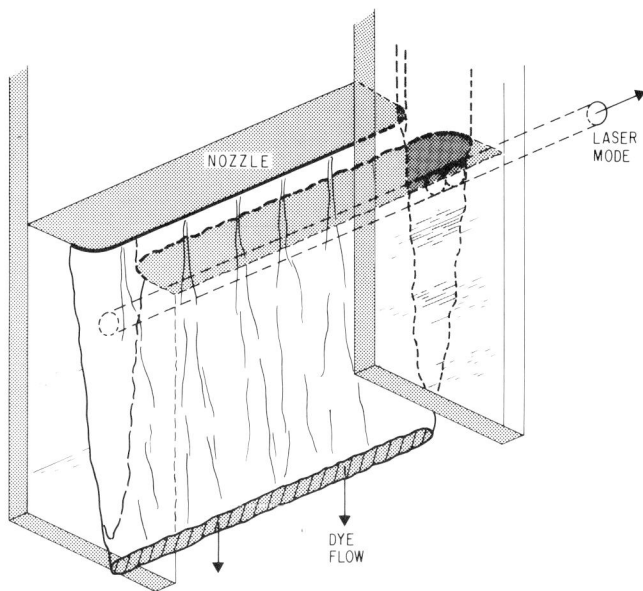

Fig. 31. Diagram of jet of dye solution for use in flashlamp-excited lasers. The view is looking up at the nozzle, which forces the jet to flow in a vertical downward direction. The specially shaped nozzle extrudes the solution into a flat sheet.[81]

nozzle there will be a distribution of fluid velocities with the maximum occurring at the center of the nozzle and the minimum occurring at the walls. In the jet the fluid must be flowing at a uniform velocity. Therefore, as the liquid breaks away from the nozzle, the outer layers of liquid must be accelerated. From simple considerations of mass flow conservation it is clear that the jet must shrink. This effect is exaggerated if the velocities are small and the flow is also subjected to gravitational acceleration. As a consequence, the optical path through all jets is narrower than would be estimated from the nozzle width.

The surface waves on the face of the jet would damp out if the flow were laminar but will increase if the solution flow rate is greater than the laminar-turbulent transition point.[82] Useful jets can be fabricated to operate in both velocity regimes. The transition point from laminar to turbulent flow occurs where the solution velocity v satisfies the relation

$$v = R_c \nu/D, \qquad (5.5.21)$$

where R_c is Reynolds number characteristic of the transition, ν the kinematic viscosity of the solvent, and D the effective diameter of the nozzle.

[82] H. Lamb, "Hydrodynamics." 6th ed. Dover, New York, 1932.

The effective diameter is calculated from

$$D = 4A/P, \qquad (5.5.22)$$

where A and P are the cross-sectional area and wetted perimeter of the nozzle, respectively. The critical value of the Reynolds number R_c for smooth pipes is approximately 2000.[82]

From Eq. (5.2.27) it is clear that the power that can be absorbed by a jet for a selected wavelength variance is

$$P = z \frac{\rho C_p}{d\eta/dT} \frac{d\lambda}{\lambda} vw, \qquad (5.5.23)$$

where w is the narrow dimension of the nozzle and v the solution flow velocity. Substituting Eqs. (5.5.21) and (5.5.22) into (5.5.23) yields

$$P = z \frac{\rho C_p \nu}{d\eta/dT} \frac{d\lambda}{\lambda} \frac{R_c}{2}, \qquad (5.5.24)$$

where w is assumed to be much less than the nozzle length. Equation (5.5.24) calculates a maximum power that a solution can absorb if its velocity is limited to maintain laminar flow conditions. The parameters in this equation that are characteristic only of the solvent can be collected together to generate a figure of merit in a manner similar to that used previously to define a thermal property figure of merit for solution[83]

$$M_L = \frac{\rho C_p \nu}{d\eta/dT}. \qquad (5.5.25)$$

This factor has been listed in Table I. It is seen that while water is superior to other solvents for closed cells and jets where turbulent flow can be tolerated, high-viscosity fluids such as ethylene glycol and glycerol are better for laminar flow jets. These materials have therefore been most frequently used in laminar-flow systems with ethylene glycol, often preferred for its ease of handling. Pure glycerol is sufficiently viscous to inhibit the mixing in the dye solution reservoir necessary to obtain thermal homogeneity of the solution. The viscosity of this solvent can be easily adjusted to an optimum value for a particular apparatus by diluting it with water. The viscosity of such mixtures has been measured and is tabulated in handbooks.[21] The higher velocities that can be used with the more viscous solvents improve the dimensional properties of the useful region of the jet. The area where the surface remains flat is greatly lengthened in the direction of the fluid flow. Further, the shrinkage of the jet thickness is extended over a greater length, which significantly improves the parallelism of the jet surfaces.

---

[83] B. Wellegehausen, H. Welling, and R. Beigang, *Appl. Phys.* **3**, 387 (1974).

For both turbulent- and laminar-flow jets, it is necessary that the liquid flow be exceptionally smooth. Pressure fluctuations introduced by the circulating pump must be removed with surge dampers and the nozzle must be backed up with a large plenum equipped with screens to make the flow into the nozzle uniform along its length and also to break up the turbulence into eddies smaller than the nozzle width. The nozzle is constructed with a long channel that is only partially shown in Fig. 31. The long narrow channel straightens out the flow of the liquid so that the exit velocities of packets of solution have a predictable direction. These channels are usually made with lengths approximately 20 times the thin dimension of the nozzle. If these precautions are not taken, the jet will have significant thickness fluctuations, which will effect the wavelength stability and efficiency of the laser device.

To maximize the average power attainable with the illustrated device, it is necessary to operate in the turbulent regime. Since surface imperfections will grow as a function of time and distance, greater surface perfection is required of the nozzle components than is desired of the jet surface. The nozzle must be fabricated of a hard material that can be polished to optical perfection. The edge where the fluid breaks away from the nozzle requires special attention. It must be polished to a minimum radius of curvature. If this edge has any significant radius, the liquid will break at different places at different times. These fluctuations in the break-away position will generate surface waves on the jet in much the same fashion that one creates waves in a blanket by shaking one edge.

Since the laser emission does not have to pass through the jet surface, a certain amount of surface imperfection can be tolerated. The small-angle scattering of the excitation light by this surface does not significantly degrade the laser performance since the absorbing region is in intimate contact with the scattering source. The most serious limitation introduced by surface imperfections is the possibility of the waves extending far enough within the liquid to scatter the laser light. A good nozzle will limit this disturbed region to less than 5% of the stream thickness.

Since jets are completely unsupported, they are susceptible to many minor perturbations in the ambient conditions. They are easily altered by mechanical vibrations and particularly by pressure fluctuations in the fluid flow system. They are also affected by air movements and acoustic waves. In the laminar regime, the surface can be affected by waves with propagation velocities greater than the flow velocity that are generated when the jet strikes a surface. The safest way to eliminate these waves is to allow the jet to continue in space until it breaks up into drops. To shorten this distance the jet can be impinged upon a surface at a very high angle of incidence but should not be allowed to splash. To achieve stable laser operation with low noise in both the amplitude and wavelength character of the emission, it is necessary to give careful consideration to all of these effects.

In both the wall-confined dye cell and the free jet, the interface region of the solution flow is lost at high levels of average power operation because of optical distortions. The closed cell is obviously easier to design and to operate. The free jet has the advantage that there is no surface wetted by the dye solution, which is exposed to the high-intensity excitation. The air–liquid interface eliminates any possibility of decomposition products adhering to closed-dye cell windows, thereby degrading the operation of the laser. This advantage may prove decisive for sustained operation at high average powers.

5.5.11. Shock Wave Effects

The other characteristic of liquid media that requires special attention are their susceptibility to shock-wave-induced optical inhomogeneities. Such shock waves are produced by all flashlamp discharges with sufficient intensity that the resulting optical inhomogeneities will terminate laser action.[84-86] It is necessary, therefore, that the acoustic wave be delayed long enough so that the laser pulse is complete before the shock wave has been transmitted to the dye cell. A small air gap is usually sufficient to accomplish this purpose as is seen in Table II, where sound velocities[21] are listed in units appropriate to this application, mm/μsec. It may also be necessary to use nitrogen gas instead of air to fill the flashlamp cavity to eliminate shock waves produced by absorption of radiation near 200 nm by the oxygen in the air.

Some of the shock wave effects have been attributed to the absorption of the excitation by the dye solution, which cannot be delayed by the air

TABLE II

Material	Sound velocity (mm/μsec)
Air	0.33 (STP)
Water	1.48 (20°C)
Pyrex glass	5.6
Fused silica	6.0
Aluminum	6.4
Stainless steel	5.8

[84] T. F. Ewanizky, R. H. Wright, Jr., and H. H. Thessing, *Appl. Phys. Lett.* **22**, 520 (1973).

[85] A. Hirth, K. Vollrath, and J. P. Fouassier, *Opt. Commun.* **9**, 139 (1973).

[86] S. Blit, A. Fisher, and U. Ganiel, *Appl. Opt.* **13**, 335 (1974).

gap.[86] To minimize this possibility, all radiation that the dye molecules cannot use for producing excited-state populations should be filtered out of the excitation. Care should also be exercised to ensure uniform absorption of the excitation.

5.5.12. Coaxial Flashlamps

The introduction of coaxial flashlamps to dye lasers was one of the first significant improvements in the technology.[87-89] It is a completely integrated system designed to obtain the greatest single-pulse efficiency possible from a flashlamp-excited dye laser. In addition to other novel attributes, it was one of the first successful attempts to match the discharge circuit impedance to the impedance of the discharge in the short time regime suitable for use with dye lasers. It involved careful assembly of the driving circuit to minimize its 0.3-Ω impedance and clever design of the coaxial discharge volume so that the discharge would approximate the impedance of the driver circuit. The system is schematically outlined in Fig. 32. The flashlamp is a thin annulus coaxially surrounding the dye cell. The thickness of the discharge channel is carefully chosen so that the electrical discharge will be close to the critically damped condition. If this condition is achieved, the discharge will uniformly fill this annular region. The annular discharge greatly reduces the inductance of the circuit as compared to what would be obtained if the discharge were filamentary. The resultant shorter discharge pulse width greatly increases the peak excitation intensity. This high intensity, of course, is ideal for dye laser applications as has been discussed. Dye lasers excited by these lamps have been very efficient. More importantly, they permit lasing output to be obtained from marginal dyes, thus significantly increasing the wavelength range of operation. Discharge pulse widths as short as 200 nsec have been obtained and laser action observed over the range of 340–900 nm with this style of lamp.

The short discharge times are critically dependent on the design of the entire discharge circuit. The circuit inductance must be maintained at an absolute minimum. This low inductance is achieved by placing the energy storage capacitor and its high-voltage switch in a cylindrical can that serves as a coaxial ground return lead. The flashlamp is fitted with a similar coaxial ground return cylinder and the two parts are connected together with a minimum length of parallel-plate transmission line. It is imperative that the energy storage capacitor be one specifically designed

[87] H. W. Furumoto and H. L. Ceccon, *Appl. Opt.* **8**, 1613 (1969).
[88] H. W. Furumoto and H. L. Ceccon, *IEEE J. Quantum Electron.* **qe-6**, 262 (1970).
[89] J. Bunkenberg, *Rev. Sci. Instrum.* **43**, 1611 (1972).

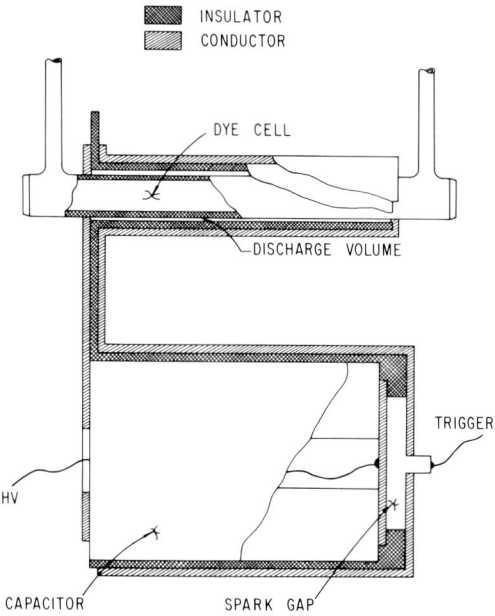

Fig. 32. Coaxial flashlamp-excited dye laser assembly including the low-inductance discharge circuit for the flashlamp. The flashlamp discharge fills the coaxial volume surrounding the dye cell. The current is carried from the low-inductance capacitor by coaxial conductors surrounding both the lamp assembly and the capacitor, which are connected together by a short section of flat strip-line transmission line.[87]

for minimum inductance. Such capacitors are often, and preferably for this application, fabricated in the form of a cylinder with electrodes on each end. One such discharge circuit that has been reported had a total inductance of approximately 30 nh of which 13 nh were due to the capacitor. The impedance $Z = (L/C)^{1/2}$ for this circuit, which included a 0.3-μf energy storage capacitor, was 0.3 Ω.

With the driver impedance reduced to an acceptable minimum it was necessary to select the discharge channel parameters in such a manner that the impedances would be closely matched. The maximum light was emitted when the damping factor as defined in Eq. (5.5.18) for linear flashlamps is between 1.4 and 1.9. To calculate the damping factor the diameter d in Eq. (5.5.16) was replaced by $d = 4A/P$, where A is the cross-sectional area and P the perimeter of the discharge volume. For the example described, the optimum thickness of the annular discharge volume was 0.33 mm for a 6-cm-long lamp and 0.45 mm for a 12-cm-long lamp when the annular region had a diameter of 11 mm and 18 kV were applied to the energy storage capacitor. The circuit was still slightly underdamped at the point of maximum light output. However, further reduction in the annular spacing apparently cooled the plasma and reduced its irradiance.

Only xenon gas was found to operate efficiently in this lamp. When the discharge is operating in the desired coaxial mode, the current has a very distinctive signature characterized by a prominent "hash" superimposed on the current waveform. Additionally, the maximum di/dt occurs at the start of the pulse, as contrasted with filamentary pulses, which rise exponentially. There is no question when the lamp discharges in a single filament instead of the desired annular discharge because such an asymmetric discharge will impose similarly asymmetric pressures on the quartz components. This pressure will usually fracture the lamp.

The spectral distribution of the lamp output indicated that the plasma temperature was in the range 21,000–25,000 K for the 6-cm-long lamp. This temperature is rather high for effective coupling of the energy into the dyes. However, the high brightness of the lamps causes the laser to operate well over threshold, which dramatically improves efficiency. The close proximity of the plasma to the active medium makes the coupling of the radiation into the dye very efficient except that most of the radiation from the outer surface of the plasma is lost. This loss is due to the opacity of the plasma. The close proximity also couples any shock waves created by the discharge into the cell and thus disturbs the optical quality of the active volume.[84-86] The compact geometry of this lamp, which is very important for achieving low inductance, makes the lamp very difficult to cool. It is therefore necessary to limit this lamp to applications with low duty cycles. This limitation is also imposed by the low-inductance capacitor, which is usually constructed with a relatively high loss dielectric.

This lamp has been very useful in the development of dye laser technology. Its fast rise times and high brightness have extended the wavelength range of flashlamp-pumped dye lasers by exciting marginal dyes to laser action. These same characteristics have made it an efficient laser using the better dyes.

5.5.13. Resonators and Tuning Elements

The resonators used for flashlamp-pumped dye lasers are in general indistinguishable from those used in other visible laser systems except for the proliferation of wavelength-selective elements contained in these resonators. Resonator designs have been strongly affected by thermal effect considerations. The thermally induced optical inhomogeneities have made it quite difficult to obtain single-mode output from dye lasers simultaneously with good output efficiency. The best output efficiency has most often been obtained with flat mirrors in a high Fresnel number resonator.[71–75,87–91] The Fresnel number is defined by $N_F = r^2/\lambda l$, where r is the radius of the smallest aperture in the system, λ the wavelength, and l the distance between the resonator mirrors. Such resonators permit the laser emission to fill very nearly the active volume and to extract the maximum amount of energy from the excited region. The high Fresnel number makes the resonator less sensitive to minor beam deflections caused by the optical inhomogeneities.

The design of single transverse-mode resonators has been discussed in many references and will not be repeated here.[58] The resonator stability criteria and beam divergence equations will be considered later as they apply to cw dye lasers.

During the development of dye lasers almost every type of wavelength-dispersive element has been put into a dye laser resonator to tune the output wavelength. Since the characteristics of most such elements are adequately described in standard optics textbooks[92] they will not be considered in detail. The choice of tuning elements depends on the gain characteristics of the laser and the desired output bandwidth. Short-pulse dye lasers that normally exhibit high transient gains require tuning elements with high dispersion, e.g., gratings.[36] On the other hand, cw lasers require only a small amount of wavelength discrimination to control their output but must be tuned with elements with very low insertion losses, e.g., prisms[93] and Lyot filters.[94] These elements will tune the output wavelength of the dye laser but yield a relatively broad-band, multimode output.

To obtain narrow-line emission from any dye laser requires the addition of interferometric devices as Fabry–Perot etalons[95] or Fox–Smith[96] adaptions of the Michelson interferometer. The number of etalons and their required reflectivity are also dependent on the laser gain characteristics. High reflectivities, e.g., 50%, may be necessary for short-pulse lasers, while uncoated ($R = 4\%$) etalons often suffice for cw devices. The free spectral range of successive elements is chosen so that only one transmission peak of a particular element falls within the passband of the element with the next larger free spectral range.

[90] B. H. Soffer and B. B. McFarland, *Appl. Phys. Lett.* **10**, 266 (1967).
[91] G. Marowsky and F. Zaraga, *IEEE J. Quantum Electron.* **qe-10**, 832 (1974).
[92] F. A. Jenkins and H. E. White, "Fundamentals of Optics." McGraw-Hill, New York, 1957.
[93] S. A. Tuccio and F. C. Strome, Jr., *Appl. Opt.* **11**, 64 (1972).
[94] G. Holtom and O. Teschke, *IEEE J. Quantum Electron.* **qe-10**, 577 (1974).
[95] G. Marowsky, *Rev. Sci. Instrum.* **44**, 890 (1973).
[96] P. W. Smith, *Proc. IEEE* **60**, 422 (1972).

The ultimate of narrow-line output is obtained when the laser has been limited to oscillation on a single longitudinal mode of the resonator. In this mode of operation the intraresonator intensity is a single standing wave unless the resonator is a unidirectional ring resonator. This standing wave produces a corresponding periodic structure within the gain medium. The excited-state population will be fixed at the threshold value at the antinodes of the standing wave, but can be much higher at the nodes of the pattern where there is no intraresonator intensity to stimulate the emission from the excited molecules. The periodic gain structure helps to discriminate against laser action on nearby modes, but allows very high gains to exist for modes that are completely out of phase with the selected mode within the excited volume. The wavelength-selective elements in the single-mode resonator therefore serve two purposes: first, to select the desired mode for oscillation by making this mode have the lowest threshold for laser action; and second, to prevent parasitic laser action on satellite modes. The suppression of the parasitic oscillation usually requires much higher discrimination from the tuning elements than the simple selection of the favored output mode. The calculation of the spacing between the desired mode and the nearest satellite mode is straightforward from the spacing between the center of the gain medium and the closest end mirror of the resonator, l_{cm}. This spacing is

$$\Delta\lambda = \lambda^2/4l_{cm}, \qquad (5.5.26)$$

where the two modes are assumed to be in phase at the end mirror. If the thickness of the active medium is small compared to the spacing l_{cm} then the gain of the satellite and central mode will be identical without further discrimination introduced by the tuning elements. If the active medium has a significant length compared to the distance to the end mirror then the satellite mode gain will be reduced as the two modes become more nearly in phase as they approach the end mirror.

There are several specialized techniques of particular utility with dye laser devices. Included among these is the simultaneous tuning of the series of intraresonator elements by changing the gas pressure in the resonator, which uniformly changes the index of refraction within the tuning elements.[41] By controlling the gas pressure a series of air-spaced etalons therefore do not have to be individually tuned as the wavelength is shifted. The wavelength tuning can be controlled electrically by the use of either acousto-optic[97] or electro-optic devices.[98] In the latter case the tuning can be sufficiently rapid that the tuning element can be oscillated in phase with the round trip transit time of the light within the resonator.

[97] P. Saltz and W. Streifer, *IEEE J. Quantum Electron.* **qe-9**, 563 (1973).
[98] J. M. Telle and C. L. Tang, *Appl. Phys. Lett.* **26**, 572 (1975).

The traveling waves within the resonator will therefore have different wavelengths at different positions. This must be the ultimate in scanning speed that can be obtained in a controlled manner. It is also possible to tune the laser using distributed-feedback techniques. The distributed feedback yields very narrow line output but presents significant problems if the output must be tuned.[99,100]

5.5.14. Dye Flow System

The dye solution handling system is conceptually simple in design. In its simplest form, it consists of a pump, a reservoir, and a dye cell with appropriate connections.[75] More sophisticated systems may include damping devices to absorb the energy in the pressure surges induced by the mechanical pump. They may also contain filters to remove particulate matter and air bubbles and may include heat exchangers.

The most important consideration in selecting components for such a system is the chemical inertness of the construction materials. The safest materials are glass, stainless steel, and simple polymers such as polyethylene, polypropylene, and Teflon™. The performance of the dye molecules was found to be degraded by the use of polymers containing plasticizers for the flow system components. Components made of polymeric material should be minimized since the dyes slowly diffuse into them and consequently can be leached into subsequent solutions with possible deleterious effects.

It is necessary to keep the pressure relatively high in the dye cell. This pressure is needed to minimize the bubble formation caused by dissolved air coming out of solution. These bubbles can cause severe losses to the laser system by scattering the laser emission. Flow-induced pressure minima can cause such bubble formation, particularly if the minima are below the ambient pressure in the reservoir where the solution and air are in equilibrium. Another source of bubbles is surface turbulence in the reservoir or surge damper, where the rapidly flowing dye solution comes into contact with air. This surface must be maintained relatively quiescent to reduce the possibility of entrapping additional air into the solution. The bubbles that are still in the system can be removed by the in-line filter as long as the pressure drop across the filter is lower than that required to force air through the pores in the filter.

The surge damper in its simplest configuration (Fig. 33) is a closed chamber containing a volume of air trapped over the solution. The air ab-

---

[99] J. E. Bjorkholm and C. V. Shank, *IEEE J. Quantum Electron.* **qe-8,** 833 (1972).
[100] Y. Aoyagi, T. Aoyagi, K. Toyoda, and S. Namba, *Appl. Phys. Lett.* **27,** 687 (1975).

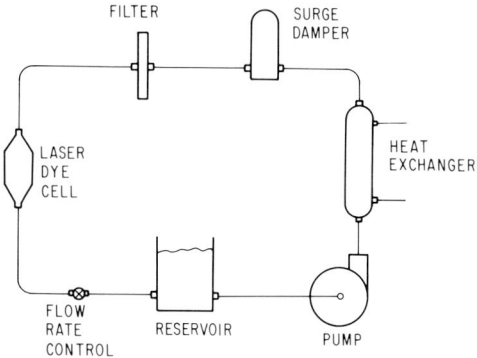

FIG. 33. Dye solution flow system schematic.

sorbs the energy in the pressure surges induced by the pump and therefore smooths out the flow. More sophisticated devices will separate the liquid and air with an elastic membrane to permit some adjustment in the air pressure independently of the fluid pressure at that point. The damper may be tuned in this manner for maximum absorption of the particular amplitude and frequency of the pressure fluctuation characteristic of the specific system.

Most of the items used in the flow system are commercially available. The choice between competing devices is dictated primarily by material compatability and ease of cleaning of the components. This latter characteristic is very important if the system is to be versatile and operate at different wavelengths using different dyes. It is particularly important to prevent contamination of a short-wavelength dye by a long-wavelength one since the long-wavelength dye can cause severe absorption losses at the lasing wavelength of the short-wavelength dye.

5.6. CW Laser

5.6.1. Analytical Description

The cw dye laser is most amenable to definitive calculations of its characteristics. Since the excitation of such devices is of necessity the output of another cw laser, the spectral and geometrical overlap functions can be accurately calculated in principle. These calculations are facilitated by the fact that the usual excitation sources are high-powered ion lasers that have well-defined output mode structure that permits precise control of the excitation volume.

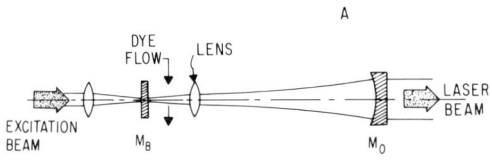

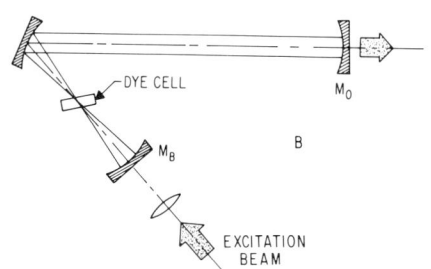

FIG. 34. cw dye laser resonators using three-element, thick mirrors to obtain the required small-mode diameters. The laser in (A) uses a lens to focus the intraresonator intensity into the dye solution while the laser in (B) uses a short radius mirror M_F for the same purpose. In both cases the excitation enters the resonator through a dichroic mirror M_B, which transmits the excitation but reflects the laser intensity. The laser output is extracted through a partially transmitting mirror M_0.

Two representative geometries are illustrated in Fig. 34.[93,101,102] The important characteristic common to both designs is a very small active volume generated by the tight focusing of the excitation and emission beams into a small dye volume. The small diameter, 20–50 μm, of the excitation region is required to achieve approximately 1 MW/cm² excitation and emission intensities needed to sustain laser action in organic dyes. The small diameter also allows the potential thermal problem to be mitigated since dye solution transit times through this volume will be a few microseconds for flow rates of several meters per second. Laser resonator stability was achieved by focusing the emission intensity into the excited volume with a thick mirror configuration composed of a long radius of curvature end mirror and a short-focal-length focusing element. The details of the resonator design will be considered later.

The equilibrium equations (5.2.1)–(5.2.5) apply to the cw laser and are listed below using the notation defined in Section 5.2.2:

---

[101] F. C. Strome, Jr. and S. A. Tuccio, *IEEE J. Quantum Electron.* **qe-9,** 230 (1973).

[102] H. W. Kogelnik, E. P. Ippen, A. Dienes, and C. V. Shank, *IEEE J. Quantum Electron.* **qe-8,** 373 (1972).

5.6. CW LASER

$$dI_X/dx = -n_0\sigma_{0X}I_X - n_S\sigma_{SX}I_X - n_T\sigma_{TX}I_X, \quad (5.6.1)$$

$$dI_M/dx = n_S\sigma_{SM}I_M - n_T\sigma_{TM}I_M - n_0\sigma_{0M}I_M, \quad (5.6.2)$$

$$dn_S/dt = n_0\sigma_{0X}I_X + n_0\sigma_{0M}I_M - n_S\sigma_{SM}I_M - k_{ST}n_S - n_S/\tau_S, \quad (5.6.3)$$

$$dn_T/dt = k_{ST}n_S - n_T/\tau_T, \quad (5.6.4)$$

$$n = n_0 + n_S + n_T. \quad (5.6.5)$$

These equations can be further simplified by applying approximations specific to the cw laser. The excited-state absorptions are assumed to have no effect on the excited-state populations because of the extremely short lifetimes of the higher excited levels. Since the laser is operating cw, all of the time derivatives will be set equal to zero. The intersystem crossing term $k_{ST}n_S$ is important for the triplet concentration equation but its effect on the excited singlet population is insignificant. Since cw dye lasers always have been longitudinally excited, it is possible to estimate the dye molecule concentration and from this concentration the relative importance of the different terms in the equation. To absorb the excitation effectively the dye concentration must be adjusted so that the optical density in Naperian units will be approximately $\sigma_{0X}n_0l = 3$ at the excitation wavelength. At the laser emission wavelength this same ground singlet state will have an absorption coefficient that can be as much as 10^4 times smaller. The term $n_0\sigma_{0M}I_M$ associated with the ground state absorption of the emission can be ignored in both Eqs. (5.6.2) and (5.6.3). For improved accuracy the absorption term in the dI_M/dx equation can be included as a correction to the resonator losses. Incorporating these approximations and substituting for the excited-state values yields

$$\frac{dI_X}{dx} = -\frac{n\sigma_{0X}I_X[\tau_S(\sigma_{SX} + k_{ST}\tau_T\sigma_{TX})I_X + \tau_S\sigma_{SM}I_M + 1]}{\tau_S(1 + k_{ST}\tau_T)\sigma_{0X}I_X + \tau_S\sigma_{SM}I_M + 1}, \quad (5.6.6)$$

$$\frac{dI_M}{dx} = \frac{n\sigma_{0X}I_X\tau_S(\sigma_{SM} - k_{ST}\tau_T\sigma_{TM})I_M}{\tau_S(1 + k_{ST}\tau_T)\sigma_{0X}I_X + \tau_S\sigma_{SM}I_M + 1}. \quad (5.6.7)$$

Useful solutions to these equations can be obtained by dividing Eq. (5.6.6) into (5.6.7):

$$\frac{dI_M}{dI_X} = -\frac{I_M(\sigma_{SM} - k_{ST}\tau_T\sigma_{TM})\tau_S}{I_X(\sigma_{SX} + k_{ST}\tau_T\sigma_{TX})\tau_S + \sigma_{SM}\tau_S I_M + 1}. \quad (5.6.8)$$

This equation can be solved analytically in a trivial manner if the laser operating conditions are such that the intraresonator intensity I_M can be assumed to be constant. These lasers are normally operated in low-loss resonators, which implies small gain per pass and therefore a very uniform intraresonator laser power. The intensity variation can be easily estimated if it is assumed that the resonator losses are predominantly at one end of the resonator.[103] If we consider the elementary laser in Fig. 3 and the intensities labeled in the figure, then after one pass through the active region we have $I_2 = I_1 e^G$, where G is the logarithmic gain. With the assumption of lossless reflection, the intensity I_3 equals I_2 and the intensity I_4 equals $I_3 e^G$. The ratio of the total intensities at each end of the amplifying medium is

$$\frac{I(T_1)}{I(T_2)} = \frac{I_1 + I_4}{I_2 + I_3} = \frac{I_1(1 + e^{2G})}{2I_1 e^G} = \cosh G. \quad (5.6.9)$$

For a 1% deviation, i.e., $\cosh G = 1.01$, the gain G equals 0.14. The ratio I_1/I_4 equals 0.75 for this value of G. The result is that double-pass resonators with losses less than 25% have less than 1% deviation in the total intraresonator intensity exclusive of diffraction and interference effects.

It will be assumed in this treatment that the laser is operating in many longitudinal modes so that interference effects will be randomized and the intensity can be assumed to be spatially uniform. In addition, the divergence of the two beams will be at a minimum in the excitation region so that the transverse spatial distributions of both beams can be assumed to be constant through the active volume. This constancy is a consequence of the fact that the active volume is placed at the resonator focus where the laser mode has minimum cross-sectional area. At this minimum point the beam divergence is also at a minimum. The point where the beam cross-sectional area has increased by a factor of two as it diverges away from the minimum point is defined as the confocal distance b, which can be calculated from the relation

$$b = \pi \omega_0^2/\lambda, \quad (5.6.10)$$

where ω_0 is the $1/e^2$ radius at the minimum point. The active region in these devices is purposely kept equal to or smaller than the confocal volume, which has a length b or $2b$ in the resonator diagrammed in Figs. 34A and B, respectively. It is reasonable, therefore, to assume I_M to be a constant through this region, which permits the above relation to be easily integrated, i.e.,

$$\Delta I_M = I_M \gamma \ln \left[\frac{(\sigma_{SX} + k_{ST}\tau_T\sigma_{TX})\tau_S I_X^i}{\sigma_{SM}\tau_S I_M + 1} + 1 \right], \quad (5.6.11)$$

---

[103] U. Ganiel, A. Hardy, G. Neumann, and D. Treves, *IEEE J. Quantum Electron.* **qe-11**, 881 (1975).

where ΔI_M is the increase in the intraresonator intensity I_M, I_X^i is the input excitation intensity, and

$$\gamma = \frac{\sigma_{SM} - k_{ST}\tau_T\sigma_{TM}}{\sigma_{SX} + k_{ST}\tau_T\sigma_{TX}}. \tag{5.6.12}$$

In evaluating the limits of integration, it was assumed that the excitation was totally absorbed. It is interesting to note that the ground-state absorption of the excitation does not enter this relation explicitly although its role is important in validating the assumptions used to evaluate the integral. It was also assumed that the excitation and emission beams are perfectly collinear. For efficient operation it is important that this be maintained since the length-to-width ratio of the excitation volume is very large, typically $b/\omega_0 = \pi\omega_0/\lambda \approx 100$. This restriction can be relaxed somewhat by increasing the dye concentration so that the excitation volume is reduced in length.

Equation (5.6.11) can be used to outline the conditions for optimum laser operation. It is again useful to convert the equation into one using dimensionless parameters:

$$\Delta\phi_M = \phi_M\gamma \ln\left(\frac{\phi_X'}{\phi_M + 1} + 1\right), \tag{5.6.13}$$

where $\Delta\phi_M$ is the incremental increase in ϕ_M due to single-pass amplification, $\phi_M = I_M\sigma_{SM}\tau_S$ as defined previously, and $\phi_X' = I_X^i(\sigma_{SX} + k_{ST}\tau_T\sigma_{TX})\tau_S$, which is different from the definition given for ϕ_X in Section 5.4.1 but close in value since σ_{TX} is expected to be small and σ_{SX} is approximately equal to σ_{0X}. This analytical expression for the gain can now be used to evaluate the optimum resonator parameters for efficient laser operation. To evaluate these we shall impose lowest order mode Gaussian spatial distributions on the excitation and emission intensities:

$$\phi_X' = \phi_{X0}' \exp(-2r^2/\omega_X^2) = \phi_{X0}' \exp(-2r^2\rho^2/\omega_M^2),$$
$$\phi_M = \phi_{M0} \exp(-2r^2/\omega_M^2),$$
$$\rho = \omega_M/\omega_X, \tag{5.6.14}$$

where ϕ_{i0} is the peak of the Gaussian curve, ω_i the $1/e^2$ radius of the distribution in the focal region, and ρ is defined as the ratio of the emission to excitation beam radii. For steady-state operation, the threshold condition Eq. (5.2.6) is applicable, where the total integrated laser intensity is required to be reproduced upon one round trip through the laser.

To perform this integration, additional physical assumptions were required. The excited-state population depends on the balance between the excitation and the sum of forward and backward directed emission beams that exist in the usual laser oscillator. To allow the two beams to be considered together, the resonator can be considered to be an infinite string of gain regions separated by loss elements that are equivalent to the total intraresonator losses. To simplify the evaluation of the equivalent loss elements the laser will be assumed to be symmetric with one-half of the output being emitted from each end.[104] Each equivalent loss element will have a transmissivity T_i of each of the intraresonator elements and the end mirror reflectivity $R_{1/2}$, where the mirror reflectivity is calculated to transmit one-half the desired output:

$$T_R = T_1 T_2 \cdots T_i \cdots T_N R_{1/2}. \tag{5.6.15}$$

If each of these elements is sufficiently close to unity the transmissivity factor becomes

$$T_R = 1 - \sum_i L_i - T_{1/2}, \tag{5.6.16}$$

where L_i is the loss of each resonator element, and $T_{1/2}$ the transmissivity of the symmetric resonator end mirror.

The transmission factor T_R will be arbitrarily divided into two components T_S and $R_{1/2}$, so that $T_R = T_S R_{1/2}$. The quantity T_S is the product of all the element transmissivities exclusive of $R_{1/2}$. The quantity T_S, which it is desirable to maximize, is limited by small scattering and absorption losses at the surfaces of the intraresonator optical elements. The output mirror reflectivity $R_{1/2}$ is an independent variable that will be adjusted to maximize the laser output. This laser is unique in that it is one of the few that contain no hard apertures that can diffract or vignette the laser beam. No allowance need be made therefore for any diffraction losses.

By using this notation, the steady-state condition where the gain in each stage exactly compensates for the losses occurs when

$$T_R \int_A (\phi_M + \Delta\phi_M)\, dA = \int_A \phi_M dA, \tag{5.6.17}$$

where the area integral is taken over the entire spatial extent of the laser intensity.

An inspection of Eq. (5.6.17) and its components shows that the steady-state equation is a function of four resonator parameters ϕ_{M0}, ϕ'_{X0}, ρ, and T_R in addition to the parameters associated with the lasing dye. Physically, ϕ_{M0} is the dependent variable and it is expected that optimum values for ρ and $R_{1/2}$ can be identified for selected values of T_S and ϕ'_{X0}.

Equation (5.6.17) has been evaluated numerically by using the cross-

---

[104] W. W. Rigrod, *J. Appl. Phys.* **36**, 2487 (1965).

5.6. CW LASER

TABLE III

$\sigma_{SX} = 2.6 \times 10^{-16}$ cm^2
$\sigma_{SM} = 1.3 \times 10^{-16}$ cm^2
$\sigma_{TX} = 0$
$\sigma_{TM} = 0.6 \times 10^{-16}$ cm^2
$k_{ST}\tau_T = 1.0$
$\tau_S = 4.8 \times 10^{-9}$ sec

section values for rhodamine 6G for absorption at 515 nm and laser emission at 590 nm as listed in Table III. The value of σ_{TX} is unknown and therefore has been neglected. These results are then used to calculate the laser output efficiency from the following equation:

$$E = \frac{P_O}{P_I} = \frac{(1 - R_{1/2})\int_A I_M dA}{\int_A I_X dA}, \qquad (5.6.18)$$

where P_I is the total input power and P_O the output power. It can be shown that this efficiency is related to the value of T_R obtained with Eq. (5.6.17) in the following way:

$$\frac{P_O}{P_I} = \frac{\phi_{M0}}{\phi'_{X0}} \frac{\sigma_{SX} + k_{ST}\tau_T\sigma_{TX}}{\sigma_{SM}} \rho^2 \left(\frac{1}{T_R} - \frac{1}{T_S}\right). \qquad (5.6.19)$$

This efficiency has been plotted in Fig. 35 as a function of the output mirror transmission T, where $T = 2T_{1/2} = 2(1 - R_{1/2})$. A value of 10 was selected for the excitation intensity ϕ_{X0} and three arbitrary values for the round trip resonator loss L were used in the evaluation, where $L = 2(1 - T_S)$. The loss values employed, 1.5, 4, and 10%, were selected to

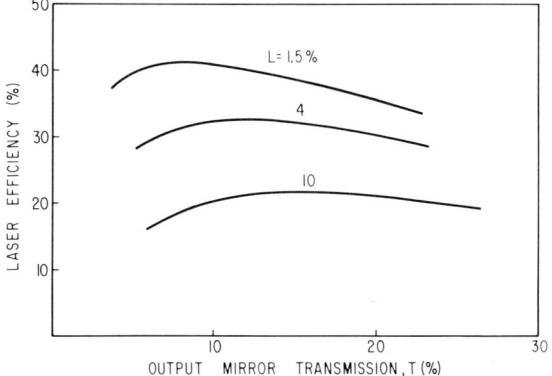

FIG. 35. cw laser efficiency as a function of output mirror transmission for three values of the resonator losses L.

span the anticipated experimental range for this parameter. The minimum value was obtained by using the rule of thumb that optical surfaces have a minimum scattering loss of 0.1% and dielectric mirrors have an additional minimum absorption of 0.1%. The curves have relatively flat maxima that occur at larger values of mirror transmission as the resonator loss increases.

These curves and the subsequent graphs are all optimized with respect to the ratio of the laser mode radius to the radius of the excitation beam $\rho = \omega_M/\omega_X$. The optimum value for this parameter is remarkably constant, as has been analytically corroborated.[104a] It was found to be within the narrow range from 1.0 to 1.2 for excitation levels of $\phi'_{X0} = 5$ or greater. For excitation levels closer to laser threshold where $\phi'_{X0} = 1$ or 2, the optimum was found to be between 0.7 and 1.0. Variations in this parameter of 0.1 altered the peak output by less than 5%.

The analysis has been extended to other values of excitation intensity and the efficiency maxima for both ρ and T are plotted in Fig. 36 as a function of the excitation intensity ϕ'_{X0} for the three previously mentioned resonator loss values. It is seen that for low values of ϕ'_{X0} the efficiency is increasing rapidly with ϕ'_{X0}. The rate of improvement diminishes rapidly after the excitation level reaches $\phi'_{X0} = 10$. The graph has been plotted for excitation values up to $\phi'_{X0} = 20$ to display the region of rapid changes in efficiency. The calculations were extended to $\phi'_{X0} = 100$, where the efficiencies were found to be 55, 37, and 27% for loss values of 1.5, 4, and 10%, respectively. These values are seen to be at most 25% higher than the efficiency at $\phi'_{X0} = 10$ but required an order of magnitude increase in excitation intensity.

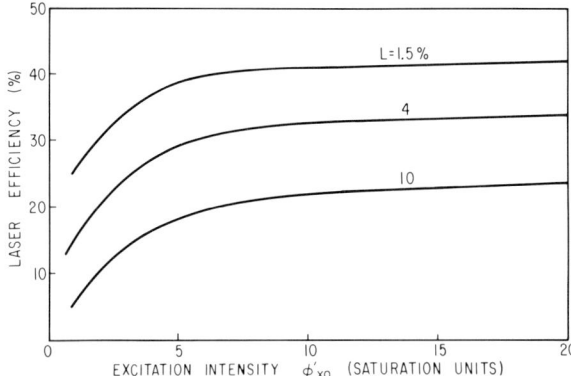

FIG. 36. cw laser efficiency at the optimum output mirror transmission as a function of excitation intensity for three values of resonator losses L.

[104a] A. Dienes, B. Couillaud, and A. Ducasse, *IEEE J. Quantum Electron.* **qe-14**, 702 (1978).

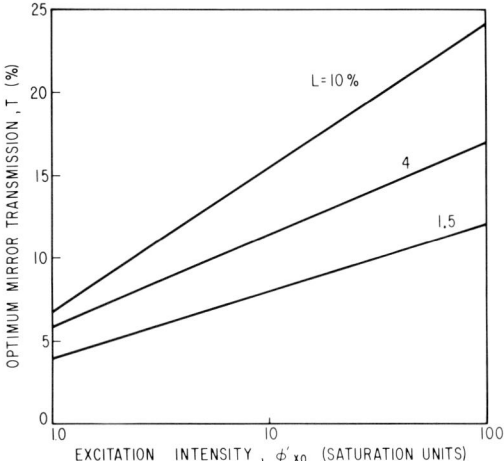

FIG. 37. Variation in the optimum output mirror transmission as a function of excitation intensity for three values of resonator losses L.

In these calculations, the output mirror reflectivity has been continuously changed to maintain optimum conditions. The value of this parameter is shown in Fig. 37 also as a function of the input intensity. It is seen that the optimum transmission of the output mirror is a monotonically increasing function of the input intensity. As was clear in Fig. 35, the mirror transmission function has a broad maximum region so that these values are to be considered only guidelines.

Additional manipulations can be performed to evaluate the distribution of the input photons into each of the loss channels, laser output, scattering losses, triplet state absorption, excited singlet absorption, and fluorescence. Figure 38 shows the fraction of the input photons that goes into each of these channels as the input intensity is varied. This figure is similar to Fig. 22 but there are significant differences between the two. It is seen that the inclusion of excited-state absorptions in this calculation has substantially altered the functional relationships. Also, in this case the beam geometries could be predicted so that fewer assumptions were required and the calculation could be performed with greater accuracy. It is instructive to note that at high excitation intensities the major loss is still the triplet absorption. This fact further emphasizes the importance of judicious selection of the organic dye to be used in the laser system.

The model that has been outlined here has been applied, with minor modification, to an experimental device to compare the predictions with actual laser performance.[9] The excited-state absorption was directly measured by monitoring the transmission of the excitation beam through

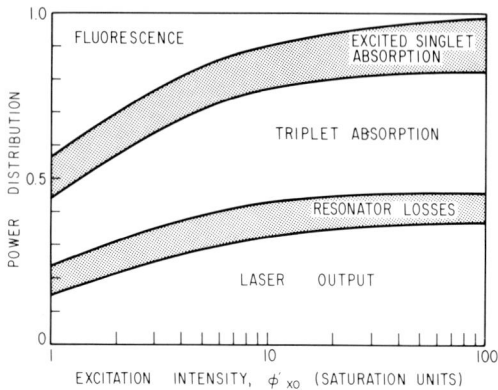

FIG. 38. Distribution of excitation photons into laser output and various cw laser losses plotted as a function of the excitation intensity.

the dye solution in the laser device. The values of this absorption in rhodamine 6G at selected wavelengths are shown in Fig. 2.

These absorption values were used to predict the laser output power. The results of the confrontation between the model and experiment are displayed in Fig. 39, where both the transmission of the excitation beam and the output power are shown as a function of input power. The transmission is seen to change significantly depending on whether or not the dye solution is lasing. The close comparison between the predicted laser output and the experimentally realized output power is most rewarding.

Machine calculations have also been performed on the cw dye laser system by assuming a constant output coupling so that the modeling prediction can be compared with an experimental device.[93,101] The experimental laser used for comparison purposes had a minimum number of optical surfaces and consequently a minimum of resonator element losses. It was similar to Fig. 34A and was composed of two mirrors and a lens. The space between the flat mirror and the lens was filled with dye solution, which helped index match one surface of the lens and made a separate dye cell unnecessary. In this configuration, only half of the confocal region is available for use as the gain medium. For this design it was necessary to extend the calculation to include the region of beam divergence in the modeling. The calculation required three-dimensional numerical integration of the gain relation and iterative solutions to match the boundary conditions. The complexity of the calculation was reduced dramatically by using Gaussian–Laguerre quadrature to evaluate the integrals. The excited singlet-state absorption was not included in this model since measurements of this quantity were not available at that time. However,

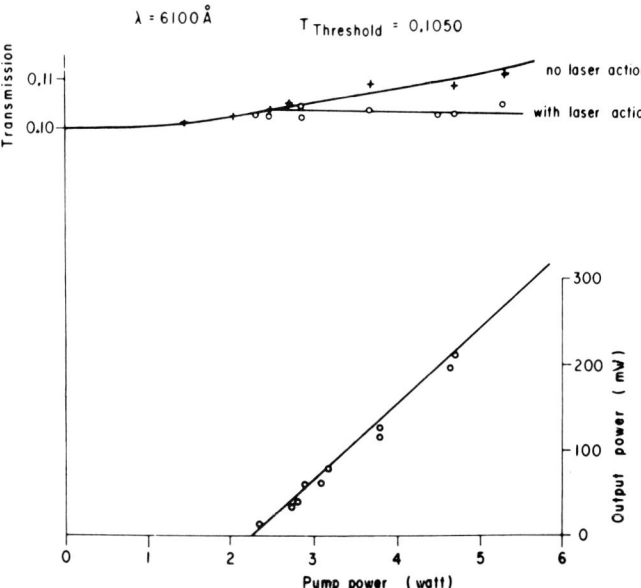

FIG. 39. Dye cell transmission and output power vs. pump power for the 514.5-nm argon laser excitation. Solid lines are theoretical curves matched to the experimental points.[9]

fluorescence quantum yield measurements were performed on the dye under simulated laser operating conditions, which partially corrected for this omission. The result was a remarkable agreement between the model prediction and the measured characteristics of the cw dye laser.

The model predictions are displayed in Fig. 40. In Fig. 40A the power going into laser output and selected losses are shown. In this case resonator losses other than output transmission were neglected. The comparison between the model and measurements on the experimental device is illustrated in Fig. 40B. In this case all resonator losses were carefully measured and included in the calculation. Again, excellent agreement was obtained between model predictions and experimental results.

In both of the model calculations discussed it was found necessary to use a fluorescence lifetime value that reflects concentration quenching of the excited molecules. The value of the lifetime at high concentrations τ_S' can be calculated from the relation[93]

$$1/\tau_S' = 1/\tau_S + K_Q n, \qquad (5.6.20)$$

where τ_S is the intrinsic lifetime and K_Q the concentration quenching rate constant that was found to be 1.35×10^{-10} cm^3 sec^{-1} for rhodamine 6G.

The excellent agreement between model predictions and experimental

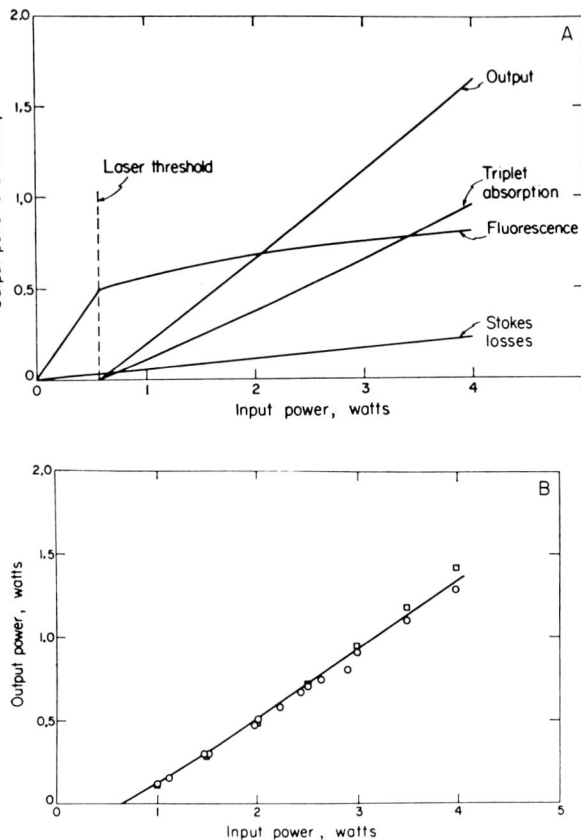

Fig. 40. (A) Calculated rhodamine 6G dye laser output at 593 nm and losses as a function of input power at 515 nm for a 1-mm dye path. Output mirror transmittance, 0.15. Scattering and reflection losses neglected. (B) Calculated and measured output of dye laser with 1-mm dye path. Rhodamine 6G, 4×10^{-4} M in water, plus 5% Ammonyx LO. Input, argon laser line at 515 nm; output, 593 nm; output mirror transmittance, 0.148; effective reflectance, 0.829; flat mirror reflectance, 0.993; circles, no COT; squares, 10^{-2} M COT; continuous curve, calculation.[101]

devices gives dramatic evidence that the operation of the dye laser is understood and its characteristics can be predicted with confidence. Accurate calculations, of course, are predicated on the measurement of the characteristics of the laser dye to be used in the system.

5.6.2. CW Resonator Geometry

In Section 5.2.3, it was shown that laser threshold for rhodamine 6G required an excitation level of approximately 85 kW/cm² of laser mode cross section. Since the large ion lasers normally used to excite cw dye lasers have output powers in the range of a few watts, it is clear that the dye laser resonator must have an active volume with a very small cross section. A first-order estimate shows that this cross section must have a diameter less than 100 μm. To obtain mode diameters of that small size necessitates special optical designs. The essential elements of such a design have been illustrated in Fig. 34. The resonator contains at least one element of high optical power or short focal length to focus the laser emission back through the amplifying region. In the schematic, a short focal length lens, e.g., a microscope objective, has been used for this purpose in one case[93] and in the other case the lens has been replaced by a concave mirror of equivalent focal length.[102] The laser active dye solution is placed at the point of minimum mode diameter where the emission intensity will be at its maximum value. The excitation must be focused into the active volume defined by the resonator mirrors.

The first successful dye laser contained only two optical elements, as is seen in Fig. 41.[105] In this case, the output mirror M_O, which had a 4.5-mm radius of curvature, also served as the focusing element. The back mirror M_B was coated on a sapphire flat to conduct the heat away from the active region. The back mirror dielectric coatings were designed to be totally reflective at the dye laser wavelength and partially transmitting at the excitation wavelength. The volume between the two mirrors was completely filled with the dye solution. The mirror spacing was adjusted by inserting shims between the components of the laser.

This design cannot contain any wavelength-selective elements because of its small size. Since the one outstanding attribute of dye lasers is the wavelength tunability of the output, it is clearly necessary to extend the resonator. The three-element design permits the resonator to be extended almost without limit. This extension is achieved by replacing the focusing mirror with a thick mirror composed of a mirror and a focusing element.[106] In this manner, an equivalent mirror is made whose focal length is often much shorter than focal lengths that can be obtained from commercially available mirrors. In the region between the two components of the thick mirror the laser beam will be nearly collimated and will have a relatively large mode diameter. Both of these conditions are necessary for efficient and precise wavelength control when using most of the common wavelength-tuning elements. In addition, the resultant short focal length of the equivalent mirror greatly increases the stability of the laser resonator. This resonator is sufficiently stable that once the fo-

---

[105] O. G. Peterson, S. A. Tuccio, and B. B. Snavely, *Appl. Phys. Lett.* **17**, 245 (1970).
[106] H. Kogelnik, *Bell Syst. Tech. J.* **44**, 455 (1965).

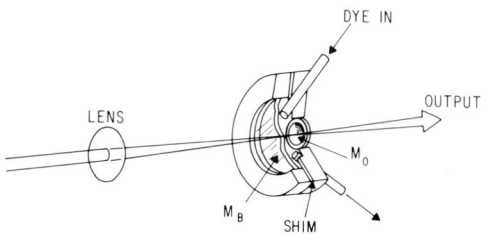

FIG. 41. Original cw dye laser schematic. The resonator had a hemispherical geometry composed of one mirror M_B coated on a sapphire flat and a curved mirror M_0 with a very short radius of curvature. The mirror spacing was adjusted with shims and the volume between the mirrors was filled with dye solution.[105]

cusing element is adjusted with respect to the back mirror, the output mirror, in some cases, can be positioned and aligned by hand without the aid of micrometer adjustments.

5.6.3. Resonator Stability Criteria

Resonators with such small-mode radii are most easily analyzed with the aid of the concept of stability circles.[107,108] Excellent discussion and derivation of the concept are available elsewhere. They state that any mirror can be conceptually replaced by a stability circle whose diameter is equal to the radius of curvature of the mirror. This circle is positioned so that it passes through both the center of curvature and the vertex of the mirror. These circles can be used to determine whether or not an assembly of two mirrors will produce a stable resonator. The stability criterion requires that the circles representing each mirror must intersect. The minimum waist of the stable mode will be located at the intersection of the optical axis and the chord connecting the two intersection points of the circles. The length of this chord is a measure of the minimum mode diameter, which can be calculated from

$$h = \omega_0^2 \pi / \lambda, \tag{5.6.21}$$

where ω_0 is the lowest-order Gaussian mode radius to the $1/e^2$ point, λ is the wavelength of the light, and h is one-half of the chord length and also equal to the confocal parameter as defined in Eq. (5.6.10).

An example of this construction is shown in Fig. 42 for the case of two mirrors of unequal radii of curvature. It is clear from the construction that small-mode radii can be obtained by two alternative schemes. One method is to select one of the mirrors with a radius of curvature suffi-

[107] D. C. Sinclair and W. E. Bell, "Gas Laser Technology." Holt, Rinehart, and Winston, New York, 1969.
[108] P. Laurès, *Appl. Opt.* **6,** 747 (1967).

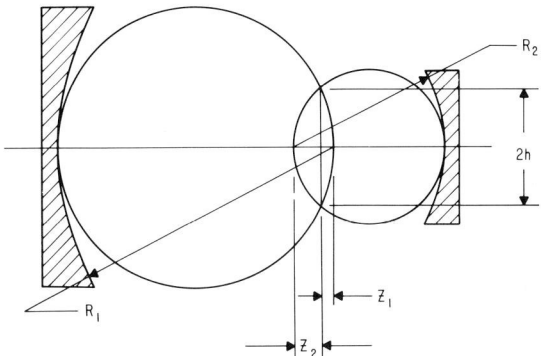

FIG. 42. Two mirror resonator including the stability circles characteristic of each mirror.

ciently small that the desired waist radius can be obtained when the stability circle for the small radius of curvature mirror is bisected. This technique is employed in designing three-element resonators where the equivalent mirrors can have radii of curvature in the millimeter range.

The second method that can be used to produce small-waist radii depends on minimizing the overlap of the stability circles. Therefore, the parameter Z must become very small. In this region of the circle the slope is approaching infinity so that very small changes in Z yield large changes in the mode radius. This method was used to control the mode radius in the device illustrated in Fig. 41.

The mode radius that is produced at this waist or minimum point in the resonator mode, can be calculated by using Eq. (5.6.21). The resulting relation is

$$\omega_0^4 = (\lambda/\pi)^2 Z_i (R_i - Z_i), \qquad (5.6.22)$$

where the indices have been included to ensure that the R and Z values for only one mirror are used in the equation. This relation is the familiar equation for the size of the mode at the flat mirror of a hemispherical resonator. The value of this mode radius in free space is plotted in Fig. 43 as a function of the distance Z for three values of mirror radii and with $\lambda = 590$ nm. The point at which the mode radius is at a maximum for a given mirror radius of curvature is also the point of maximum positional stability. This point corresponds to the intersection of the stability circle at its midpoint. It is obvious from the graph that extreme precision is required in the resonator dimension adjustments to maintain a small-radius mode waist with mirrors of radius of curvature equal to 10 mm or more. Again it is seen that a small radius of curvature mirror or equivalent mirror is not only desirable but necessary.

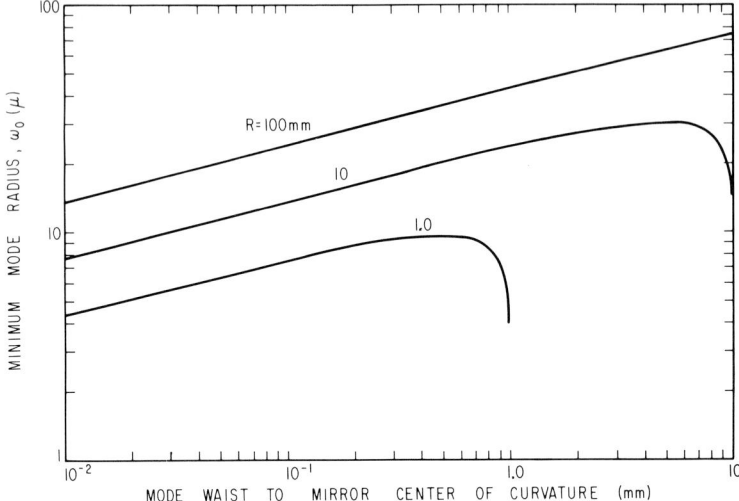

FIG. 43. Minimum resonator mode radius as a function of relative position of the minimum waist and a curved mirror for three values of the radius of curvature R of the curved mirror. The relative position is measured from the center of curvature of the curved mirror to the position of the minimum waist as fixed by the resonator geometry.

Once the position and size of the mode waist are determined the $1/e^2$ radius ω of the mode and the radius of curvature R_c of the wavefront can be calculated at any point a distance Z away from the waist. This calculation can be performed using the usual diffraction formulas[58]

$$\omega^2 = \omega_0^2 \left[1 + \left(\frac{\lambda Z}{\pi \omega_0^2}\right)^2\right], \quad (5.6.23)$$

$$R_c = Z \left[1 + \left(\frac{\pi \omega_0^2}{\lambda Z}\right)^2\right], \quad (5.6.24)$$

$$\theta_{1/2} = \lambda/\pi\omega_0, \quad (5.6.25)$$

$$b = \pi\omega_0^2/\lambda. \quad (5.6.26)$$

The quantity $\theta_{1/2}$ is the half-angle that the diverging beam approaches asymptotically and b is the confocal parameter, which is the distance to the point where the beam cross-sectional area has doubled from its value at ω_0. This point is also that at which R_c is at a minimum. It is to be noted that R_c is infinite both for Z equal to zero and infinity. The geometric relationships between these quantities are illustrated in Fig. 44.

It has been mentioned previously that the ideal resonator will include a thick mirror that can produce an equivalent mirror of very small radius of

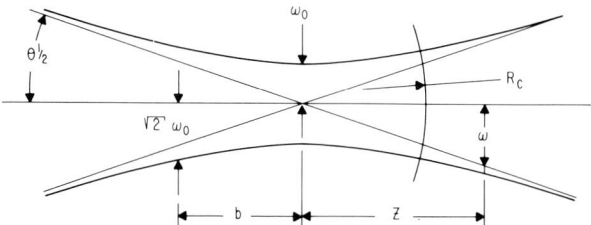

FIG. 44. Geometric relationships between parameters characteristic of the divergence of a Gaussian beam.

curvature. The thick mirror consists of a mirror in combination with a focusing element. It has been shown that such a combination can be conceptually replaced with an equivalent mirror whose position and radius of curvature can be determined by simple imaging techniques. It was shown that the centers of curvature and vertices of the real and equivalent mirrors are images of each other.[106] This situation makes the calculation of the equivalent mirror radius of curvature R_e a trivial matter. As an illustration, the combination of a flat mirror and a lens has been shown schematically in Fig. 45. The image of the center of curvature of the real mirror will be at the lens focal point, a distance f from the lens. The image of the vertex will be at a distance

$$S_v' = S_v f/(S_v - f'), \qquad (5.6.27)$$

where the prime indicates a distance in image space. The equivalent mirror radius of curvature is the difference between these two:

$$R_e = f^2/(S_v - f). \qquad (5.6.28)$$

The generalization of this case to that of a curved output mirror requires some additional though trivial algebra.

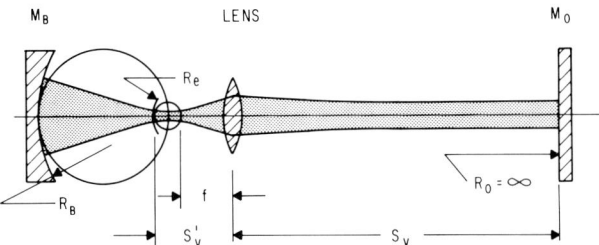

FIG. 45. Three-element, thick mirror resonator schematic including stability circles for the back mirror M_B and the equivalent mirror of radius R_e. The equivalent mirror is representative of the thick mirror composed of the lens of focal length f and the output mirror M_0 positioned a distance S_V from the lens.

A stable resonator configuration can now be assembled by forcing the stability circle for the back mirror to intersect the stability circle for the equivalent mirror. This configuration is also shown in Fig. 45. The resonator end mirrors have been defined according to common usage at the present time. It is seen in the diagram and from Eq. (5.6.28) that the equivalent mirror can have a very small radius of curvature. The equivalent mirror therefore becomes the dominant element in fixing the mode waist radius. The preferred scheme for obtaining a desired waist radius then involves estimating the probable value of S_v and selecting the focal length of the focusing element so that the desired waist radius will be obtained when the resonator is aligned in its most stable configuration. As discussed previously, maximum stability will be achieved when the back mirror is positioned with respect to the equivalent mirror such that the stability circle for the equivalent mirror is exactly bisected by the corresponding circle for the back mirror. In this configuration the waist radius can be calculated from

$$\omega_0^2 = \frac{\lambda}{2\pi} \frac{f^2}{S_v - f}, \qquad (5.6.29)$$

$$= \frac{\lambda}{2\pi} \frac{R_0 f^2}{(S_v - f)(S_v - R_0 - f)}, \qquad (5.6.30)$$

where Eq. (5.6.29) relates to Fig. 45, where the output mirror is flat, and Eq. (5.6.30) is the more general relation for a system with a curved output mirror of radius of curvature R_0. If necessary, fine tuning of the mode size can be achieved by adjusting S_v.

This resonator will remain stable with minor changes in the waist radius for variations in the focusing element to back mirror separation that do not exceed $R_e/2$. The waist radius will decrease as the focusing element to back mirror spacing is changed by a small amount $\pm \delta$ according to the following relation:

$$\omega_0^4 = \omega_{0c}^4 [1 - (2\delta/R_e)^2], \qquad (5.6.31)$$

where ω_{0c} is defined for this equation as the confocal or maximum value for ω_0 that can be obtained for a selected f and S_v.

The beam size can be determined at other points in the resonator by using the divergence relations, Eqs. (5.6.23)–(5.6.26), and normal lens-imaging formulas. The radius of curvature of the wavefront must match the radius of curvature of each end mirror. There will be beam waists on each side of the focusing element but these will not be images of each other. For the case where the output mirror is flat, the second waist must be at this plane mirror. There are more sophisticated and precise methods for propagating the beam through optical elements by using complex beam parameters and ray matrices.[106] However, the simple concepts and formulas presented here are sufficient for most design calculations.

5.6.4. Astigmatic Resonator

An important special case of the three-element resonator class uses a short radius of curvature spherical mirror in place of the internal lens.[102,109] The resonators are exactly equivalent with one important exception—the focusing mirror must be used slightly off axis as illustrated in Fig. 46. This small deflection angle introduces astigmatism into the focus of the resonator. This astigmatism can be used to advantage to compensate astigmatism that would otherwise be introduced into the resonator by the use of a dye cell with Brewster's-angle windows. The use of a mirror for this element can reduce the resonator losses since the transmission of an antireflection-coated lens is usually less than the reflection of a high-reflectivity mirror. Also, the focal length of the mirror is always achromatic in contrast to the simple low-loss lenses that would otherwise be used for this element.

The small deviation angle, where the half-angle is defined as θ, gives the focusing mirror two focal lengths: f_T for the tangential bundle of rays that

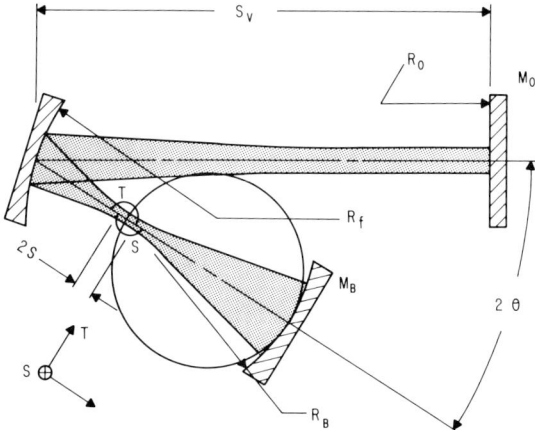

FIG. 46. Schematic of three-element resonator using curved mirror as the focusing element. The stability circle for the back mirror M_B is shown intersecting the stability circles for the equivalent mirrors produced by both the sagittal S and tangential T focus of the folding mirror.

[109] D. C. Hanna, *IEEE J. Quantum Electron.* **qe-5**, 483 (1969).

lie in the plane of Fig. 46 and f_S for the sagittal bundle of rays that lie outside the plane of the illustration. These focal lengths are given by the following relations[92]:

$$f_T = \tfrac{1}{2} R_f \cos \theta, \qquad (5.6.32)$$

$$f_S = \frac{R_f}{2 \cos \theta}, \qquad (5.6.33)$$

where R_f is the radius of curvature of the focusing mirror. Each of these foci will produce an equivalent mirror that will have a corresponding stability circle. For stable laser operation, both sets of rays must form stable configurations. Therefore, it is necessary that the stability circle for the back mirror intersect the stability circles for both foci. For the illustrated resonator, therefore, the sagittal and tangential stability circles must overlap each other. This condition is illustrated in Fig. 46 with an exaggerated geometry. Only half of each stability circle has been shown. The diagram shows the desired overlap of the three circles of stability and demonstrates that the range of stability for positioning the mirrors can become severely limited if the overlap of the sagittal and tangential stability circles is reduced. For a particular set of resonator elements there will be a maximum value of θ for which a stable resonator can be assembled without other compensation. It is clear from the figure and equations that maximum stability can be achieved for a resonator with minimal compensation with a deflection angle θ as close to zero as is consistent with fabrication restraints.

The magnitude of the stability range $2S$ can be calculated from the following relations, which are applicable to the two regions in the parameter space where the thick mirror can be used to advantage to produce small mode waists:

$$2S = \begin{cases} \dfrac{(S_v - R_0) f_T}{S_v - R_0 - f_T} - \dfrac{S_v f_S}{S_v - f_S} & \text{for } S_v - R_0 - f_S > 0, \quad (5.6.34) \\[1em] \dfrac{S_v f_T}{S_v - f_T} - \dfrac{(S_v - R_0) f_S}{S_v - R_0 - f_S} & \text{for } \begin{array}{l} S_v - R_0 < 0, \\ \text{and } S_v - f_S > 0. \end{array} \quad (5.6.35) \end{cases}$$

For the special case of a flat output mirror, the stability range becomes

$$2S = f_S \left(\frac{f_T - S_v \sin^2 \theta}{S_v - f_T} \right). \qquad (5.6.36)$$

5.6.5. Dye Cell Astigmatism

As was previously mentioned, it is possible to use a dye cell with Brewster's-angle windows at the mode waist and to adjust the focusing mirror deviation angle to compensate exactly for the astigmatism introduced by the dye cell. The dye cell, or more probably, the jet of dye solution introduces additional optical pathlength as compared to the free-space optical pathlength. As was previously discussed, the spacing between the back and focusing mirrors is extremely critical for determining the laser characteristics. Therefore, any optical elements inserted in this region can have a dramatic impact on the laser operation. The additional effective pathlength will be different for the rays lying in the plane of incidence than for those outside the plane of incidence. The cell geometry is illustrated in Fig. 47. The cell or jet thickness is labeled w, the index of refraction η, the incident angle ψ_B, and the refracted angle ψ_η. The distance that the light travels in the cell is defined as l, and the distance in free space that the cell displaced is defined as U. These parameters are related to each other by Snell's law of refraction and the definition of Brewster's angle:

$$\sin \psi_B = \eta \sin \psi_\eta, \tag{5.6.37}$$

$$\tan \psi_B = \eta, \tag{5.6.38}$$

$$l = w(\eta^2 + 1)^{1/2}/\eta, \tag{5.6.39}$$

$$U = 2w/(\eta^2 + 1)^{1/2}. \tag{5.6.40}$$

To calculate the effect of the cell, it will be assumed that the mode waist occurs at the incident surface. As the beam passes into the cell, it will be magnified in the T dimension and remain constant in the S dimension:

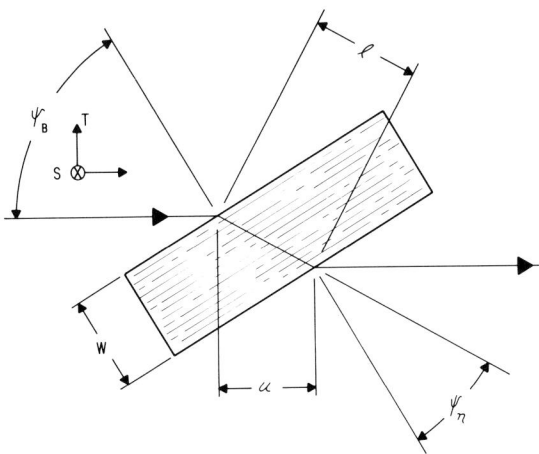

FIG. 47. Geometric relationships between parameters used for calculation of astigmatism of Brewster's-angle dye cell.[102]

$$\omega_{0S} = \omega_0, \tag{5.6.41}$$

$$\omega_{0T} = \omega_0 \frac{\cos \psi_\eta}{\cos \psi_B} = \eta\omega_0. \tag{5.6.42}$$

The beam will expand at different rates in the two dimensions according to Eq. (5.6.23) so that after the beam has traversed the optical pathlength through the cell, the beam radii become

$$\omega_{0S}^2 = \omega_0^2 \left[1 + \left(\frac{\lambda l}{\eta \pi \omega_0^2} \right)^2 \right], \tag{5.6.43}$$

$$\omega_{0T}^2 = \eta^2 \omega_0^2 \left[1 + \left(\frac{\lambda l}{\eta \pi \eta^2 \omega_0^2} \right)^2 \right], \tag{5.6.44}$$

where the reduced wavelength λ/η in the cell is also included. After leaving the cell ω_{0T} will be demagnified by $1/\eta$. It is seen by referring these two equations to Eq. (5.6.23) that the effective distance U_i the rays have traversed in the cell is

$$U_S = l/\eta = w(\eta^2 + 1)/\eta^2, \tag{5.6.45}$$

$$U_T = l/\eta^3 = w(\eta^2 + 1)^{1/2}/\eta^4. \tag{5.6.46}$$

These relations can be derived by alternative techniques using Snell's law or Huygen's principle.[109] It is seen that the optical pathlength for the rays reflected from the focusing mirror in the tangential plane has been shortened with respect to the pathlength for the sagittal rays. The optical pathlength change is equivalent to moving the back mirror of the resonator closer to the focusing mirror selectively for the tangential rays as is desired to compensate for the astigmatism introduced by the focusing mirror. The point of compensation occurs where the dye cell decreases the optical pathlength for the tangential rays by the same amount that the mirror astigmatism has decreased the focal length for these rays. This is easily calculated from the following equation:

$$U_S - U_T = f_S - f_T, \tag{5.6.47}$$

which becomes after substitution

$$2N_A w = R_f \tan \theta \sin \theta, \tag{5.6.48}$$

where the factors dependent only on the index of refraction have been collected together:

$$N_A = \frac{(\eta^2 - 1)(\eta^2 + 1)^{1/2}}{\eta^4}. \tag{5.6.49}$$

This function varies slowly with η, whose values are listed below.

η	1.3	1.4	1.5	1.6	1.7
N_A	0.396	0.430	0.445	0.449	0.446

The compensation angle has been calculated and displayed in Fig. 48 as a function of the resonator parameters w/R_f for two values of N_A corresponding to an indices of refraction of $\eta = 1.335$ and 1.76, which yield values of N_A of 0.411 and 0.442, respectively. These indices were selected to represent the extremes in range of index of refraction for commonly used materials. The low index is characteristic of dye solvents and the high index is for sapphire (αAl_2O_3), which is often used as a window or substrate material because of its high thermal conductivity. It is seen in the figure that commonly used resonator parameters, $w = 0.25$ mm and $R_f = 100$ mm, require very small compensation angles $\theta = 3°$.

5.6.6. Mode Area

It is necessary to calculate the cross-sectional area of the mode within the dye cell. This area is the critical parameter for the calculation of laser characteristics. It will be elliptical in shape and to a good approximation is equal to

$$A = \pi \omega_{0S} \omega_{0T} = \pi \eta \omega_0^2, \quad (5.6.50)$$

where ω_0 is the free-space mode radius calculated at the confocal resonator configuration as diagrammed in Fig. 46.

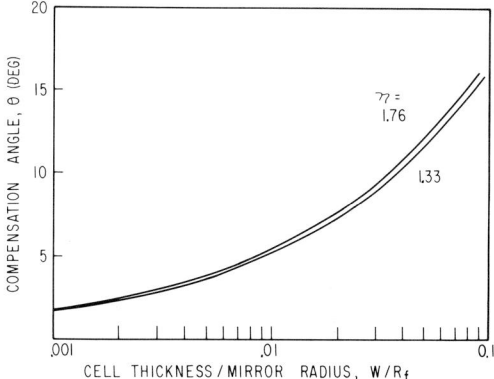

FIG. 48. Compensation angle θ required for ratio of cell thickness w to folding mirror radius of curvature R_f. At the compensation angle the astigmatism introduced by the folding mirror is exactly compensated by the astigmatism of the Brewster's-angle dye cell. The curves are plotted for two values of the dye cell index of refraction η.

For a more accurate determination it is necessary to find the location of the waists for both the sagittal and tangential bundle of rays within the dye cell and to calculate the area as a function of position. To find this position it will be assumed that the intraresonator beam is nearly circular in cross section throughout the resonator except for the region within the dye cell. This approximation is adequate for a completely compensated resonator. The effective distance from the back mirror to each waist therefore must be identical, which means that the cell surface to waist effective distances also must be identical. The relationship between the effective distances and the cell thickness has been presented in Eqs. (5.6.45) and (5.6.46). It will be assumed in this case that the waists are symmetrically located with respect to the center of the cell. It can then be shown that the waists will be located at positions l_T and l_S along the propagation direction, where[102]

$$l_T = w\eta/(\eta^2 + 1)^{1/2}, \qquad (5.6.51)$$

$$l_S = w/\eta(\eta^2 + 1)^{1/2}. \qquad (5.6.52)$$

Evaluation of l_T and l_S over the normal range of refractive indices shows that the symmetrically positioned waists are at a distance from the cell center that is between one-sixth and one-fourth of the total distance through the cell.

Using Eqs. (5.6.50)–(5.6.52), the general expression for the cross-sectional area within the cell becomes

$$A(l) = \pi\eta\omega_0^2 \left[1 + \left(\frac{\lambda}{\pi\eta\omega_0^2}\right)^2 (l' - l_S)^2\right]^{1/2}$$
$$\times \left[1 + \left(\frac{\lambda}{\pi\eta^3\omega_0^2}\right)^2 (l' - l_T)^2\right]^{1/2} \qquad (5.6.53)$$
$$= \pi\eta\omega_0^2 \left[1 + \left(\frac{w}{b}\right)^2 \left(\frac{\eta^2 + 1}{\eta^4}\right)\left(x - \frac{1}{\eta^2 + 1}\right)^2\right]^{1/2}$$
$$\times \left[1 + \left(\frac{w}{b}\right)^2 \left(\frac{\eta^2 + 1}{\eta^8}\right)\left(x - \frac{\eta^2}{\eta^2 + 1}\right)^2\right]^{1/2}, \qquad (5.6.54)$$

where l' is defined as the length variable along the propagation direction, x is the fractional distance through the cell, and b is the confocal parameter for the free-space mode-waist radius, $b = \omega_0^2\pi/\lambda$. This area has been plotted in Fig. 49 together with the free-space area, as a function of the fractional distance through the cell. The areas have been normalized to the minimum free-space value $\pi\omega_0^2$ and the cell thickness has been chosen to be twice the confocal parameter. An index of refraction of 1.33 was selected for the example. It is seen that the minimum area within the cell

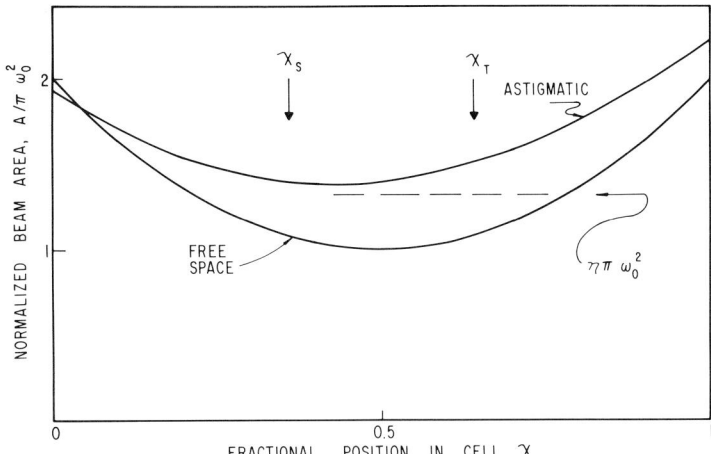

FIG. 49. Cross-sectional area of beam within dye cell normalized with respect to the free space value $\pi\omega_0^2$. The position of the sagittal waist is labeled x_S and the position of the tangential waist is labeled x_T. The curve was generated for the special condition of the cell thickness being equal to twice the confocal parameter b for the mirror. The cell index of refraction was assumed to be 1.33.

is larger than the free-space value by approximately a factor of η. However, the divergence of the area is less because of the separation between the sagittal and tangential waists.

It is clear from inspection of Eqs. (5.6.54) and (5.6.53) and Fig. 49 that the minimum area occurs very near to $l = l_s$ or, equivalently, $x = 1/(\eta^2 + 1)$. The area at this point is

$$A_{\min} = \pi\eta\omega_0^2 \left[1 + \left(\frac{w}{b}\frac{\eta^2 - 1}{\eta^4}\right)^2 \frac{1}{\eta^2 + 1}\right]^{1/2} \quad (5.6.55)$$

$$= \pi\eta\omega_0^2 \left[1 + \left(\frac{\lambda w}{\pi\omega_0^2}\frac{\eta^2 - 1}{\eta^4}\right)^2 \frac{1}{\eta^2 + 1}\right]^{1/2} \quad (5.6.56)$$

As the cell thickness is increased, this minimum area asymptotically approaches

$$A_{\min} \longrightarrow \frac{\lambda w}{\eta^3} \frac{\eta^2 - 1}{(\eta^2 + 1)^{1/2}}. \quad (5.6.57)$$

In this limit, the minimum area has become completely independent of the focusing elements in the resonator. Thus, even though the astigmatism of the focusing mirror is compensated by the cell, the primary objective of small mode cross sections for which these resonators are designed cannot be achieved.

Equation (5.6.55) can be rearranged so that it can be expressed in terms of normalized parameters:

$$\frac{A}{\eta \pi \omega_0^2} = \left[1 + \left(\frac{w}{bN_t}\right)^2\right]^{1/2}, \qquad (5.6.58)$$

where N_t is a parameter dependent only on the index of refraction:

$$N_t = \frac{\eta^4(\eta^2 + 1)^{1/2}}{\eta^2 - 1}. \qquad (5.6.59)$$

Values of N_t are given in the following tabulation:

η	1.3	1.4	1.5	1.6	1.7	1.8
N_T	6.79	6.88	7.30	7.93	8.72	9.65

Equation (5.6.58) shows the increase in the minimum mode area over that which would be obtained for an infinitely thin cell $\eta \pi \omega_0^2$ as a function of the cell thickness to confocal parameter ratio w/b. This function has been plotted in Fig. 50 for the two previously selected indices of refraction, 1.335 and 1.76. It is seen that relatively large cell thicknesses are needed to enlarge significantly the minimum beam area. In most applications, this enlargement will not be of concern. The active volume is usually restricted to the confocal region around the minimum waist. Ex-

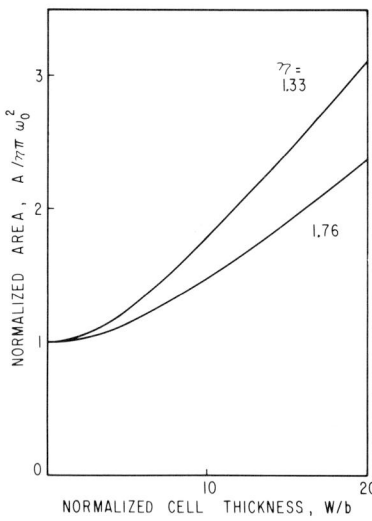

FIG. 50. Increase in minimum beam cross-sectional area as a function of the ratio of cell thickness to thick mirror confocal parameter w/b.

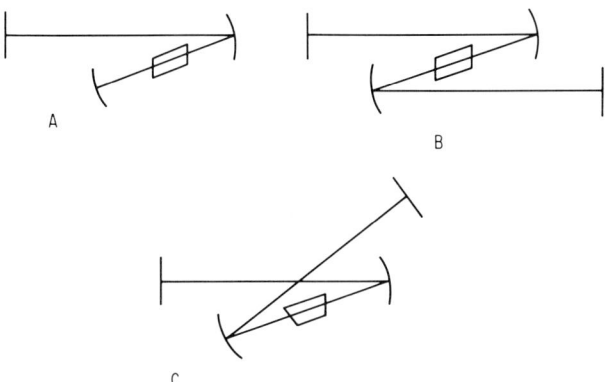

FIG. 51. Alternative geometries for astigmatic resonators displaying geometries that yield minimum coma.

cited dye molecules outside this region are relatively ineffectual because the beam intensity is rapidly decreasing due to divergence. Therefore, the useful cell thickness is equal to $2b$. Any additional material will be nonactive such as cell windows or extra dye solution included either for astigmatic compensation or for facilitation of handling of the dye solution.

5.6.7. Alternative Geometries

Alternative geometries have been suggested for more complete compensation of the aberrations introduced by the off-axis mirrors and Brewster's-angle surfaces.[110] Some of these are schematically shown in Fig. 51. The resonator in Fig. 51A is the one that has been previously analyzed in which the astigmatism introduced by the focusing mirror is compensated by passage of the beam through two Brewster's angle surfaces. Equation (5.6.48) defines the conditions at which the resonator is astigmatically compensated. Even when the resonator is compensated, the focus within the active volume is asymmetric. By adding an additional off-axis mirror to the resonator, a condition can be found where each mirror is compensated by a single Brewster's-angle surface. In this geometry both the resonator and the focal volume can be compensated simultaneously. The conditions for compensation can be calculated from relation (5.6.48) if the cell thickness is replaced by $1/2w$, thus

$$N_A w = R_f \tan \theta \sin \theta. \tag{5.6.60}$$

Configurations B and C also permit compensation for coma. The astigmatism of the mirror is compensated by either a positive or negative

---

[110] M. H. Dunn and A. I. Ferguson, *Opt. Commun.* **20**, 214 (1977).

Brewster's angle. This is not true for the coma compensation. Only those angles illustrated will allow the contribution from the two elements to cancel rather than to add. It is seen that the Brewster's angle illustrated in Fig. 51A as the angle yielding the minimum effect of coma is of the opposite sign from the commonly used angle illustrated in Fig. 34B.

The increase in the mode waist δ_y and the optical pathlength δ_z due to tangential coma are given by the following relations:

$$\delta_y = \frac{3\eta}{8} \left[R_f \sin \theta - \frac{2w(1 + \eta^2)^{1/2}(\eta^4 - 1)}{\eta^7} \right] \left(\frac{\lambda}{\pi \omega_0}\right)^2, \quad (5.6.61)$$

$$\delta_z = -\frac{1}{4} \left[R_f \sin \theta - \frac{2w(1 + \eta^2)^{1/2}(\eta^4 - 1)}{\eta^7} \right] \left(\frac{\lambda}{\pi \omega_0}\right)^3, \quad (5.6.62)$$

where the terms are the same as used in the astigmatism calculation.

It is clear from these relations that both effects are compensated when

$$R_f \sin \theta = 2w \cdot \frac{(\eta^2 + 1)^{1/2}(\eta^4 - 1)}{\eta^7}. \quad (5.6.63)$$

Equation (5.6.63) has the same form as the condition for astigmatism compensation [Eq. (5.6.48)] so it is seen that there is a specific angle at which both aberrations can be compensated simultaneously. This angle can be evaluated from

$$\tan \theta_{ac} = \frac{1}{2} \frac{\eta^3}{\eta^2 + 1}. \quad (5.6.64)$$

The variation of this angle with index of refraction is given in the following tabulation:

η	1.2	1.3	1.4	1.5	1.6	1.7	1.8	1.9
θ_{ac}	19.5	22.2	24.9	27.4	29.9	32.3	34.5	36.6

By inserting commonly used values characteristic of dye lasers for the parameters in Eqs. (5.6.61) and (5.6.62) it can be shown that coma can be ignored in most situations. Coma becomes significant when the divergence angle is very large, particularly when the mirror radius of curvature or the cell thickness is unusually large. These conditions are more probable in resonators that focus the laser intensity into a nonlinear element than into the dye-laser-active volume.

Coma could become an important consideration for special-purpose resonators. An example of such a resonator would be a four-element resonator with unequal focusing mirror radii of curvature. Such a configuration could yield a greatly expanded intraresonator beam in one of its arms.

This expanded beam could be used to advantage for obtaining much improved precision and efficiency in the wavelength tuning of the laser. One arm of such a device would be used for selecting the wavelength of the laser and the other arm would control the mode waist dimensions.

An additional example of a device that would require careful consideration of the effects of coma would be one that would purposely produce an astigmatic mode waist. Such an asymmetric mode waist should allow the laser to operate at significantly higher power levels than symmetric waist devices by reducing the dye solution transit time through the active volume. Obviously the dye would be transported through the active region in the short direction. In this case the sagittal and tangential modes must be stabilized separately. This stabilization may require a three-dimensional configuration with an arm of the device in each of the orthogonal x, y, and z directions. Alternatively, it is conceivable that a single mirror can simultaneously satisfy the stability criteria for two equivalent mirrors well separated in space. In this case, particularly, coma considerations would be important.

5.6.8. Dye System for cw Lasers

Early cw dye lasers contained the rapidly flowing dye solution in dye cells. It was realized from the start that the optical element through which the excitation entered the dye solution would need to be fabricated from a high-thermal-conductivity material. Crystalline quartz and sapphire were the commonly used materials, with the latter being the preferred one because of its unusually high conductivity. It was soon discovered, however, that the liquid–solid interface through which this excitation passed collected decomposition products from the organic materials. These materials accumulated and rather quickly burned this surface.

Fortunately, it was found that the surface of a small free-flowing jet of dye solution had sufficiently good optical characteristics so that it could be used in place of the dye cell.[111] The excitation could now be made incident on the dye solution through a liquid–air interface. The problems of decomposition product accumulation and heat storage in the window material were totally eliminated. In exchange, problems with the surface quality of the jet required careful attention. If the solution flow rate could be kept below the transition from laminar to turbulent flow, surface waves introduced by the nozzle would damp out as the function of distance away from the nozzle. Above this transition the surface imperfection would grow as a function of distance. Nozzles can be operated in both regimes.

[111] W. D. Johnston and P. K. Runge, *IEEE J. Quantum Electron.* **qe-8,** 724 (1972).

354   5. DYE LASERS

However, it is obvious that a nozzle that is to be used for producing a high-velocity jet where the fluid flow is turbulent must be much more carefully designed and fabricated.

Many of the characteristics of liquid jets were discussed in Section 5.5.10 for flashlamp-excited laser applications. In cw laser designs, however, the laser emission must pass through the liquid–air interface, so that the optical perfection of this surface is of the absolute essence. The surfaces through which the beam is to pass must be optically flat over the illuminated region as well as being free of any optical imperfections.

Two examples of nozzle designs are illustrated in Fig. 52. The diagrams are cross-sectional views through the short dimension of long, narrow nozzles. In both designs the edge where the fluid breaks away from the nozzle requires special attention. It must be polished to a minimum radius of curvature.

The nozzle in Fig. 52A gives the jet stability by rapidly accelerating the fluid as it is breaking free from the nozzle. The direction that the jet assumes after leaving the nozzle depends on the relative orientation of the knife (razor) blade edges of the nozzle. Any small change in this orientation can change the jet flow direction significantly. The polishing of such knife edges to a flawless minimum radius is very difficult in practice. For these reasons, this geometry is not often employed.

The nozzle in Fig. 52B is easier to fabricate and yields more predictable results. As discussed in Section 5.5.10 this nozzle straightens out the so-

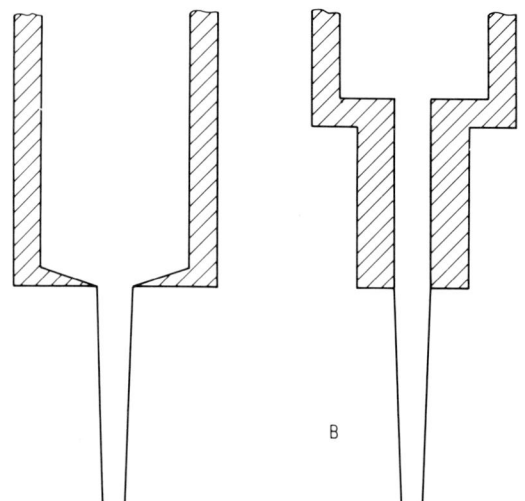

FIG. 52. Nozzles for dye solution jets. Diagram (A) uses knife edges to form the flat jet and diagram (B) uses a long narrow channel to generate the jet.

lution flow velocities by using a long-flow channel, which should be approximately 20 times the narrow dimension of the nozzle for optimum results. For operation in the laminar flow regime, these nozzles have successfully been fabricated by flattening small-diameter, 3-mm seamless tubing and carefully cutting and polishing a flat exit surface normal to the tube axis on the flattened section. For high-velocity operation, the inside surface of the nozzle must be polished to greater perfection than is desired of the jet surface. This flawless surface is necessary because any imperfection induced into the flow by the nozzle will grow as a function of time and distance. Such perfection usually can be achieved only by assembling the nozzle from separate components, each being carefully honed and polished.[83]

The only stable geometry for a free-flowing jet of liquid is a circular cross-sectional cylinder. The flat surfaces produced on these jets therefore generate a metastable condition that will only exist for a short distance away from the nozzle. The length of the flat region depends on the aspect ratio (length/width) of the nozzle orifice and the solution velocity. The parallelism of the flat surfaces of the jet is also improved as a function of velocity. These facts make high-viscosity solvents very attractive. The comparisons of different solvents are given in Table I.

As also discussed in Section 5.5.10, these jets are extremely sensitive to any disturbances in the ambient conditions. They require total isolation from pressure surges in the solution, vibrations in the mechanical components, and acoustic waves or drafts in the air.

5.7. Laser Dyes

One of the most outstanding attributes of dye lasers is the wide selection of dye molecules that are available and can be used in laser devices. Good dyes are available that will permit efficient laser operation between 450 and 900 nm.[112] At reduced efficiency, this wavelength range can be extended to approximate limits of 320 and 1200 nm.[112] The less efficient dyes can be excited only with high-intensity, fast-risetime laser excitation. The extent of the wavelength range covered by efficient dyes is defined by the range over which continuous laser operation is achievable. Only the dyes that have a minimum of internal losses will operate cw and therefore are the dyes of choice for all applications within their range of operation. A partial list of such dyes is illustrated in Fig. 53, which pictorally shows the wavelength range and relative outputs for cw operation

[112] K. H. Drexhage, in "Dye Lasers" (F. P. Schafer, ed.), p. 144. Springer-Verlag, Berlin and New York, 1973.

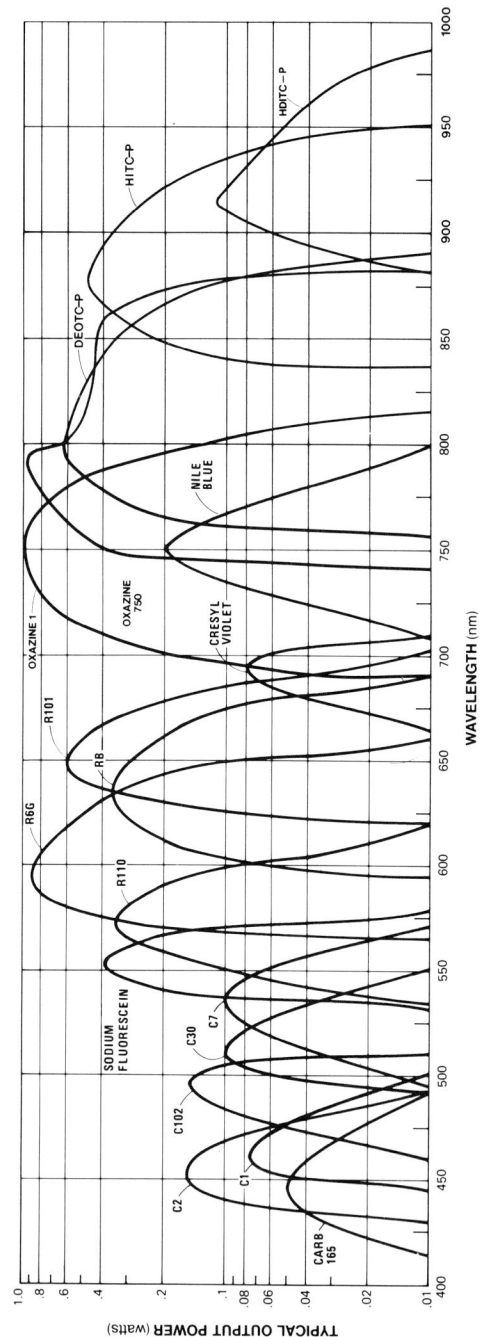

FIG. 53. Relative outputs of various organic dyes used in cw dye lasers excited by ion lasers. Rhodamine dyes are labeled **R** and coumarin dyes are labeled **C** (diagram courtesy of Spectra Physics, Inc., Mountain View, CA).

using commercial ion-laser-excited dye lasers. This list is a very small fraction of the organic dyes that have been found to be capable of laser action. The dyes have been chosen to satisfy the criteria of high efficiency, commercial availability, and completeness in filling the entire wavelength range of operation.

On the ultraviolet side of the list in Fig. 53 are several dyes that were synthesized for use in scintillators for detecting nuclear particles. Some of the better dyes in this class are listed in Table IV together with approximate wavelength ranges of operation.

On the infrared side of Fig. 53 list are several cyanine dyes of which representatives are listed in Table IV.

The correlation between dye structure and fluorescence characteristics has been carefully studied and is quite well established. Following these principles, many new dyes have been synthesized since the invention of the dye laser, thereby extending the wavelength range over which high-efficiency dyes are available. These high-efficiency dyes have come almost exclusively from four classes of dyes: xanthenes, coumarins, oxazines, and cyanines.

Although the wavelength range has been enlarged, the synthetic efforts have not been successful at significantly improving the efficiency of the laser dyes. Consequently rhodamine 6G, which was among the first to be

TABLE IV. Laser Dyes

Dye	Solvent	Lasing wavelength (nm)
p-Terphenyl	DMF[a]	341
PPD: 2,5-diphenyl-1,3,4-oxadiazole	Dioxane	348
PPF: 2,5-diphenylfuran	Dioxane & DMF	371
p-Quaterphenyl	DMF	374
PPO: 2,5-diphenyloxazole	Dioxane	381
α-NPO: 2-(1-naphthyl)-5-phenyloxazole	Ethanol	400
BBO: 2,5-bis(4-biphenyl)oxazole	Benzene	410
POPOP: 2,2'-p-phenylenebis(5-phenyloxazole), 1,4-bis[2-(5-phenyloxazolyl)]benzene	Toluene	419
bis-MSB: p-bis(O-methylstyryl)benzene	Toluene	424
1,1'-Diethyl-2,2'-tricarbocyanine iodide	Acetone	898
3,3'-Diethyl-12-ethylthiatetracarbocyanaine iodide	Ethanol	920
1,1'-Diethyl-4,4'-quinotricarbocyanine iodide	Acetone	1000
3,3'-Diethyl-12-ethylthiatetracarbocyanine carbocyanine iodide	Methanol	1100
Pentacarbocyanine dye	Methanol	1120

[a] DMF, dimethylformamide.

discovered, is still one of the best dyes available for laser application. It is at this time the most thoroughly studied and characterized of the laser dyes and therefore has been used throughout this discussion as the example material. Although most of the characteristics of the other dyes have not been measured, a close similarity between the values of the parameters for all the high-efficiency dyes can be implied from the similarity in output efficiency observed during cw operation. It is expected that the excited singlet state absorption will have an increased effect in dyes operating in the short-wavelength portion of the spectrum. In the long-wavelength region the dyes are expected to have increased radiationless transitions competing with the fluorescence and thereby reducing the quantum yield for fluorescence. The triplet absorption is expected to be a universal problem in all regions of operation. There is a tendency for the shorter-wavelength dyes to have a greater breadth in wavelength of absorption and emission bands as compared with the dyes fluorescing in the long-wavelength region. The cross sections for absorption and emission are inversely proportional to wavelength extent of these bands. Therefore, it is usually necessary to use higher concentrations of short-wavelength dyes than long-wavelength ones in lasing solutions in a specific device.

Essentially all dyes are easily dissolved in alcohol solutions and usually have their highest fluorescence quantum yields in these solutions. The alcohols most often used are simplest ones, methanol and ethanol, which have the greatest ultraviolet transparency, and ethylene glycol which is used for its high viscosity. The fluorinated alcohols, especially hexafluoro isopropanol, have been used to shift the wavelength of some dyes.[112] The solvent dimethyl sulfoxide has been found to improve the quantum yield of several dyes, particularly the long-wavelength cyanines. This solvent is seldom used, however, because of its adverse biological effects. Water with surfactants to solubilize the dyes has superior thermal properties as compared to the alcohols when it is used in turbulent flow systems. These thermal properties are discussed in more detail in Section 5.2.5.

Measurable improvements have been made in laser solution performance by the addition of materials to quench populations of triplet-state molecules.[113–115] These have been most effective in dye lasers driven by monochromatic excitation where the absorption of the additives does not reduce the excitation. Two triplet-state quenchers have been identified

[113] R. Pappalardo, H. Samelson, and A. Lempicki, *Appl. Phys. Lett.* **16**, 267 (1970).
[114] J. B. Marling, L. L. Wood, and D. W. Gregg, *IEEE J. Quantum Electron.* **qe-7**, 498 (1971).
[115] F. P. Schäfer and L. Ringwelski, *Z. Naturforsch., Teil A* **28**, 792 (1973).

5.7. LASER DYES

that are effective in flash-lamp-excited systems. These two are oxygen and cyclo-octatetraene (COT). The latter material has excellent UV transparency in the few percent solution concentrations in which it is used and is more effective than oxygen. However, COT is quickly decomposed under laser operating conditions. Fortunately, dissolved oxygen is a quite satisfactory triplet-state quenching material that is not subject to depletion. Also it has been found to be in the optimum concentration for rhodamine 6G when it is dissolved into the alcohol solution at the concentration characteristic of equilibrium solution for air at normal ambient conditions.[115]

AUTHOR INDEX FOR PART A

Numbers in parentheses are footnote reference numbers and indicate that an author's work is referred to although his name is not cited in the text.

A

Ablekov, V. K., 32
Abraham, A., 2, 20(9)
Abrahams, M. S., 229
Abrams, R. L., 113
Abrosimov, G. V., 45
Adams, A., Jr., 59, 73
Agheev, V. P., 72, 82(188), 83(188)
Aiki, K., 248, 249(53)
Aisenberg, S., 57
Akirtava, O. S., 42, 45, 91
Alaev, M. A., 43, 45
Aleinkov, V. S., 126, 134
Aleksandrov, E. B., 58
Aleksandrov, I. S., 45, 46
Alferov, G. N., 73, 93, 94
Alferov, Zh. I., 209, 215(3)
Anderson, R. S., 38, 45
Andrade, O., 51, 52, 53, 54, 55
Andreev, V. M., 209, 215(3)
Aoyagi, T., 324
Aoyagi, Y., 324
Armand, M., 60
Armstrong, D. R., 57, 58
Arrathoon, R., 101, 108, 110
Asmus, J. F., 43
Aspnes, D. E., 232
Astheimer, R. W., 205(19), 266, 267
Auzel, F., 194

B

Babaev, A. A., 45, 46
Babeiko, Yu. A., 45, 46
Ball, J. T., 187
Ballick, E. A., 72, 84(209, 212)
Balucani, U., 264

Banse, K., 99
Baranov, A. I., 43, 45
Barker, G. C., 96
Baron, K. U., 71
Barthel, K., 73(225), 74
Bartram, R. H., 193
Basov, N. G., 1, 6(1), 244
Batenin, V. M., 47
Beck, R., 31, 56(3), 61
Begley, R. F., 118(350), 119
Beigang, R., 316, 355(83)
Bell, W. E., 37, 40, 60, 61, 73, 80, 90(206),
    91(206), 101, 104(313), 122(367), 123,
    124(24, 367), 126(372), 127(24), 128(367),
    129, 149, 152, 153, 338
Bennett, W. R., Jr., 1, 15, 36, 52, 56, 60,
    73(113), 82, 83, 84, 85(209, 212), 87, 92,
    97, 115(309), 116, 117(340), 127(113), 132,
    143, 144(434), 145(434)
Berezin, A. K., 92
Bergou, J., 123(481), 124(481), 159, 161,
    162(479), 165, 166
Berolo, O., 232
Bessarab, Ya. Ya., 92
Bigio, I. J., 118(350), 119
Birnbaum, M., 75
Bjorkholm, J. E., 281, 287(48), 324
Bliss, E. S., 188
Blit, S., 301, 318, 319(86), 321(86)
Bloembergen, N., 1, 187
Bloom, A. L., 37, 40, 61, 73, 101, 103(311),
    104(311, 313), 122(367), 123, 124(367),
    127, 128(367), 129, 131, 143, 149, 152,
    153(465)
Bobroff, D. L., 57, 59
Bockasten, K., 51, 52, 53(71), 54, 55(71)
Bockasten, K. B., 37, 40
Boersch, H., 72, 93, 94, 99
Bogdonova, I. P., 126

361

Bogdanovich, A. V., 92
Bogus, A. M., 91
Bokhan, P. A., 37, 39, 40, 42, 43, 45, 46, 48, 71(63)
Bolotin, L. I., 92
Bonner, W. A., 198
Boot, H. A. H., 101, 115
Boscher, J., 73(223, 224, 225), 74, 89, 93(221), 94, 99
Boyd, G., 167
Brau, C. A., 2
Brecher, C., 177
Breton, J., 122
Breton, L., 121
Brewer, R. G., 2
Bricks, B. G., 38, 45
Bridges, W. B., 31, 51, 53, 56(1), 57, 59, 60, 61, 63, 66, 67, 68 (1), 69(1), 70(1), 7(114), 72(114, 132, 136, 144, 160), 77, 78(176), 80, 83, 84, 86, 88(116), 93, 94, 95(132, 144), 97(144), 99, 113, 122(365), 123, 124, 126(371), 127, 157(132, 175)
Brown, D. C., 137, 191, 200(62b)
Brown, R. M., 190
Browne, P. G., 133
Browness, C., 310
Buczek, C. J., 281, 287(50)
Buettner, A. V., 256
Buiocchi, C. J., 229
Burkhard, P., 97
Burlakov, V. D., 40
Bunkenberg, J., 319, 322(89)
Burlamacchi, P., 311
Burmakin, V. A., 47
Burnham, R. D., 227
Burns, G., 214
Burrus, C. A., 192, 198
Buser, R. G., 92, 95(294), 100
Butler, J. K., 215, 220, 221, 222, 238(20), 239, 240, 242(20), 246, 247(20)
Buzhinskii, O. I., 45, 46
Byer, R. L., 73, 122(367), 123, 124, 128, 129

C

Cabezas, A. Y., 191, 200(62a)
Cahuzac, Ph., 36, 37, 39, 48, 71(20, 60, 61)
Calawa, A. R., 233
Chapovsky, P. L., 50, 51, 117
Carlson, R. O., 214

Carlstein, J. L., 276, 281(39)
Casey, H. C., Jr., 216
Ceccon, H. L., 319, 320(87), 322(87, 88)
Chebotayev, V. P., 73, 101, 103
Chebotaev, V. P., 74
Chen, B. U., 200
Chen, C. C., 71, 74(159), 81(239)
Chen, C. J., 40, 42, 45
Cheo, P. K., 61, 70(119), 75, 76, 88(248), 89
Cherezov, V. M., 40, 47
Cherkasov, E. M., 117
Chernen'kii, Yu. N., 92
Chernock, J. P., 189
Chesler, R. B., 167
Chester, A. N., 31, 51, 53, 56(1), 61, 63, 68(1), 69(1), 70(1), 71(169, 170), 72(160), 78(160), 82, 83(160), 88(116), 93(160)
Chihara, M., 152
Chinn, S. R., 203
Chkuaseli, Z. D., 96
Clark, P. O., 53, 54, 57, 72, 77(176), 78(176), 94
Cline, C. F., 193
Clunie, D. M., 51, 52, 101, 115(314)
Collins, G. J., 121, 126, 132, 133, 143, 144(434), 145(434), 147, 154, 155, 157, 158, 160(475), 161(475, 476), 162(475, 476), 163, 164(478, 484, 485, 486), 165(476, 486)
Colombo, J., 91, 124, 129
Condon, E. V., 55, 62(86), 64(86)
Convert, G., 60
Cooper, H. G., 61, 70(119), 75, 76, 88(248), 89
Cormier, A. J., 199
Cottrell, T. H. E., 90
Creedon, J. E., 91
Crow, T., 188
Csillag, L., 31, 122(364), 123(481), 124(481), 133, 134, 158, 159, 161, 165, 166
Cunningham, F., 276, 281(38)

D

Dahlquist, J. A., 70, 95(139)
Damen, T. C., 203
Dana, J., 121, 123, 124(358)
Dana, L., 122(366), 123
Danch, F. F., 85, 86(264)
Danielmeyer, H. G., 168, 180, 190, 197, 203, 281, 287(48)
Dapkus, P. D., 227

Daschenko, A. I., 85, 86(264)
Dauger, A. B., 116
Davis, C. C., 61, 70, 71, 90, 95(141), 96
Davis, W. C., 289
Decker, C. D., 270
Deech, J. S., 36, 38, 39, 71(19)
de Hoog, F. J., 162
DeLuca, J. A., 177, 186
deMaria, A. J., 28
deMars, G., 17, 19, 73, 202
Detch, J. L., 60, 73(113), 84(113), 86(113), 92(113), 127(113)
DeTemple, T. A., 117
Deutschbein, O., 193, 200(71)
Devlin, G. E., 185
Dexter, D. L., 180, 181(28)
Dianov, E. M., 198
Dieke, G. H., 174
Dielis, J. W. H., 85
Dienes, A., 255, 256, 257(9), 269, 270(9), 326, 333(9), 335(9), 337(102), 343(102), 345(102), 348(102)
Dill, F. H., Jr., 214
Dirac, P. A. M., 2, 20(9)
Dishington, R. H., 302
Dixon, R. W., 243
Dolan, G., 270
Donin, V. I., 74, 93(247), 94
Drexhage, K. H., 355, 358(112)
Druyvesteyn, M. J., 50, 78, 112(68)
Dubrovin, A. N., 37, 40
Duffendack, O. S., 120, 154, 155, 157(354)
Dujardin, G., 276
Dumke, W. P., 214
Dunn, M. H., 70, 133, 351
Dunning, F. B., 276, 278(42), 286(42), 287(42)
Dyatlov, M. K., 133
Dyson, D. J., 125
Dzhikiya, V. L., 42, 45, 91

E

Eberly, J. H., 255, 256, 257(8), 258(3), 260(8), 261(8), 263(8), 264(8)
Efimov, A. V., 45, 46
Eichler, H. J., 73(226), 74
Eisenberg, H., 265(16, 20), 266(16), 267
Eletskii, A. V., 47
Elliot, C. J., 73

Emmett, J. L., 302
Englisch, W., 31, 56(3), 61(3)
Eppers, W., 56
Erez, G., 43, 45, 46
Erickson, L. E., 281, 287(49)
Ettenberg, M., 215, 218, 223(9), 224(9), 225(9), 226, 228, 230, 231(31), 238, 243, 244(42)
Evtuhov, V., 168, 185(9), 202(9)
Evtyunin, A. I., 47
Ewanizky, T. F., 318, 321(84)
Ewing, J. J., 2

F

Fabelinskii, I. L., 32
Fahlen, T. S., 43, 45, 47
Fainberg, Ya. B., 92
Falkenstein, W., 255
Faulstich, M., 193, 200(71)
Faust, W. L., 54, 55(85), 56, 57, 116, 117(340)
Fedorov, L. S., 72
Felt'san, P. V., 87
Fendley, J. R., Jr., 73, 94, 99, 134, 135, 137, 149(417)
Fenner, G. E., 214
Ferguson, A. I., 351
Ferrar, C. M., 42, 43, 305, 307, 308(75), 309(75), 322(75), 324(75)
Ferrario, A., 124, 126, 127
Feucht, D. L., 216
Fgorov, V. G., 43, 46(45a)
Field, R. L., Jr., 114
Finzel, R., 73(224), 74, 90(224), 93(224)
Fisher, A., 318, 319(86), 321(86)
Fisher, C. L., 91, 124, 129
Flamant, P., 276, 281, 287(52)
Flint, G., 188
Florin, A. E., 118(349), 119
Foley, R. J., 314, 315(81)
Fouassier, J. P., 318, 321(85)
Fountain, W., 188
Fowles, G. R., 36, 37, 38, 40, 42(13), 70, 128, 130, 131, 143, 146, 154(387), 155(387)
Fox, A. G., 11
Foy, P. W., 209, 215(2), 216
Franken, P. A., 1
Franzen, D. L., 158, 159, 160(475), 161(475, 476), 162(475, 476), 165(476)
Frapard, C., 121, 122(366), 123, 124(358)

Freiberg, R. J., 281, 287(50)
Frez, G., 42
Fridrikhov, S. A., 115
Friedman, H. A., 187
Fujimoto, T., 133
Fukuda, K., 70, 92(149), 133
Fukuda, S., 141
Furomoto, H. W., 319, 320(87), 322(87, 88)

G

Gabay, S., 42, 46
Gadetskii, N. P., 92
Gallardo, M., 51, 52, 53(71), 54, 55(71), 70, 95(140, 142)
Gallego Lluesma, E., 70, 95(140, 142)
Galvan, L., 54, 55
Ganiel, U., 271, 275(34), 276, 281, 287(51), 301, 318, 319(86), 321(86), 328
Gannon, J. J., 230, 231(31)
Garavaglia, M., 37, 40, 70, 95(140, 142)
Garrett, C. G. B., 56, 57(89)
Gassman, M. H., 274
George, E. V., 83
Gerasimov, V. A., 40
Gerritsen, H. J., 62, 97
Gerry, E. T., 46, 50, 51
Gerstenberger, D. C., 161, 163, 164(478, 485), 165
Geusic, J. E., 167, 168, 194
Gill, P., 144, 145, 146
Ginsburg, N., 137
Glaze, J. A., 199
Glenn, W. H., 307, 308(74, 75), 309(74, 75), 322(73, 74, 75), 324(75)
Gnedin, I. N., 43, 45
Goedertier, P. V., 61, 97
Gol'dinov, L. L., 96
Goldsborough, J. P., 73, 90(206), 91, 122(368), 123, 124, 129, 131, 134(391), 135, 143, 149(416), 152, 153(465)
Goldschmidt, C. R., 270
Goldstein, R., 305
Goldstein, Y., 227
Gomolka, S., 186
Goncz, J. H., 302, 303, 304(66)
Goodwin, A. R., 219, 225(18), 226
Gordon, E. I., 62, 71(114, 164, 166), 72(114), 75(166), 78, 80, 81(163), 82(164), 83(164), 84(163), 86(163), 88(166), 95(163), 106, 108, 110, 111, 112(326), 113, 157(163)
Gordon, F. B., 43, 46(45a)

Gordon, J. P., 1, 6(1)
Gorog, I., 73, 88, 89, 134, 135(417), 137(417), 149(417)
Goto, T., 124, 127(375), 133, 134(408a, 408b), 147, 148, 151
Gould, G., 32, 34, 36, 40(10), 41(10), 42(10), 45, 46(10), 50(10), 71(10)
Graef, W. P. M., 85
Graham, W. J., 70
Gran, W. H., 120, 154(354b), 155
Green, J. M., 121, 132, 145(401), 151(401), 152, 154(401), 155, 156, 162
Gregg, D. W., 358
Greskovich, C., 189
Griffith, J. S., 171
Grove, A., 234
Grover, N. B., 227
Groves, S. H., 233
Grum, F., 310
Guggenheim, H. J., 179, 185, 194
Gumeiner, I. M., 100
Gundersen, M., 96, 100
Gunmeiner, I. M., 92, 95(294)
Gürs, K., 31, 56(3), 61(3)

H

Hagen, W. F., 199
Hall, R. N., 214
Halsted, A. S., 65, 67, 71, 72(160), 77(176), 78(176), 79(132), 83(132), 84, 86, 88, 93, 94, 95(132), 157(132, 175)
Hammer, J. M., 72, 86
Hammond, E. C., 270
Hanna, D. C., 276, 278(44), 279(44), 343, 346(109)
Hänsch, T. W., 275, 276(36), 280(36), 281(41), 322(36), 323(41)
Hardwick, D. L., 103
Hardy, A., 271, 275(34), 281, 287(51), 328
Harman, T. C., 233
Harper, C. D., 96, 100
Harris, S. E., 197
Hartman, R. L., 243
Hashino, Y., 70, 92
Hattori, S., 124, 126, 127(375), 133, 134(408a, 408b), 147, 148, 151
Haug, H., 247
Hawkes, J. B., 265(19), 266, 267
Hawrylo, F. Z., 215, 231
Hayashi, I., 209, 215(2)
Hayward, J. S., 269

Heard, H. G., 124, 127(373)
Heavens, O. S., 146, 157, 164
Heckscher, H., 203
Heil, H., 105, 106, 107
Heiman, D., 193
Hellwarth, R. W., 27, 193
Hendorfer, B. W., 276
Hernqvist, K. G., 72, 134, 135(417),
  137(417), 149(417), 150, 151, 158, 159, 162
Herriott, D. R., 97, 115(309)
Herriott, E. R., 1
Herz, A. H., 269
Herziger, G., 73, 74, 82, 83, 94, 99
Heynau, H., 28
Hilberg, R. P., 302
Hill, A. E., 1
Hirth, A., 318, 321(86)
Hocker, L. O., 118(351), 119
Hoder, D., 72, 92(221), 94, 99
Hodges, D. T., 134, 154
Hodges, E., 122(367), 123, 124(367),
  128(367), 129
Hodges, E. B., 73, 90(206), 91(206)
Hoffman, V., 100
Hoffmann, V., 70, 75, 95(143), 96, 100
Holonyak, N., 227
Holstein, T., 35, 60(11), 87(11), 88
Holt, D. B., 227
Holtom, G., 322
Holton, W. C., 190
Honig, R. E., 42
Hook, W. R., 302
Hopkins, B. D., 130, 131, 154(387), 155(387)
Hopper, R. W., 193
Horrigan, F. A., 64, 66, 72(129), 73(129), 85
Hunt, J. T., 188
Hyman, H., 73, 87(210)

I

Iijima, T., 131, 138(395, 396), 143(395, 396)
Ikegami, T., 246
Illingworth, R., 92
Ill'yushko, V. G., 156
Imre, A. I., 85, 86(264)
Ippen, E. P., 326, 337(102), 343(102),
  345(102), 348(102)
Irwin, D. J. G., 85
Isaev, A. A., 36, 37, 38, 39, 40, 41, 42, 45,
  46, 47, 48, 51, 52, 71(21), 124, 127
Ishchenko, P. I., 41, 48(31)
Isjikawa, M., 147

Itagi, V. V., 133
Itzkan, I., 276, 281(38)
Ivanov, I. G., 133, 155

J

Jacobs, R. R., 186, 192, 200(69)
Jacobs, S., 32
Jacobs, S. D., 191, 200(62b)
Jánossy, M., 31, 122(364), 123(481),
  124(481), 133, 134(409), 138, 158(4), 159,
  161, 162(479), 165, 166
Jarrett, S. M., 96
Javan, A., 1, 32, 97, 110, 115
Jeffers, W. Q., 118(348), 119
Jenkins, F. A., 322, 344(92)
Jensen, R. C., 70, 132, 143, 144, 145, 146
Jensen, R. D., 143.A, 145
Jensen, R. J., 118(349), 119
Jenssen, H. P., 185, 205(40)
Jethwa, J., 305
John, W., 193, 200(71)
Johnson, A. M., 65, 66, 71(171), 72, 93
Johnson, L. F., 167, 179, 185, 194
Johnson, W. L., 163, 164(484, 486), 165(486)
Johnston, T. F., Jr., 138, 139, 140, 141, 142
Johnston, W. D., 355
Judd, B. R., 177
Juramy, P., 281, 287(52)

K

Kagen, Yu. M., 81, 89
Kaiser, W., 255
Kamiide, N., 147
Kaminskii, A., 181, 188
Kaminskii, A. A., 181, 188, 194, 203
Kano, H., 124, 126, 127, 147, 148, 151
Kántor, K., 133, 134(409)
Karabut, E. K., 131, 138(397), 157, 162, 163, 164
Kärkkäinen, P. A., 276, 278(44), 279(44)
Karras, T. W., 38, 45
Kaslin, V. M., 51, 53
Kato, I., 91, 122, 123
Katsurai, M., 91
Katsuyama, Y., 70, 92(149)
Katzenstein, J., 92
Kavasik, A. Va., 198
Kawahara, A., 133
Kazaryan, M. A., 36, 38, 39, 40, 41, 42, 45,
  46, 47(16), 71(21)

Keefe, W. M., 70
Keidan, V. F., 70, 150, 151
Kel'man, V. A., 85, 86(264)
Kemp, J. D., 265(22), 266, 267
Kerman, A., 43, 45
Keune, D. L., 227
Keyes, R. J., 214
Kimiya, T., 247
Kindlemann, P. J. 52, 73, 87(210)
Kindt, T., 73(223), 74
King, T. A., 62, 70, 71(89), 95(141), 96
Kingsley, J. D., 214
Kirkby, P. A., 215, 216
Kisliuk, P., 186
Kitaeva, V. F., 71, 72(158), 82, 83, 85(183)
Kiumira, T., 203
Klauminzer, G. K., 276, 278(45), 279(45), 280(45)
Klein, M. B., 88, 89, 136, 149, 150, 151, 152
Klement'ev, V. M., 74
Klimkin, V. M., 37, 48, 71(62, 63)
Klimovskii, I. I., 47
Klüver, J. W., 59
Knutson, J. W., Jr., 60, 73(113), 84(113), 87(113), 92(113), 127(113)
Knyazev, I. N., 52, 115
Kobiyama, M., 142
Kocherga, L. F., 72
Koechner, W., 169, 170, 188
Kogelnik, H., 59, 248, 337, 341(106), 343(106)
Kogelnik, H. W., 326, 337(102), 343(102), 345(102), 348(102)
Kolb, W. P., 138, 139, 140, 141, 142
Kolpakova, I. V., 116
Koozekanani, S. H., 64, 66, 72(129), 73(129), 85(129), 87
Kopf, L., 198
Korolev, F. A., 85
Koster, G. F., 54, 64, 66, 72(129), 73(129), 85(129)
Kovacs, M. A., 118, 119
Krag, W. E., 214
Kravchenko, V. F., 162, 163, 164(483)
Kressel, H., 209, 215(1), 216, 218, 220, 221, 222, 223(9), 224(9), 225(9), 226, 227, 228, 231, 232, 233, 234, 235, 238(20), 239, 240, 242(20), 243(37), 244(42), 245(20), 246, 247(20)
Krolla, G., 193, 200(71)
Kruithof, A. A., 120
Krupke, W. F., 186
Krysanov, S. I., 45, 46
Kuklo, T. C., 314, 315(81)
Kulagin, S. G., 91
Kulyasov, V. N., 58
Kuroda, K., 134
Kushida, T., 174, 178(15), 179(15), 181

L

Labuda, E. F., 61, 65, 71(114, 164, 166, 167, 171), 72(114), 75(166), 78, 80, 81, 82, 83, 84, 86, 88(166), 95(163), 109, 110, 112(326), 113, 157(163)
Ladany, I., 216, 231, 234, 235, 243(37), 244(42)
Laegrid, N., 165, 166(490)
Lamb, H., 315, 316(82)
Lamb, W. E., Jr., 15
Landry, R. J., 193
Langmuir, I., 78, 81
Lankard, J. R., 270
Larionov, N. P., 115
Lasher, G., 214
Lasher, G. J., 246
Latimer, I. D., 73, 87, 99
LaTourette, J. T., 59
Latush, E. L., 71, 156
Laurès, P., 121, 122(366), 123, 124(358), 338
Layne, C. B., 179, 193(25)
Lax, B., 214
Lebedeva, V. V., 85
Lee N., 191, 200(62b)
Leheny, R. F., 270
Lemmerman, G. Yu., 42, 45
Lempicki, A., 190, 358
Lengyel, B. A., 37, 40
Leonard, D. A., 46, 49, 50, 51, 52
Lesnoi, M. A., 47
Levin, L. A., 42, 43, 45, 46
Levinson, G. R., 89
Li, T., 11
Likhachev, V. M., 91, 92
Lin, C., 255
Lin, S.-C., 71, 74(159), 93(20), 94, 99
Linford, G. J., 51, 53, 54
Ling, H., 91
Lisitsyn, V. N., 50, 51, 117(66)
Littlewood, I. M., 47, 126, 154
Liu, C. S., 36, 42

AUTHOR INDEX FOR PART A

Livingston, A. E., 85
Lobsiger, W., 74, 81(233), 93(233), 99
Lockwood, H. F., 215, 225, 243, 244(42)
Long, E. A., 265(22), 266, 267
Lopez, F. O., 37, 40, 61, 124(24), 127(24)
Loree, T. R., 118(352), 120
Lovberg, R. H., 92
Lowdermilk, W. H., 179, 193(25)
Luckey, G., 310
Lundholm, T., 37, 40, 53, 54
Luo, H. H., 74, 93(240), 99
Lüthi, H. R., 74, 81(228), 82(227), 90, 93, 97, 99
Lynch, D. W., 232

M

McClure, D. S., 176
McColgin, W. C., 255, 256, 257(8), 258(3), 260(8), 261(8), 263(8), 264(8), 269
McCumber, D. E., 185
McFarland, B. B., 322
McFarlane, R. A., 54, 55(85), 56, 57(89), 62, 70(117), 122, 116, 117(340)
McIlrath, T. J., 276, 281(39)
Mack, M. E., 307, 308(71), 322(71, 72)
McKenzie, A. L., 151
McNeil, J. R., 158, 159, 160, 161, 162, 163, 164(478, 484, 485, 486), 165(476, 486)
McWhorter, A. L., 214
Magda, I. I., 92
Maiman, T. H., 1, 8(2), 167
Maloney, P. J., 58
Mamyrin, A. B., 58
Manley, J. H., 120
Many, A., 227
Marantz, H., 64, 66, 67, 72(130, 138), 86(173)
Marcos, H., 168
Markiewicz, J. P., 302
Markova, S. V., 36, 39, 47
Markuzon, E. V., 91
Marling, J., 197
Marling, J. B., 61, 95, 100, 358
Marowsky, G., 276, 289, 297, 322
Marshak, I. S., 302
Martinot-Lagarde, P., 60
Marusin, V. D., 126
Mash, L. D., 142
Maslowski, S., 247, 248
Massone, C. A., 70, 95(142)

Mastrup, F. N., 305
Mathews, J. W., 229
Maupin, R. T., 109
Mercer, G. N., 60, 70, 72(144), 73(113), 84(113), 85(209, 212), 87(113, 210), 92(113), 95(144), 97(144), 99, 127(113), 157(175)
Mertenat, R., 74
Meyer, Y. H., 281, 287(52)
Mikhalevskii, V. S., 70, 131, 138(97), 149, 150, 157(397)
Milam, D., 188, 193
Miller, R. C., 71(164, 166, 167), 72, 75(166), 78(163), 80(163), 81(163), 82(164), 83(164), 84(163, 164), 86(164), 88(166), 95(163), 157(163)
Milnes, A. G., 216
Minnhagen, L., 64, 69(128)
Miya, M., 141
Miyakawa, T., 181
Miyazaki, K., 133
Mizeraczyk, J. K., 148
Moffett, J., 304
Moncur, N. K., 43
Moore, C. A., 186
Moore, C. E., 67, 68, 69, 70(135), 123
Morey, W. W., 307, 308(74, 75), 309(74, 75), 322(73, 74, 75), 324(75)
Mori, M., 133, 134(408a, 408b)
Morradian, A., 186
Morris, R. C., 185, 205(40)
Morruzzi, A. L., 270
Moskalenko, V. F., 43, 45, 95
Moulton, P. F., 186
Müller, G., 192
Murayama, M., 134
Mutegi, N., 180
Myers, R. A., 124, 128, 129
Myers, S. A., 276, 278(43)

N

Nakamura, M., 248
Nakaya, M., 122(363), 123
Namba, S., 142, 324
Nassau, K., 167, 188, 190
Nathan, M. I., 214
Neal, R. A., 50, 51
Neeland, J. K., 168, 185(9), 202(9)
Neely, D. F., 159, 162
Nelson, H., 209, 215(1)

Nerheim, N. M., 42, 43, 45
Neumann, G., 271, 275(34), 276, 328
Neuroth, N., 192, 193, 200(71)
Neusel, R. H., 70
Neustraev, V. B., 198
Nikolaev, G. N., 45, 46
Nikolaev, V. N., 45
North, D. O., 242
Novik, A. E., 85
Novikov, M. A., 276, 278(46)
Nuese, C. J., 230, 231, 232, 233(35)

O

O'Dell, E. W., 185, 205(40)
Odintsov, A. I., 71, 72(158), 85
Ofelt, G. S., 177
Ogata, Y., 133
O'Grady, J. J., 66, 73
Ogura, I., 134, 152
Oleinik, Yu. M., 42, 45, 91
Olsen, G. H., 228, 229, 230, 231(31), 232, 233(35)
Olson, C. G., 232
Orlov, V. K., 46
Osiko, V., 181, 188(34)
Osipov, Yu. I., 72, 82(177, 178, 187, 188), 83(178, 188), 85(183)
Ostapchenko, E. P., 96, 133
Ostravskaya, L. Ya., 72
Otsuka, K., 190, 203

P

Paananen, R. A., 57, 59, 65, 66,73, 95(201)
Pacheva, Y., 122, 123
Padovani, F. A., 132
Paik, S. F., 91
Palenius, H. P., 70
Panish, M. B., 209, 215(2), 216
Papakin, V. F., 131, 138(397), 157(397), 162, 163, 164(483)
Papayoanou, A., 92, 95, 100
Pappalardo, R., 358
Papulovskiy, V. F., 88
Parke, S., 186
Parker, J. V., 71, 72(160), 78(160), 83(160), 93(160)
Patek, K., 191
Patel, C. K. N., 56, 57(89), 111, 116, 117(340)

Pavlenko, V. S., 43, 46(45a)
Pavlova, L. S., 72
Peacock, R. D., 177
Pearson, W. M., 264, 281, 287, 288(53), 291(53), 294(10)
Pease, A. A., 264, 281, 287, 288(53), 291(53), 294(10)
Pechurina, S. V., 96
Penning, F. M., 50, 112(68), 120
Penzkofer, A., 255
Perchanok, T. M., 115
Perel', V. I., 81, 89
Persson, K. B., 158, 159, 160(475), 161(475, 476), 162(475, 476), 163, 164(478), 165(476)
Pesin, M. S., 32
Peters, C. W., 1
Peters, J. R., 219, 225(18), 226(18)
Peterson, J., 124, 127(373)
Peterson, O. G., 185, 205(40), 255, 256, 257(8), 258(3, 6), 260(8), 261(8), 262(8), 263(8), 264(8), 269, 287, 288(53), 291(53), 294(10), 314, 315(81), 337, 338(105)
Petersson, B., 64, 69(128)
Petrash, G. G., 35, 36, 37, 38, 39, 40, 41(12), 42, 45, 46, 47(16), 48(31), 51, 52, 53, 71(21), 115, 124, 127
Phelps, A. V., 108
Phi, T. B., 118(351), 119
Picus, G. S., 59
Pierce, J. W., 203
Piltch, M., 34, 36, 40(10), 41(10), 42(10), 45, 46(10), 50(10), 71(10)
Pinnington, E. H., 85
Pion, M., 219, 225(18), 226 (18)
Piper, J. A., 124, 126, 128, 129, 138, 144, 145, 146, 147, 148, 151, 153, 154, 159, 162
Pivirotto, T. J., 43
Pleasance, L. D., 83
Pole, R. V., 128, 129
Pollini, R., 289, 294, 297(57)
Polyakov, V. M., 72
Portnoi, E. L., 209, 215(3)
Povch, M. M., 87
Powell, C. G., 124, 129
Pramaturov, P., 122, 123
Pratesi, R., 264, 311
Prokhorov, A. M., 1, 6(1), 198
Prokop'ev, V. E., 37, 48, 71(63)
Pugnin, V. I., 147
Pultorak, D. C., 135, 150, 151
Purohit, R. K., 216

Q

Quist, T. M., 214

R

Rabin, H., 1
Rabinovich, M. S., 91
Rabinowitz, P., 32
Rabkin, B. M., 142
Racah, G., 50, 101
Rautian, S. G., 32, 46
Rawson, H., 192
Razmadze, N. A., 96
Redaelli, G., 89
Rediker, R. H., 214
Redko, T. P., 116
Reed, T. B., 186
Reid, R. D., 161, 163, 164(478), 165
Reisfeld, R., 193
Reisler, E., 265(20), 266, 267
Remer, D. S., 57, 134(98)
Rempel, R. C., 101, 104(313)
Renard, P. A., 188
Rigden, J. D., 101, 103(312), 104(312)
Rigrod, W. W., 330
Ringwelski, L., 358, 359
Riseberg, L. A., 145, 177, 178, 179(24), 190
Rock, N. H., 115
Rode, D. L., 219
Rodin, A. V., 47
Röss, D., 310, 311(77), 312(77)
Ronchi, L., 264
Rosenberger, D., 51, 52
Ross, J. N., 71, 86
Rózsa, K., 31, 122(364), 123, 124, 133, 134(409), 158(4), 159, 161, 162, 165, 166
Rubin, P. L., 72, 85
Rudelev, S. A., 147
Rudko, R. I., 64, 65(131), 66, 67, 72(131, 138), 85, 86
Runge, P. K., 355
Russell, G. R., 42, 43, 45
Russov, V. M., 115

S

Sahar, E., 270
Salamon, T., 31, 122(364), 123 133, 134(409), 158(4)
Salimbeni, R., 311
Salimov, V. M., 85
Salk, J., 73(224, 225), 74, 90(224), 93(224)
Saltz, P., 323
Samelson, H., 358
Sanders, J. H., 36, 39, 71(19)
Sandoe, J. N., 186
Saruwatari, M., 203
Sasnett, M. W., 117
Satake, T., 122(363), 123
Schacter, H., 90
Schäfer, F. P., 251, 305, 358, 359(115)
Schäfer, G., 73(223, 224, 225), 74, 90(224), 93(221, 224), 94, 99
Schawlow, A. L., 1, 6(1), 185
Schearer, L. D., 95, 100, 132, 145
Scherbakov, I. A., 198
Schlie, L. A., 103
Schuebel, W. K., 124, 128, 129, 131, 138, 166
Schulman, J. H., 180
Schulz, G. J., 107
Schwarz, S. E., 117
Seelig, W., 73, 83, 94, 97(220), 99
Seelig, W. H., 74, 81(228, 233), 82(227), 90, 93(233)
Seiden, M., 73
Seki, N., 122
Sekiguchi, T., 91
Selezneva, L. A., 47
Sém, M. F., 70, 131, 133, 138(397), 149, 150, 155, 156, 157(397)
Shah, J., 270
Shair, F. H., 57, 134(98)
Shank, C. V., 248, 324, 326, 337(102), 343(102), 345(102), 348(102)
Shapiro, S. L., 2
Sharma, B. L., 216
Shay, T., 126, 147
Shcheglov, V. B., 42, 43, 45, 46
Shelekhov, A. L., 72, 82(188), 83(188)
Shen, Y. R., 2
Shevtsov, M. K., 37, 40
Shimizu, T., 90, 122(363), 123
Shimoda, K., 1
Shionoya, S., 180, 181
Shortley, G. H., 55, 62(86), 64(86)
Shtyrkov, Ye. I., 115
Sibbett, W., 199
Siegman, A. E., 46, 133, 295, 322(58), 340(58)
Silfvast, W. T., 36, 37, 38, 40, 42(13), 70, 128, 130, 131, 132, 134(389), 135, 137, 138, 143, 144, 149, 150, 151, 152, 154, 155

Simmons, W. W., 95, 100, 199
Simons, W., 100
Sinclair, D. C., 61, 338
Singh, S., 198
Skolnick, M. L., 281, 287(50)
Skurnick, E., 89
Slivitskii, A. A., 45, 46
Smilanski, I., 42, 43, 45, 46
Smith, P. W., 28, 58, 113, 114, 322
Smith, R. G., 197, 198
Smith, W. L., 188, 193
Smith, W. V., 245
Snavely, B. B., 251, 255, 256, 258(6), 269(2), 337, 338(105)
Snitzer, E., 167, 191, 193
Sobelman, I. I., 32
Sobolev, N. N., 71, 72(158), 82(177, 178, 181, 188), 83(178, 188), 85(183)
Soden, R., 167
Soffer, B. H., 322
Sokolov, A. V., 45, 46
Solanki, R., 161, 163, 164(478)
Solimene, N., 34, 36, 40(10), 41(10), 42(10), 45, 46(10), 50(10), 71(10)
Solomonov, V. I., 37, 39, 40, 45, 48, 71(63)
Soltys, T. J., 214
Sommargren, G. E., 188
Sommers, H. S., Jr., 215, 242
Sorokin, A. R., 50, 51, 117(66)
Sorokin, P. P., 9(10), 167, 245, 270
Sosnowski, T. P., 135, 136, 149(418)
Spaeth, M. L., 287, 288(53), 291(53)
Spong, F. W., 73, 88, 89
Springer, L., 45
Springer, L. W., 38
Stadler, A., 97
Stadler, B., 71
Stafsud, O. M., 116
Starostin, A. N., 47
Statz, H., 17, 19, 54, 59, 64, 66, 72(129), 73(129), 85, 202
Stebbings, R. F., 276, 278(42), 286(42), 287(42)
Steckel, F., 265(18), 266, 267
Steele, E. L., 289
Stefanova, M., 122, 123
Steinberg, H., 59
Steinger, J., 99
Steinger, J. H., 74, 81(233), 93(233)
Stepanov, A. F., 147
Stepanov, V. A., 96, 133

Stern, F., 222, 223
Stetser, D. H., 28
Stevenson, M. J., 9(10), 167
St. John, R. M., 87
Stockman, D. L., 255, 258(3)
Stokes, E. D., 276, 278(42), 286(42), 287(42)
Stone, J., 192, 198
Streifer, W., 323
Strihed, H., 64, 69(128)
Strome, F. C., Jr., 322, 326(93), 334(93, 101), 335(93), 336(101), 337(93)
Subbes, E. V., 115
Suchard, S. N., 54, 55
Sucov, E. W., 36, 42
Suematsu, Y., 246
Sugawara, Y., 131, 138(395, 396), 143
Sumski, S., 215, 216
Sun, C., 134, 135(417), 137(417), 149(417)
Sunderland, J., 73
Sutovskii, V. M., 91
Sutovskiy, V. M., 92
Sutton, D. G., 54, 55
Suzuki, N., 125
Szabo, A., 281, 287(49)
Szapiro, S., 265(18), 266, 267
Sze, R. C., 73, 82, 83, 118(352), 120
Szeto, L. H., 135, 137, 138

T

Tagliaferri, A. A., 70, 95(140, 142)
Takasu, K., 133, 134(408b)
Tang, C. L., 1, 17, 54, 64, 65(131), 66, 67, 72(129), 131, 138), 73(129), 85(129), 86, 200, 323
Targ, R., 117
Tatarintsev, L. V., 45, 46
Taylor, J. R., 199
Telle, J. M., 323
Tereschenkov, V. S., 45, 46
Tertyshnik, A. D., 276, 278(46)
Teschke, O., 256, 257(9), 269, 270(9), 322, 333(9), 335(9)
Theissing, H. H., 318, 321(84)
Thomas, R. A., 185
Thompson, G. H. B., 215, 216, 219, 225(18), 226(18)
Thomson, K., 120, 157(354)
Thorn, R. S. A., 51, 52, 115
Thornton, J., 188
Tibilov, A. S., 37, 40

AUTHOR INDEX FOR PART A 371

Timmermans, J., 265, 266
Tio, T. K., 99
Tittel, F. K., 276
Tkach, Yu. V., 92
Tognetti, V., 264
Tokiwa, Y., 131, 138(395, 396), 143(395, 396)
Tokutome, K., 147
Tolkachev, V. A., 51
Tolmachev, Yu. A., 147
Tolman, R. C., 2, 20(9)
Tonks, L., 81
Topfield, B. C., 203
Toschek, P. E., 70, 75, 95(143), 96, 100
Townes, C. H., 1, 6(1)
Toyoda, K., 75, 142, 324
Treat, R. P., 191, 200(62a)
Treuthart, L., 289
Treves, D., 270, 271, 275(34), 281, 287(51), 328
Trezise, K. E., 52
Trofimov, A. N., 42, 45
Trukan, M. K., 209, 215(3)
Tsukada, T., 236
Tsukanov, Yu. M., 43, 45, 96
Tuccio, S. A., 322, 326(93), 334(93, 101), 335(93), 336(101), 337(93), 338(105)
Tunitskii, L. N., 117
Turner-Smith, A. R., 145(400), 151, 154
Tychinskiy, V. P., 89

U

Uhlmann, D. R., 193
Ultee, C. J., 118, 119
Umeda, J., 248, 249(53)
Ushakov, V. V., 126, 134

V

Vainshtein, I. A., 87
Valenzuela, P. R., 54, 55
Van der Sijde, B., 85
van der Ziel, J. P., 198
Vang, K. H., 177
Van Uitert, L. G., 168, 193, 194, 198
Vasilenko, L. S., 101, 103
Vasil'ev, L. A., 45, 46
Vasil'eva, A. N., 92
Vasil'tsov, V. V., 45
Verdeyn, J. T., 103

Vereshchagin, N. M., 43, 45
Vernyi, E. A., 142
Vinogradov, A., 87
Vokhmin, P. A., 47
Vollrath, K., 318, 321(85)
von Engel, Cf. A., 112
Vukstich, V. S., 85, 86(264)

W

Wada, J. Y., 105, 106, 107
Wallace, R. W., 197
Wallenstein, R., 276, 281(37, 41), 323(41)
Walling, J. C., 185, 205(40)
Walpole, J. N., 233
Walter, H., 1
Walter, W. T., 34, 36, 37, 40, 41, 42, 45, 46, 47, 48(56), 50, 71(10)
Wang, C. P., 74, 94
Wang, S., 249
Wang, S. C., 133
Warner, B. E., 161, 163, 164
Watanabe, S., 134, 152
Watts, R. K., 190
Waynant, R. W., 70
Weaver, H. J., 188
Weaver, L. A., 36, 42
Webb, C., 147
Webb, C. E., 47, 71(164, 165, 166, 167, 168), 72, 75(166), 78(168), 82(164), 83(164, 168), 84(164), 85, 86(164), 88(166), 120, 121, 124, 126, 128, 129, 132, 138, 145, 146, 151(401), 152, 153(466), 154(401), 155, 156, 162
Webb, C. F., 47, 120, 154
Webb, J. P., 255, 256, 257(8), 258(3), 260(8), 261(8), 262(8), 263(8), 264(8)
Weber, H. P., 203, 274
Weber, M. J., 177, 178, 179(24), 180, 181, 182, 183, 186, 188, 192, 193(25), 194, 200(69), 203
Wehner, G. K., 165, 166(490)
Weinreich, G., 1
Wellegehausen, B., 316, 355(83)
Welling, H., 316, 355(83)
Wen, C. P., 73, 87
Wexler, B., 73, 87(210)
Whinnery, J. R., 256, 257(9), 270(9), 333(9), 335(9)
White, A. D., 101, 103(312), 104(312), 106, 108, 110, 111, 113

White, H. E., 322, 344(92)
Whiteaway, J. E. A., 219, 225(18), 226(18)
Wieder, H., 124, 128, 129
Wieder, I., 270
Willett, C. S., 31, 33, 56(1, 2), 61, 70, 101, 109, 110, 112, 126, 127, 146, 157, 164
Williams, C. K., 90
Wilson, D. T., 73
Wiswall, C. E., 118(348), 119
Witte, R. S., 95, 100
Wittke, J. P., 241
Wood, L. L., 358
Wooley, J. C., 232
Wright, J. H., Jr., 318, 321(84)
Wyatt, R., 276, 278(44), 279(44)
Wybourne, B. G., 173, 177(14)

Y

Yakhontova, V. E., 126
Yamada, N. S., 181
Yamada, T., 190
Yamanaka, C., 75
Yanai, H., 247
Yang, K. H., 186
Yano, M., 247
Yariv, A., 59, 248, 249(53)
Yoshino, N., 148
Young, C. G., 191
Young, G., 167
Young, R. T., 109
Yurshin, B. Ya., 74, 93(247), 94

Z

Zamerowski, T. J., 229, 230, 231(31)
Zapesochnyi, I. P., 85, 86
Zaraga, F., 322
Zarowin, C. B., 90
Zeiger, H. J., 1, 6(1)
Zemskov, K. I., 46
Zemstov, Yu. K., 47
Zherebtsov, Yu. P., 43, 45
Zhukov, V. V., 156
Ziegler, H. J., 214
Zitter, R. N., 101
Zoroofchi, J., 240
Zwicker, H. R., 227
Zwicker, W. K., 203

SUBJECT INDEX FOR PART A

A

Aluminum ion laser, 164–166
Ammonium beam maser, 6–7
Amplified spontaneous emission
  for short-pulse dye lasers, 271–276
  excitation intensity and, 275
  and gain of amplifying medium, 274
  geometry for, 273
  optical feedback and, 274
  for single-pass short-pulse amplifier, 283
Anisotype heterojunctions, 217
Argon
  double ionized, 68
  first spectrum of, 60
Argon–chlorine laser, 116
Argon ion, laser transitions of, 65
Argon ion laser, 61
Argon–oxygen laser, 117
Argon II laser, 92–95
Argon III laser, 92–95
Arsenic ion laser, 153–154
ASE, see Amplified spontaneous emission
Astigmatic resonator, for cw dye lasers, 343–344
Atom
  expectation value for, 20–21
  state of, 20–21
Atomic coherence, 20
Atomic collisions
  neutral atom lasers excited by, 116–120
  noble-gas ion lasers excited by, 121–124

B

Beryllium ion laser, 156–157
Blue–green laser lines, in Argon II lasers, 93–94
Blue–green pulsed xenon ion lasers, performance of, 100
Blumlein laser electrodes, 119
Brewster's angle surface, in cw dye lasers, 351

C

Cadmium ion laser, 128–143
  cataphoresis in, 134–137
  excitation mechanisms in, 131–134
  noise in, 137–138
  spectroscopy of, 128–131
Cataphoretic flow, in cadmium ion lasers, 134–137
Coaxial flashlamps, for short-pulse dye lasers, 319–321
Coma, in special-purpose resonators, 352–353
Continuous wave dye laser(s), 325–355
  alternative geometries for, 351–353
  astigmatic resonator for, 343–347
  coma in, 352
  defined, 325–326
  dye cell astigmatism in, 333–347
  dye system for, 353–355
  efficiency of, 331–332
  geometries of, 326
  input photons and, 333
  mode area for, 341–351
  mode radius or waist for, 339–340
  resonator for, 326, 336–338
Continuous wave ion laser, 71–74
  operating pressures for, 89
  output power variation in, 77
Copper ion laser, 158–162
Copper laser lines, singly ionized, 159
Copper vapor laser, performance of, 44–45
COT, see Cyclo-octatetraene
Crystals, for solid state lasers, 188–191
Current-dependent destruction mechanism, 107
cw laser, see Continuous wave laser
Cyclic laser, 34–55
  ideal, 34
  metal vapor, see Cyclic metal vapor lasers
  noble gas, 48–53
Cyclic metal vapor lasers
  first oscillations of, 38–40
  properties of, 36

373

Cyclo-octatetraene, in flashlamp-excited laser systems, 359

D

Degradation, in semiconductor diode lasers, 242–244
Density matrix, 20–26
 defined, 20–22
 time dependency of, 22–26
Density matrix elements, relaxation times for, 24
Density matrix equation, 22–26
 steady-state solution of, 26
Density operator, 21
Depopulation, of lower laser levels, 109
DFB (distributed feedback) laser, 247–249
Dye cell
 cross-sectional area of beam within, 349–351
 for flashlamp excitation, 312–314
 free jet, 314–318
Dye cell astigmatism, in cw dye lasers, 344–347
Dye flow system, for short-pulse lasers, 324–325
Dye laser(s), 251–375, see also Steady-state laser(s)
 continuous wave, see Continuous wave dye laser(s)
 dye cell length for, 265
 electronic energy levels and transitions in, 253–254
 first successful, 337
 flashlamp-excited, 297–298
 flashlamp plasma temperature as efficiency factor in, 300
 as four-level system, 255
 free-jet dye cell in, 314–318
 heat stored in, 313
 intense excitation from, 297–298
 laser devices and, 269
 laser dyes in, 355–359
 laser threshold relation in, 258–263
 linear flashlamps and, 302–305
 most outstanding characteristic of, 264
 oscillation conditions for, 257–259
 output wavelength selectivity of, 264–265
 rate equations for, 257–258, 269–271
 resonators for, 321–324
 resonator stability criteria for, 338–343

 rhodamine 6G type, 256–261, 331, 336, 357
 self-absorption in, 255
 short-pulse, see Short-pulse dye laser(s)
 short-pulse oscillator and, 276–281
 simulated emission cross section in, 263–264
 solvent characteristics and, 265–267
 specialized techniques with, 323–324
 Spectra-Physics model 356, 375
 steady-state laser as, 293–325
 thermal limitations in, 264–269
 threshold equation for, 259
 two classes of, 252
 uses of, 251
 water as solvent for, 268–269
 wavelength range and intensity maximum for, 258–259
Dye laser levels, optical transitions between, 253–254
Dye laser operation, high discharge plasma temperatures in, 298–300
Dye laser oscillator–amplifier system, 286
Dye laser performance, self-absorption factors in, 255–256
Dye molecules
 energy states of, 253–257
 steady-state behavior of, 294–297
Dye solution jets, 314–318
 nozzles for, 354

E

Electric dipole interaction, of linearly polarized monochromatic wave, 25
Electron collision
 ions excited by, 59–97
 neutral atom lasers excited by, 33–59
 helium metastable destruction by, 108–109
 in noble gas ion lasers, 82–83
Electron temperature, optimizing of, 112
Electrooptic shutter, 27
Energy levels
 in solid state lasers, 171–176
 in thermal equilibrium, 32
Evanescent coupler, short-pulse dye lasers and, 289

F

Fabry–Perot cavities, 10–11, 27, 210
Fabry–Perot etalons, 322

SUBJECT INDEX FOR PART A 375

Fabry–Perot interferometer, 9
Facet damage, in semiconductor dye lasers, 242
Flashlamp(s)
 ablating, 305–307
 coaxial, 319–321
 ionized gas in, 304
 linear, 302–305
 xenon-filled, 302
Flashlamp arc, wall stabilization of, 307
Flashlamp excitation, dye cell for, 312–314
Flashlamp-excited dye lasers, 297–298
 dye solution jets in, 314–318
 emission spectra in, 301
Flashlamp-excited systems, oxygen and cyclo-octatetraene in, 359
Flashlamp plasma temperature, 298–302
 laser efficiency and, 300
Fox–Smith adaptations, of Michelson interferometer, 322
Fundamental transverse mode operation, in heterojunction lasers, 238–239

G

Gas lasers, atomic and ionic, 31–166, *see also* Ion lasers; Neutral-atom lasers
Glasses, for solid state lasers, 191–193
Gold ion laser, 164

H

Helium–cadmium laser, 128, *see also* Cadmium ion laser; Ion lasers
 noise in, 137–138
 other excitation methods for, 138–142
Helium–fluorine laser, 118–120
Helium–iodine laser, 146–149
Helium–mercury laser, 125–126
Helium metastables
 creation of, 104–105
 destruction of, 108
Helium–neon lasers
 excitation processes in, 104–110
 noise in, 137–138
 number of, 31, 97–116
 performance of, 110–114
 pulsed operation of, 114–116
 small-signal gain of, 113
 spectroscopy of, 101–104
 transitions of, 103

Helium–zinc laser, *see* Zinc ion laser
Helium–zinc pulsed discharge, 143
Heterojunction lasers, 215–217
 carrier confinement in, 219–220
 emission range for, 232–233
 fundamental transverse mode operation in, 238–239
 material used in, 230–234
 operation of, 217–218
 radiation confinement in, 220–222
Hallow-cathode discharge, in cadmium-ion lasers, 129–131
Holstein trapping factor, 88
Host materials, in solid state lasers, 169, 187–193

I

Infrared emission lasers, 232–234
Intraresonator, short-pulse dye laser and, 290–292
Intraresonator laser intensity, for steady-state laser, 295
Iodine, in hollow-cathode discharge, 148
Iodine ion laser, 146–149
Ion, parallel radial electrical field of, 81
Ion–ion interactions, in solid state lasers, 179–181
Ionized argon, laser transitions of, 65, *see also* Argon
Ionized noble gases, laser lines in, 62–64, *see also* Noble-gas laser(s)
Ion laser(s)
 continuous wave, 71–74, 77, 89
 exciting of by atom or ion collisons, 120–166
 noble gas, *see* Noble-gas ion laser(s)
 sputtered metal vapor, 157–166
Ion laser discharge, energy distributions of, 81
Ion laser excitation, 90–93
 sudden perturbation excitation model of, 92
Ion laser oscillation, beam-generated plasmas for, 92
Ion laser plasmas
 inductive or B-field coupling in, 90
 radio-frequency excitation for production of, 90
 self-compressed or Z-pinched discharges in, 91
 excitation of by electron collision, 59–97

SUBJECT INDEX FOR PART A

Ion temperatures, in noble-gas ion lasers, 82
Iron group ions, in solid state lasers, 171–173
Iron transition metal ions, 184–186
Isotype heterojunctions, 216–217

K

Kerr cell, 27
Krypton, singly ionized, 67–68
Krypton ion lasers, outputs for, 93, 96

L

Lanthanide series ions, 173
Large-optical cavity laser, 215
Laser(s), *see also* Semiconductor diode lasers; Solid state lasers
  argon–chloride, 116
  argon ion, 61
  argon–oxygen, 177
  atomic and ionic gas, *see* Gas lasers
  basic frequency selectivity and, 3
  continuous wave, *see* Continuous wave dye laser(s)
  copper vapor, 44–45
  crystalline solid state, 16–17
  cyclic, *see* Cyclic laser
  density matrix formalism, 2–3
  depopulation of lower levels of, 109
  disturbed-feedback, 247–249
  dye, *see* Dye laser(s)
  features common to, 2
  four-level system and, 9
  gain medium of, 12
  helium–neon, *see* Helium–neon laser
  high pulsed output powers of, 117
  infrared, 232–234
  instabilities in, 18
  intensities of, 1, 28
  ion, *see* Ion laser; Aluminum ion laser; Beryllium ion laser; Copper ion laser; Lead ion laser
  as light amplification by stimulated emission, 4–9
  mercury ion, 124–128
  mirrors used with, 10–11
  near-infrared emission, 230–231
  neon–oxygen, 116
  neutral atom, 33–59, 97–120
  noble-gas *see* Noble-gas ion laser(s)
  noble-gas as reactant in, 120
  optical cavity in, 3–4
  optical resonator as feedback mechanism of, 9–12
  population inversion and, 6–9, 20
  properties of, 1–3
  pumping parameters and, 19
  red helium–neon, 31
  semiconductor diode, *see* Semiconductor diode lasers
  simplest form of, 3
  solid state, *see* Solid state lasers
  spectral characteristics of, 16–18
  spectral width of, 1–2, 11
  split-cylinder hollow-cathode, 43
  steady-state, *see* Steady-state laser
  stimulated emission cross section of, 5–6
  tellurium ion, 152–153
  thallium ion, 155–156
  transient, 35
  tunable, 205
  types of, 2
  visible emission, 231–232
Laser diodes, *see* Semiconductor diode lasers
Laser dyes, list of, 357, *see also* Dye laser(s)
Laser instabilities, reasons for, 18–19
Laser intensity, 1, 28
Laser ions, in solid state lasers, 169, 181–187
Laser levels, excitation mechanism models for, 84–87
Laser materials, in solid state lasers, 181–194
Laser oscillator, 12–20
  electron-excited, 70
  saturation characteristics of, 12–14
  threshold condition for, 12
Laser output
  mode-locked, 28–30
  spectral width of, 16–17
Laser output characteristics, modification of, 26
Laser rate equations, stability considerations and, 18–20
Laser solution performance, improvements in, 358–359
Laser threshold
  defined, 291
  for dye lasers, 258–263
Laser transitions
  homogeneous broadening of, 13–14, 17
  inhomogeneous, 17
  inhomogenous broadening of, 14–16
Lead-ion laser, 155
LED, *see* Light-emitting diodes

SUBJECT INDEX FOR PART A 377

Light amplification by stimulated emission, 4–9, see also Laser(s)
Light-emitting diodes, 195
Linear flashlamp, 302–305
　ablating, 305–307
　transverse dye flow in, 311
Linear flashlamp reflection cavities, 310–312
LOC, see Large optical cavity laser
Lower laser levels, depopulation of, 109

M

Magnesium ion laser, 154
Maser, ammonia beam, 6
Mercury ion laser, 60, 124–128
Metal vapor cyclic laser, 35–48
Metastable density curve, 107
Michelson interferometer, Fox–Smith adaptations of, 322
Mirrors
　Fabry–Perot cavities and, 10–11
　lasers and, 10–11
Mode area, for cw dye lasers, 347–351
Mode locked lasers, 28–30
　short pulses from, 29

N

Nd:glass lasers, 168
　properties of, 199–202
Nd:YAG (neodymium/yttrium aluminum garnet) crystals, 198
Nd:YAG lasers, 168, 195
　properties of, 196–198
Nd:YLF (neodymium/yttrium lithium fluoride) lasers, 204–205
Near-infrared emission lasers, 230–231
Neodymium glass lasers, see Nd:glass lasers
Neon–oxygen laser, 116
Neutral atom density, in noble-gas lasers, 83
Neutral atom lasers, excited by collision with atoms, 97–120, see also Argon–chlorine laser; Helium–neon lasers
Neutral atom temperature, in noble-gas ion lasers, 82
Noble-gas atom, outer electron shell of, 62
Noble-gas energy levels, transient laser lines in, 51
Noble-gas ion laser(s)
　cascade contribution to, 86
　characteristics of, 71–81
　mechanisms of, 81–93

　continuous neutral, 55–59
　electric field of, 83–84
　electron temperature in, 82–83
　excitation pathway in, 86
　exciting by atomic collisions, 121–124
　exciting mechanism models of, 84–87
　ion density in, 83
　ion temperatures in, 82
　neutral atom density in, 83
　neutral-atom temperature in, 82
　parallel or perpendicular fields of, 81–82
　performance of, 92–97
　plasma properties of, 81–82
　pulsed, 74–76, 95
　radiation trapping in, 87–90
　single-step excitation model of, 87
　spatial development in output of, 75–76
　spectroscopy of, 62–71
　transient, 53–55
　ultraviolet, 98–99
Noble-gas mixtures, high-pressure, 117–118

O

Optical cavity, feedback from, 3–4, 210
Optical resonator, laser and, 9
Oscillation, basic threshold condition for, 12
Oscillators, regenerative, see Regenerative oscillators
Oxygen, in flashlamp-excited laser systems, 359

P

Paramagnetic ions, 181–182
Penning collisions, 151–152
Penning electrons, 147
Penning lines, 159
Penning process, 133, 144–145, 159–160
Polycrystalline materials, as hosts, 189
Population inversion
　laser and, 6–9, 20
　Q-switching and, 27
Pulsed ion lasers, operating pressure for, 89, see also Laser(s)
Pulsed noble-gas ion lasers, see also Noble-gas ion lasers
　characteristics of, 74–76
　discovery and disappearance of, 95
　longitudinal dc electric field in, 80–81
　oscillation of, 75
Pumping, Q-switching and, 27

Q

Q-switching, for laser pulses, 27–28

R

Radiation patterns, of semiconductor diode lasers, 237–242
Radiation trapping, in noble-gas ion lasers, 87–90
Rare earth ions, 182–184
  in solid state lasers, 173–176
Rare earth lasers, optical pumping efficiency and output power of, 193–194, *see also* Laser(s)
Rate equation description, for short-pulse dye lasers, 269–271
Regenerative oscillators, short-pulse dye lasers and, 287–293
Relaxation processes, atomic forces and, 23
Relaxation times, for different density matrix elements, 24
Resonator(s)
  astigmatic, 343–344
  for flashlamp-pumped dye lasers, 321–324
  modifications of characteristics in, 26–30
  two-mirror, 287–290
Resonator stability criteria, for cw dye lasers, 338–343
Rhodamine 6G dye lasers, 331, 357
  characteristics of, 256–261
  laser threshold for, 336
  output of, 336
Ring geometry regenerative oscillator, 291–292
Ruby lasers, 202–203

S

Selenium-ion laser, 149–151
  operating parameters of, 150
Self-compressed discharges, 91
Self-sustaining oscillations, in semiconductor diode lasers, 246–247
Semiconductor diode lasers, 209–249, *see also* Heterojunction lasers
  carrier confinement in, 219–220
  defined, 209–211
  degradation of, 242–244
  distributed-feedback lasers and, 247–249
  facet damage in, 242
  gain coefficient and threshold condition in, 222–224
  heterojunction, *see* Heterojunction lasers
  homojuction, 214
  injection in, 216–219
  internal defect formation process in, 242
  laser topology for, 211–213
  materials used in, 226–230
  modulation characteristics in, 244–247
  performance of, 234–237
  radiation patterns for, 237–242
  self-sustaining oscillations in, 246–247
  temperature dependence of threshold current density in, 224–226
  threshold condition for, 222–224
  vertical geometry of, 213–216
Shock wave effects, in short-pulse dye lasers, 318–319
Short-pulse amplifiers, 281–287
Short-pulse dye lasers, 269–293, *see also* Dye lasers; Laser(s)
  amplified spontaneous emission from, 271–276
  coaxial flashlamps in, 319–321
  dye flow system in, 324–325
  excitation volume for, 280
  intraresonator and, 290–292
  laser threshold in, 291
  resonators and tuning elements in, 321–324
  ring oscillator parameters and, 291–292
  short-pulsed amplifiers and, 281–287
  thermal limitations of, 293
  two-mirror resonator and, 287–290
  wavelength-tunable oscillator and, 276–278
Short-pulse single-pass amplifier, output and point-by-point efficiency of, 284
Short-pulse wavelength-tunable oscillator, 276–278
Silver ion laser, 162–164
Single-electron collision process, 109
Singly ionized krypton, energy level of, 68–69
Solid state lasers, 167–207
  characteristics of, 169–170
  crystals for, 188–191
  energy levels in, 171–176
  excitation of, 169
  fluorescence sensitization in, 193–194
  glasses for, 191–193

hazards associated with, 205–207
history of, 167–168
host materials in, 169, 187–193
ion–ion interactions in, 179–181
laser materials in, 181–194
nonradiative transitions in, 177–179
Nd:glass, 199–202
Nd:YAG, 195–198
optical radiation hazards from, 206
physical processes in, 170–181
properties and comparison of, 195–205
radiative transitions in, 176–177
size of, 169
stoichiometric, 203–204
structure of, 168–169
transition probabilities for, 176–179
upconversion in, 194
Sputtered metal vapor ion lasers, 157–166
Steady-state lasers, 292–325
   flashlamps and, 297–307
   general description of, 293–294
   vortex stabilized lamps and, 307–310
Steady-state power balance, 294–298
Stimulated emission cross section, for dye lasers, 5–6, 263–264
Stoichiometric lasers, 203–204
Stoichiometric materials, as host materials, 190–191
Sudden perturbation model, of ion laser excitation, 91

T

Tellurium ion laser, 152–153
Thallium ion laser, 155–156
Threshold, laser, *see* Laser threshold
Threshold current density, temperature dependence of in semiconductor diode lasers, 224–226
Threshold relationships, in dye lasers, 258–263

Tin ion laser, 154–155
Transient laser lines, between lowest neutral noble-gas energy levels, 51
Transient noble-gas lasers, 53–55, *see also* Noble-gas ion laser(s)
Tunable lasers, 205
Tunable short-pulse oscillator, 276

U

Ultraviolet ion lasers, 93
Ultraviolet noble-gas ion lasers, 98–99, *see also* Noble-gas ion laser(s)
Upconversion, concept of, 194

V

Visible emission lasers, 231–232
Vortex-stabilized loops, in steady-state lasers, 307–310

W

Wavelength-tunable oscillator, 276
Wigner spin rule, 104, 121

X

Xenon flashlamps, Nd:glass lasers and, 199
Xenon gas, in coaxial flashlamps, 321
Xenon ion lasers, 100

Z

Zinc ion lasers, 143–146
   excitation mechanisms in, 144–145
   performance of, 145–146
   spectroscopy of, 143–144
Z-pinched discharges, 91